Urban Economy

Urban Economy: Real Estate Economics and Public Policy analyses urban economic change and public policy in a more practical way than a typical urban economics book. The book has a distinctive framework that considers the underlying reasons, and the consequences of, urban change for real estate investors and policy makers.

Part 1 covers the basics of urban economics and real estate markets, including housing and commercial. Part 2 looks at the reformulation of urban systems and the reasons why. It then considers the consequences for real estate markets and investment of decentralisation forces and emerging technology. The issues that arise for urban public policy are then discussed, notably transport policies, public finance and sustainability, before a chapter examining housing neighbourhood and housing market dynamics and a shift from spatial change to regeneration. Part 3 reverses the dominant perspective of Part 2 to assess the effectiveness of how property-led policies can positively influence a local economy and urban regeneration. The chapters consider several important policy questions and constraints and draw on a number of case studies that illustrate the benefits and drawbacks.

The book includes chapter objectives, self-assessment questions, chapter summaries, learning outcomes, case studies, global data and statistics and is a new textbook for core courses in urban economics and real estate economics on global Real Estate, Planning and related degree courses.

Colin Jones is an urban economist who has been a professor at Heriot-Watt University since 1998. He formerly worked at the Universities of Manchester, Glasgow and the West of Scotland. His research interests span commercial, industrial and housing market economics, investment and policy, the macroeconomy and local economic development. Colin has edited or authored eight books, published more than seventy papers in academic journals and has taught urban economics and real estate investment at undergraduate and postgraduate levels for more than 40 years.

Urban Economy
Real Estate Economics and Public Policy

Colin Jones

LONDON AND NEW YORK

First published 2022
by Routledge
2 Park Square, Milton Park, Abingdon, Oxon OX14 4RN

and by Routledge
605 Third Avenue, New York, NY 10158

Routledge is an imprint of the Taylor & Francis Group, an informa business

© 2022 Colin Jones

British Library Cataloguing-in-Publication Data
A catalogue record for this book is available from the British Library

Library of Congress Cataloging-in-Publication Data
Names: Jones, Colin, 1949 January 13– author.
Title: Urban economy : real estate economics and public policy / Colin Jones.
Description: Abingdon, Oxon ; New York, NY : Routledge, [2022] |
Includes bibliographical references and index.
Subjects: LCSH: Real estate development—Government policy—Great Britain. |
Real estate development—Government policy—United States. |
Urbanization—Great Britain. | Urbanization—United States.
Classification: LCC HD593 .J65 2022 (print) |
LCC HD593 (ebook) | DDC 307.76/0941—dc23
LC record available at https://lccn.loc.gov/2021020265
LC ebook record available at https://lccn.loc.gov/2021020266

ISBN: 978-0-367-46197-3 (hbk)
ISBN: 978-0-367-46194-2 (pbk)
ISBN: 978-1-003-02751-5 (ebk)

DOI: 10.1201/9781003027515

Typeset in Times New Roman
by codeMantra

Access the Support Material: www.routledge.com/9780367461942

Contents

Preface

The number of people living in cities around the world is growing, and at the same time as dynamic entities, they continue to evolve. It means that there is an increasing need to plan and manage the issues that arise. In turn, it requires an understanding of their underlying economics and why they are changing. This book seeks to contribute to the training and decision-making of real estate and planning professionals who are crucial to shaping cities.

It is based on my teaching of urban economics to real estate, planning and geography students over decades. Many of these students find traditional economics texts difficult and inaccessible. To address these students' needs, the book therefore has no equations and arguments are all expressed in words.

The book also draws on my research on the subject over my career. This research has benefitted from the inputs of many colleagues and PhD students along the way. I particularly wish to thank John Parr who was a strong supporter at the beginning of my career.

1 Introduction

The first ancient cities, thousands of years ago, were located in what is now Iraq and were much smaller than what would be regarded as a city today. They were usually centres of commerce and public administration with often their own sanitation systems. They were surrounded by walls for protection but these were eventually discontinued partly because they were ineffective against the emergence of artillery and partly because population growth meant expansion beyond the boundaries. Some cities such as Rome and Athens can trace a continuous history back to ancient time but most cities today are much younger. Cities as we know them now evolved with the industrial revolution.

Today the definition of a city will vary from one country to another. What is the difference between a city, a town and an urban area? The book does not concern itself with these issues, for example whether a town has been designated a city by a national government. A city is taken to be a spatial concentration of linked economic activities, namely of industries and housing.

Households live and work in a city and consume services produced there. The term city and urban area will be used interchangeably. The book also recognises that cities are not entities in a vacuum but part of a national urban system. Within this system of urban areas, there are functional specialisms, competition for businesses and a hierarchy of services provision in terms of shopping and personal/business services.

Often a city has an administrative boundary that is different from the functional boundaries linked to its economic activities. There could then be substantial differences between the populations given by a narrow administrative area and the actual city. In this case, the city boundary will be too narrow, by excluding say suburban or surrounding areas that are linked not least by commuting to the urban cores.

This underbounding could occur when the boundary was set historically, and the city has subsequently grown, so the administrative area will underestimate the true population. In some countries, the standard administrative definition will encompass the surrounding region of a city and hence exaggerate the size of cities.

Land-use patterns in cities across the world have many common features. There is always one traditional central locality associated with the administration of the city that also acts as a focus of religious activities and retailing. Retailing began life originally as open-air markets, and these still occur to a greater or lesser extent partly depending on the degree of economic development. The emergence of offices in the late 1900s also changed the complexion of this central area, and as a result, it is often referred to as the central business district.

DOI: 10.1201/9781003027515-1

A key influence on the location of land uses within an urban area is the role of accessibility. The primary religious building needs to be accessible to as many people as possible. The city chambers or a town hall requires to be at the most central location. Retailers want to have the greatest access to customers.

Locations within a city have different degrees of accessibility to the centre (although as we will see accessibility is a more complex concept). In terms of the urban economy, one of the key issues is how much is each land user in the form of firms and households prepared to pay for accessibility. This is a key determinant of relative land and real estate values within a city. It also determines the spatial pattern of land uses.

Housing is the land use that covers the largest area of a city. Housing itself is differentiated by tenure, namely rented and owner occupation. In some parts of the world, there are also forms of social housing provided by the state or a public agency that allocates such housing on an administrative basis. The internal operations of the other housing tenures are underpinned by the interaction of household choices and constraints.

There are many constraints on household choices, not least income but for house purchase, the credit rules set by banks are also very important. Ultimately, households can adjust their housing circumstances by moving or improving an existing home. When households move, then they choose a location with a (new) set of accessibility relationships and by participating in the market can influence house prices.

Over time, the spatial structure of the housing market has changed as many people are able to travel greater distances to work. This trend reflects the fact that cities are dynamic entities driven not least by technological change. This is most evident in the changing face of transport over the centuries from horse-drawn carriages, canals, trams, trains to cars.

These developments have in their time transformed both the movement of people and freight. It is not only transport that has been subject to innovations, but also gas, electricity, telecommunications, new building technologies and now the Internet. To take one example, the invention of the electric lift permitted the construction of office skyscrapers.

It was not just new buildings that have been developed as a result of technological change but the nature of urban form has also adapted. Urban tramways, railways and motorways, for example, have been built tearing up historic townscapes and changing accessibility relationships in the city.

Technology has also meant individual buildings became obsolete such as large analogue telephone exchanges in the centres of towns, even telephone boxes on streets. In their place came a demand to locate mobile phone transmission masks. All these forces have shaped cities.

Other related factors are at work in changing the face of cities. Long-standing traditional industries can become no longer viable. This can occur either because their product is no longer wanted, replaced by a modern invention, or because the product can be made much cheaper somewhere else.

New industries may take their place providing employment to the city's population, but the scars of the original industry can linger on in the shape of vacant land and forsaken industrial landscapes awaiting redevelopment/regeneration. In some cases, it can take decades, if at all, to find alternative uses for what often become eyesores representing vestiges of an industrial past.

The density and dynamic nature of cities has the potential to create negative consequences for existing land users as neighbouring plots are demolished or redeveloped, in some case by the city government for the common good. Partly as a result, town planning emerged at the beginning of the twentieth century to address these issues and exists virtually everywhere in the world in some form. It can take a strategic approach to planning the city encompassing infrastructure, public welfare and the location of different land uses, as well as micro-moderating individual developments to address any negative consequences for neighbours.

Planning in the broader sense has also brought redevelopment of substandard housing, slums, that no longer meet society's norms. It has contributed to the regeneration of areas, often ravaged by industrial decline with a coordinated revival plan in some cases supported by government subsidies.

Urban regeneration is not normally achieved simply by state actions or indeed by the private sector alone. There may be serious site contamination problems. It normally requires a form of public/private partnership and can take a very long time to be successful, especially if the dereliction is on a large scale.

Unused warehousing and abandoned factory buildings cannot be physically redeveloped or converted overnight. Finding and creating new uses is often a slow business. Ultimately, success is dependent on discovering a suitable market. It could be warehousing converted to flats, the sites of factories replaced by distribution warehousing or a shopping centre.

There are patterns to urban land use, but they are created by a myriad of individual decisions set within a framework of any planning rules. Each individual firm, investor, developer, public agency and household is making periodic decisions about real estate locations (moving or staying put), buying/creating new buildings or improving/renovating instead on the basis of what is optimum for themselves.

In addition, households are making decisions about how to travel, where to shop and what leisure activity to undertake. Planners at the same time may be seeking to moderate travel patterns and control development to address the congestion and pollution problems of a city.

The greenhouse gases created, particularly by motor traffic in cities, also have implications for climate change. Sustainability of cities is becoming a necessary imperative for the world as it faces an increasing climate emergency. How this is addressed will almost certainly have major consequences for urban form, for example if cars are banned from city centres.

In fact, there are arguments that urban form should be modified to improve the sustainability of cities. However, such a goal can only be achieved in a market economy by a solution that ensures the viability of new development.

The book examines these issues and more in a three-part framework. Part 1 is entitled 'Spatial Pattern of Economic Activity' and considers, first, the basic economics of cities beginning with why cities exist, what are their functions and why they are located where they are. Chapters 3–5 explain the concept of urban agglomeration economies and the spatial structure of cities. These chapters represent the heart of urban economic theory.

In Chapter 6, the economic logic of planning and its implications for real estate markets is set out. The remainder of Part 1 focuses on the housing and commercial real estate markets, examining the specifics of how these markets work and how cycles occur. These chapters also encompass the role of investors and banks in shaping these markets.

Part 2 of the book, 'Spatial Change and Public Policy', uses the theoretical base given in Part 1 to understand urban economic change and policy questions that arise. First it considers the dynamics of change in terms of the growth, decline and revival of cities around the world. The theories presented in Part 1 are revisited to explain decentralisation of cities. Chapter 11 considers how and why national urban systems are changing, with a particular emphasis on transport innovations.

Chapter 12 focuses on particular land uses by analysing the detail of the evolution of retail, office and warehousing forms. It identifies new types of retail centres, a changing retail hierarchy and the reasons for the decline of high/main streets. It also examines the implications for offices of information communication technologies.

In Chapter 13, the attention turns to the response of real estate investors and planning to these changes and the green agenda. It examines how investors react to new real estate forms requiring a potential rethink to their existing portfolios. The chapter similarly asks to what extent planning can thwart urban decentralisation, namely sprawl.

Chapter 14 assesses the challenges of urban public finance to support local government services given the consequences of suburbanisation with declining urban cores and the implications for its tax base. The economic theory supporting policies to reduce congestion through road tolls and its alternatives is evaluated in Chapter 15. Together with low-emission zones of cities, their impact on urban land-use patterns and property values is also assessed.

A more broad view of urban sustainability is reviewed in Chapter 16 drawing on the policy debate that has concentrated on the arguments for and against the concept of the 'compact city'. It also identifies that any solution must be linked to real estate market viability.

The final chapter in Part 2, Chapter 17, relates to neighbourhood and housing market dynamics. It considers the economics of neighbourhood decay and revitalisation. It identifies neighbourhood succession as a key issue, namely the replacement of one predominant type of household by another as an area changes.

Chapter 17 leads into Part 3 whose subject is 'Regeneration and Urban Growth Policies'. The thread to this part is the way public policy can influence a city's development and regeneration where necessary. Chapter 18 sets the scene by chronicling the development of urban policy approaches to address this issue in the UK and the United States. It highlights the role and place of real estate in promoting urban change.

The next chapters in turn examine how to influence urban competitiveness by policies linked to commercial real estate markets, and physical and housing-led urban regeneration. In Chapter 20, there is an evaluation of the use of enterprise zones, and this is followed by a chapter that similarly assesses the scope and success of urban development corporations. Part 3 is completed by considering how we test that real-estate-led urban regeneration is successful in promoting new uses in secondary and tertiary localities laid waste by spatial change.

To summarise, the book begins by examining the underlying economics of cities. It then focuses on why cities and the urban system are changing with its implications for real estate markets. The book then addresses the urban policy issues that arise from this change. The final part of the book reverses the causality. Instead of looking at how urban changes have brought implications for real estate and policy, it assesses how intervention in real estate markets can mould and change cities.

Urban economics is itself a relatively young discipline emanating from the 1960s although it draws on geographical thought that has a longer pedigree. The seminal research papers were written in the 1960s. These developed models of cities with simplifying assumptions to develop our understanding especially of urban spatial structure. These are presented in Part 1. Urban economics is also eclectic, drawing on sociological writings as well as different branches of economics. However, it is essentially a branch of applied economics, and this is reflected in the tone of the book. To make the subject matter as accessible as possible, the book contains no algebra.

Part I

Spatial pattern of economic activity

2 Location of economic activity

Objectives

Cities have been around since ancient times as centres of commerce and public administration. But the functions of today's cities are very different from their historic roles, not least because of the industrial revolution. Indeed, the roles of individual cities have evolved over time, even over just a few decades.

The location of economic activity is also changing with new technologies and transport improvements. It leads to questions about why and where cities are located? There are also questions about the definition of a city in terms of economic activity, population size and the internal spatial structure.

The aim of this chapter is to introduce the idea of what a city is, ask why cities exist and consider why they locate where they do. This discussion opens up into a wider perspective of an examination of theories of industrial location theory. The contents of the chapter are as follows:

- Definitions of a city
- Why cities are located where they are
- Industrial location theory

Definitions of a city

The statistical definitions of a city will vary from country to country around the world and are usually measured in terms of population. But cities are more than simply concentrations of population, they are also places where industrial production occurs and a range of services are provided. The balance of manufacturing and services will vary from city to city.

Cities in countries with advanced economies have generally shifted from having a predominantly manufacturing base to be more service oriented. Indeed, the vast majority of workers in cities of advanced economies are now employed in services. Cities are therefore clusters of economic activity encompassing similar and dissimilar businesses.

As services centres, cities not only have a concentration of shops of different kinds but also have a variety of personal services/amenities such as health care/hospitals. Within cities, there are also located a range of services for firms such as lawyers, printers, hotels and consultants offering advice, including on real estate decisions. Some

DOI: 10.1201/9781003027515-3

of the markets for services overlap with demands from households. But in general, a distinction can be made between business and personal services.

Cities can offer personal services that rural areas cannot because their size makes it possible for such amenities as cinemas, theatres and sports to be viable. Cities also operate as tourist centres, with often people attracted to the scale of the leisure facilities including the range of shops or by historic townscape and buildings.

Why are there concentrations? First, there are simple social arguments that humans in general enjoy living in relatively close proximity. But clearly, the population of cities is so large that people can interact at a social level with only a miniscule minority. Furthermore, large cities are criticised as lacking humanity, even causing society to breakdown. The answer must lie elsewhere.

Historically, cities offered the opportunity for common defences, and many cities were surrounded by walls. The history of city walls can be traced back thousands of years. There are many examples of cities where walls survive such as at York in the UK, Carcassonne in France, Harar in Ethiopia, Pingyao in China and Taoudant in Morocco.

At one time fortifications were the norm but in most cases they have not been preserved. To some extent, this reflected they were no longer effective to artillery but also in more peaceful times they were no longer needed. The key point is that settlements continued to exist after the obsolescence of walls and grow beyond them so that defence *per se* was never the reason for their existence.

The dominant reasons for cities and towns are economic as places for markets or production of goods and services. Initially, markets were places for local agricultural produce, and this still occurs today although its importance has been much reduced in advanced economies. Similarly, manufacturing goods were produced locally for nearby needs. The first shops often sold goods made on the premises.

The industrial revolution dramatically changed these activities with the emergence of the factory and the enabling of intercity trade in manufacturing products, but the fundamental functions of cities did not change. Further understanding of the underlying economics is considered in the next chapter.

Cities are not just simply defined by concentrations of people and economic activity but also by their physical characteristics that are encapsulated in the umbrella term, urban form. Two cities with the same population could be very different in terms of physical layout and land-use patterns. Spatial structure is a key characteristic of a city, and density and the extent of the area covered are important dimensions.

The characteristics of urban form range from micro-level neighbourhood attributes through to city-wide perspectives. The significance of individual characteristics to urban form could vary in different parts of the world. Nevertheless, it is useful to look in detail at the most important.

Density in particular can be seen as a key characteristic of a city. Cities at their core exhibit an absence of space between people and firms. Their existence stems from people locating close to each other. Urban density has a number of different interrelated dimensions.

Gross population density at its simplest is calculated by dividing the city's population by its area whereas net residential density again starts with the population but is based on the area taken up by housing. Commercial and industrial employment densities measure the spatial concentration of employment. Density can be seen as an input into the quality of life, particularly through the availability of public and private space.

In general, there is a close association between city size and density. A very-high-density city such as Manila in the Philippines functions very differently from a low-density provincial city in Europe or the Midwest of the United States. There is also a cultural dimension to density as acceptable densities vary around the world as indicated by government planning guidance on new house building in different countries. What is an unacceptably high density in one part of the world could be seen as low in say Asian cities, and an appropriate planning density could vary by the order of ten across the globe.

Densities also vary significantly within cities. The density of an inner city neighbourhood can be substantially higher than the suburbs, so a city average may hide significant variations. A concept linked to density is sprawl. Falling densities with distance from a city centre have led to concerns about unnecessary urban sprawl, especially in the United States.

This decline in density from the city centre does not hold when there are informal residential settlements on the urban periphery. These can occur in cities in developing countries. Examples include the favelas of Brazil. Approximately 25% of the world's urban population live in these very dense settlements, not complying with planning and building regulations.

Linked to density is the spatial pattern of land use in a city and how compact or sprawling it is. While the primary land use of a city is residential, a functional urban area also requires industrial, retail, office and leisure uses. Linkages between land uses and ease of travel are based on the city's transport infrastructure. It is closely associated with accessibility as transport infrastructure determines the ease with which buildings, spaces and places can be reached.

A key accessibility relationship is between home and the city centre. Accessibility is a multilayered concept and varies with mode of travel. Cities vary by their approach to the use of the car, the scale and cost of their public transport system and how cycle-friendly they are. In general, car usage promotes urban sprawl, but the issue is more complicated and discussed in Chapter 16.

To summarise, cities are defined by more than just their population or their scale of economic activity. This is important for when Part 2 considers urban change it involves more than just the nature of economic activity or the population size, but also the internal locational pattern of land uses and the shape of cities.

Location of cities

An important aspect of the location of economic activity is why cities are located where they are. There is not one reason for the locations of cities. Cities essentially grow up around places that have locational advantages that could include the following:

• *Crossover points of transport linkages* These points could be historically where trade routes or hunting trails meet and an urban centre would grow up. Historically, waterways or seas have played an extremely important role in the location of cities. The transportation of goods was originally much cheaper over water than by land.

In (relatively) modern times, this could be at the junctions of main roads/motorways, canals and railways. Cities would be located at these crossover points as places to maximise trade (markets) or minimise the distribution cost of output.

- **Closeness of mineral resources** Industries which require large inputs of raw materials that are difficult/expensive to transport could minimise costs by locating nearby, creating a focal point for a city.
- **Proximity to power** At the time of the industrial revolution, this would take the form of waterfalls enabling the powering of mills. Cities would be established at these microlocations although today their existence has no direct relationship when most of the world is on an electricity grid.

Sometimes locations can be explained as a combination of the two, namely crossover points leading to a regional hinterland where there are natural resources. Some ports are an example of such a combination.

In practice, cities do not necessarily have one reason for their existence or the reason could change over time. In many cases, the original reason is no longer of any significance, for example if a city exploited mineral resources that have now been exhausted. The city's continued existence has been achieved by successfully restructuring the economy, often more than once. However, where this has not happened, the urban area has suffered terminal decline, for example the abandonment of some coal-mining towns.

It is useful to consider some examples of the origins of cities to illustrate this.

- Pittsburgh in the United States originated in the 1700s because of its location at the convergence of three rivers, including the Ohio River. At that time, the rivers could be used as transporting furs. It was still a village then. In the 1800s, the city became an iron and then steel-making centre with the benefit of iron ore and coal deposits nearby, with railways along the river valleys increasing access to markets.
- Manchester in the UK was settled by the Romans who built a fort almost two thousand years ago, but with the demise of the Roman Empire, it remained a small centre for centuries. By the 1500s, it had become a centre for the production of linen and wool. Later, these goods were replaced in popularity by cotton in the 1700s, and Manchester became a major marketplace for its sale.

  The city's economy took off because of its transport links. It benefitted from the first canal in the UK giving it access to the sea followed by a canal network that provided the city with access to the rest of the country. Similarly, the world's first passenger railway was opened to Liverpool with more train links ensuing. Manchester's economy boomed through the 1800s on this locational base provided by the accessibility of the railways and canals.

These examples illustrate how cities' functions and fortunes were shaped by location and the evolution of transport technologies. In these cases, their origins are linked with rivers and horse-drawn carriages through canals and railways. In the case of Manchester, the most significant growth occurred because of its location relative to transport linkages, whereas for Pittsburgh, it was linked to its closeness to mineral deposits. Today, the economies of these cities are now services based and have very little if at all any relationship to their origins.

Industrial location theory

From looking at where and why cities are located, this section seeks to broaden the perspective and examine the optimum locations of industrial activity. Insights can be achieved by looking in more theoretical terms. A theory of location decisions for manufacturing firms is therefore developed, concentrating at first on the influence of transport costs.

It can be viewed as analysis of why firms or industries choose between interurban locations or places to set up in business. This industrial location theory is closely associated with the work of Alfred Weber developed from the early 1900s. This section starts with a simple location decision-making problem and gradually makes it more complex, reflecting real-life realities.

In the Weberian world, the best location is the one at which costs for a manufacturing firm or an industry are minimised. The revenues received by the firm are not influenced by its location. Stripped to its basics, the firm chooses a location that minimises the transport costs in its production and distribution process, *assuming other costs do not vary by location*. This is a very strong assumption that we return to later in the section. To reiterate, the firm therefore chooses a location to make its product that minimises its transport costs, because all its other costs are the same wherever it locates.

This means that the choice of location is one of balancing the costs of transporting the raw materials or inputs to the industrial plant and then the finished product to its marketplace. The simplest case is when there is one marketplace (say a city) and one location from where the raw materials come from.

This means the optimum location must be somewhere on the straight line between the two points. Given transport cost is the only factor in the production process that varies with distance, then the best location for the plant will be close to the source of raw materials/inputs, at the marketplace or on the straight line transport link between the two.

The actual choice of location will depend on the nature of the firm's production process. It could be that as a result of the manufacturing process, there is significant weight loss so the final product weighs a lot less than the original materials. This would be the case with a steel manufacturing plant.

In these circumstances, it is much cheaper per unit distance to transport the final product rather than the inputs so the industrial plant should logically be located near to the raw materials. On the other hand, if the final product is made by assembling smaller components, say for example in a car plant, it may mean that the cost minimisation location is near the market.

Obviously, this is an oversimplification of the industrial location decision. Its simplicity can be stressed simply by assuming two locations for raw materials rather than one. The optimising problem then becomes choosing a location within a triangle as set out in Figure 2.1. Two points of the triangle, M1 and M2, are where the different raw materials are, and the third, MK, the market for the finished product.

The cost minimisation point lies within the triangle at T. This location is determined by the transport costs of moving the raw materials and the finished product taking into account that they each of the inputs and outputs have different weights/sizes and freight rates. This is taken account of in the diagram by the weights' axes.

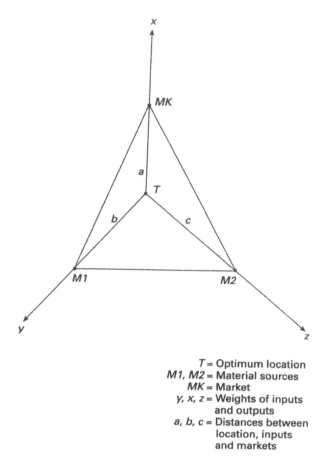

Figure 2.1 Weber's locational triangle.
Source: Balchin et al. (2000).

The optimum location in the triangle is still the point of minimum transport costs. This is because revenue is assumed not to vary with location. In other words, it does not matter where the industrial plant is situated, it can still sell the same output for the same price at the marketplace. This means that the point of cost minimisation is also the location of profit maximisation.

The theory is based on very strong assumptions. First, it assumes that manufacturers have full information on which to base their location decisions. Second, the significance of transport costs is emphasised rather than the costs of other factors of production, namely land and labour. The price of these factors, land and labour, as noted earlier is assumed to be constant at all locations.

Notwithstanding these assumptions, the practical application of the theory for a firm becomes arguably unmanageable if a firm serves more than one marketplace and has more than two locations from which it derives its inputs. Similarly, there is added complexity if a firm makes more than one product or has more than one plant that has linked production processes.

The information required to find the best location is very difficult to collect. It may be therefore problematic to work out where the optimum location is. In some cases, a firm may be located where it is simply because of proximity to a founder's home. More generally, it seems instead of optimising firms choose from locations that make a satisfactory profit, not necessarily the one that maximises profits. In this way, firms are what is called 'satisficing', a concept developed by Herbert Simon in the 1950s. While satisficing makes sense for firms, it undermines the basis of the Weberian theory.

Given these limitations, the theory has significant drawbacks to its application to a modern economy. While industrial location theory focuses on the optimum location of individual firms, it is probably best to apply it to interpret about where and why particular industries have located. This in turn as seen earlier in the chapter explains where towns and cities with an industrial specialism have grown up. The focus on transport costs within the theory in determining location fits well with why historically many industries located where they did and how this led to the creation of cities. But even from this perspective, there are exceptions such as the need to be near a source of power as discussed earlier.

Improvements in transportation such as through the building of motorway networks have reduced the relative importance of transport costs as a factor input to the production process. Accessibility to a particular location is no longer as significant as a century ago. The result is other factors have risen in importance to determine manufacturing location patterns.

In particular, the role of labour has become significant in industrial location choices. The availability and the differential price of labour at localities are now key influences on industrial location decisions. At its extreme, firms choose from a range of locations based primarily on labour rather than transport costs. Land price is logically also a factor, but there is a strong correlation between areas with low wage rates and low land prices.

This emphasis on labour is seen in moves of firms and industrial production from high-wage to low-wage areas. This occurred for example in the UK particularly during the 1970s with production moving from the prosperous south-east of the country to the poorer peripheral regions. There is also an international dimension to this process.

The same logic means that firms have moved production to low-cost countries in some cases on the other side of the world. There are innumerable examples, so Asia is the main global producer of clothes and electronic goods. Many cars for the market in the United States are now built in Mexico rather than say Detroit. Energy costs also vary between countries, and if this is an important contribution to costs for an industry, then it too is an important determinant of location.

These trends have been supported by the restructuring of many manufacturing firms from local to national and then international ownership beginning in the 1970s. Whereas a small firm is likely to remain at its original location, international conglomerates are more footloose, and less tied to traditional locations. They often set different production plants against each other in negotiations for the location of future investment in order to reduce labour costs.

This has encouraged the shift away of manufacturing from existing production centres with high wages and resulted in the decline of some cities. This is discussed more fully in Chapter 9. At the same time, location decisions are not as simple as portrayed here, and again the issue is revisited in Chapter 19.

This analysis applies to just manufacturing, and as discussed at the beginning of the chapter, the economic activities of cities in developed countries are mainly services. Services have displaced manufacturing as the economic base of cities in these countries, and the shift to low-wage economies of services activities is not so straightforward although it occurs.

As an example, the transfer of call centre employment has not been a total success because of language difficulties even with countries that share the same language. It is also possible to move some elements of services, including 'back-office' activities such as accounting, to low-wage locations.

The shift to services has been partly brought about by a redefining of the line between manufacturing and services. Historically, manufacturing combined the processes of design, testing and development, the making of the product, its marketing and distribution all 'under one roof', or at least from one site.

However, the restructuring of the industrial process over time has wrought a breakdown of this chain of activities. Instead, these functions can occur in different locations, even in different continents, with each having separate locational determinants. It means, for example, that textiles can be designed and marketed in Europe and made in Asia, before being shipped back for local distribution.

Summary

Cities are concentrations of economic activity that encompass manufacturing and services. Services provision meets the demands of local firms and households and can attract tourists. Although city size is often thought of in terms of population size, the urban economy is also reflected in physical form, namely in terms of density, land-use patterns and spatial extent, particularly sprawl.

Density is a key characteristic and varies across the city. While in many cities there is a negative density gradient from the city to the low-rise suburbs, this is not a ubiquitous pattern. In many developing countries, there are high-density informal settlements at the urban periphery.

Where cities are located can be explained by the locational advantages of individual sites. These advantages can be seen in terms of accessibility to mineral resources or to trade routes/transport links. The original reason often no longer explains the current economic activities as successful cities have adapted over time to the changing face of economic needs and demands. They will have also successfully responded to competition from other places that may be better placed to produce the goods that were the original basis for their primary economic activity. This phenomenon is demonstrated by the experience of Manchester and Pittsburgh.

Industrial location theory offers a way to think more strategically about the location of manufacturing industry and hence urban areas. Least-cost solutions based on transport costs demonstrate some insights especially into the spatial pattern of towns and cities that have origins in the immediate postindustrial period.

However, with the decline in the importance of transport costs to the industrial process, the location and price of labour has become important. Manufacturing has tended to move to low-wage locations in many cases to the other side of the world. Low transport costs mean that it can be cheaper to make goods in low-wage economies and send them around the globe to customers in high-wage economies.

The focus of this location theory is on manufacturing, but services are now the dominant element of urban economies. Part of the reason is that the production process has been decomposed so that the services required and the manufacturing activities needed to produce and sell a given product can occur in different locations. Transport or labour cost minimisation strategies are less applicable to the location of services.

Learning outcomes

Cities are a combination of production and services centres and their roles interact.

Cities can be defined not simply by size but also by their spatial form that incorporates density and urban structure.

The locations of cities have been determined by a combination of the transport network and the spatial distribution of natural resources.

The economic basis of manufacturing location was first seen in terms of minimising transport costs but is now primarily dependent on labour costs.

Bibliography

Balchin P, Isaac D and Chen J (2000) *Urban Economics: A Global Perspective*, Palgrave, London.
Button K (1976) *Urban Economics: Theory and Policy*, Macmillan, London.
Lever W F (ed) (1987) *Industrial Change in the United Kingdom*, Longman, Harlow.

3 Spatial agglomeration

Objectives

The existence of cities begs questions about the underlying reasons for the spatial agglomeration of economic activity. It implies that there are economic benefits to firms forming spatial clusters. Similarly, there must be advantages to people living in urban areas. Addressing these subjects leads to the reasons for the continuing rise of cities and urban growth across the world.

The aim of this chapter is to consider why cities exist and grow by looking at the underpinning of an urban economy. In particular, the chapter examines:

- Role of agglomeration economies
- Types of urban agglomeration economies for firms
- Urban agglomeration benefits for households
- Why cities grow?
- Scale of urban development.

Role of agglomeration economies

As cities grow, there is by definition a spatial clustering of firms and households. This clustering is important to the economics of cities as it brings benefits through what are called agglomeration economies. Indeed, it is best to see these agglomeration economies as driving the clustering rather than the other way round (but see Chapter 19).

This section focusses on the nature of agglomeration economies for firms while the household perspective is considered later in the chapter. The overarching benefit of agglomeration economies for manufacturing firms is cost savings. In other words, it is cheaper for firms to produce goods within an urban agglomeration, i.e. a city, than say in a rural area. It is also true for services. Similarly, sales revenues of retailers are higher in urban areas than in low-density areas. Agglomeration economies are therefore at the heart of the reason for cities.

Agglomeration economies as identified above through the clustering of firms are really economies external to a firm as opposed to internal economies. In the latter, a firm could generate internal economies in a number of ways. First, a large plant could be built generating economies of scale by mass production. This is known as a horizontal economy.

Alternatively, a firm could generate economies by locating plants producing different goods next to each other. This is known as economies of scope, namely one firm

DOI: 10.1201/9781003027515-4

Table 3.1 Types of Internal and External Cost Economies

Types of Cost Savings	Internal Economies	Agglomeration Economies
Scale	Horizontal	Localisation
Scope	Lateral	Urbanisation
Complexity	Vertical	Complex economies

could produce two related goods cheaper than two separate firms producing the same individual goods. This could occur because the larger firm could make managerial/coordination savings or use design expertise common to both products. Large plants may also bring together the various stages of a complex production process together at one location, a process known as vertical integration.

To recap, there are three types of internal economies for a firm:

- Horizontal economies of scale – *costs falling with output*
- Economies of scope – costs falling through a diversity of products
- Economies of complexity – costs falling by combining processes vertically in production.

These types of economies occur where there are large individual industrial conglomerates at one location and are summarised in Table 3.1.

Although there are internal economies from a firm locating at one site, there are only limited examples of 'company towns' where an urban area has grown up around one large plant. Most of these are linked to mining, steel manufacture, aluminium smelting and car manufacture. An example is Wolfsburg in Germany, the home of Volkswagen. These towns often have only a relatively small population.

There are numerous examples of cities with one dominant employer but invariably these cities grew up not entirely reliant on this one employer, but rather evolved to this state. Furthermore, this dominance tends to be transitory even if it lasts for decades, for example the electronics company, Philips, at Eindhoven in the Netherlands. These internal economies in general on their own do not support the establishment of a large urban area.

To look for the primary reasons for cities, it is therefore necessary to look to external or agglomeration economies. The internal economies outlined above have parallels with different types of agglomeration economies. The concept can be decomposed into three inter-related elements: localisation economies, urbanisation economies and complex economies. These are now explained in the next section, with their broad relationships to internal economies as shown in Table 3.1.

Types of agglomeration economies for firms

Before looking at the detail of different types, it is useful to formally spell out what an agglomeration economy is, namely:

- a lower unit **cost** and higher **efficiency** for a firm because of the concentration of economic activity

    or
- higher **revenues** for a firm consequent on clustering, urban size or the concentration of economic activity.

The analysis here focusses only on cost economies as the relationships between revenues and clustering and size are considered in Chapter 4.

Localisation economies

These agglomeration economies are associated with the concentration of like firms or firms from the same broad economic sector. This clustering of firms in the same business area enables them to draw upon firms that offer specialist services that reduce costs. Specialised services include repair, maintenance and auxiliary trades.

These specialist subcontractors are in turn viable because of the scale and concentration of customers. It means that these services can also be provided at a lower cost because of the economies of scale in these subcontractors.

The concentration of firms in the same industry is likely to enable it to draw on a large specialist-skilled workforce pool. These skilled workers maybe have been attracted by the range of job opportunities in their locality. Alternatively, this concentration of businesses may have stimulated the provision of specialist education in local colleges, reinforcing and improving this pool of skilled labour.

The concentration of firms in the same sector is strengthened by offering a spatial focus for clients and interrelated industries and possibly advantages for industry-wide marketing. There may be specialist logistics carriers offering lower freight rates on outputs and inputs. There are also possibilities for cooperation between firms in the same sector in terms of research and development activity.

Even if there is no cooperation, physical concentration and personal contacts ensure the rapid transfer of information within the industry. This is known as knowledge spillovers across the whole production process and crucially encompasses the transfer of product innovations, by for example recruitment from competitors.

Agglomeration economies of this type can be realised in cities with dominant industries. Historically, it can be seen in textile production in the Manchester area of the UK and iron and steel in Pittsburgh in the United States. Well-known modern examples include hi-tech software in Silicon Valley and film-making in Hollywood. In South Korea, the city of Ulsan has the world's largest car assembly plant and the world's largest shipyard. Perth in Australia has a focus on mineral mining. But, these are just a few examples of cities benefitting from localisation economies.

Urbanisation economies

These economies relate to cost savings from the general concentration of economic activity. Individual firms from many industries benefit from sharing a general business infrastructure within an urban area. These include the availability of specialist business services such as accountants, lawyers, property consultants and even basic services like hiring photocopying machines. The shared use of inputs with other diverse firms who are also unrelated reduces the cost of these services.

A city also offers a large labour pool to draw upon when hiring staff, compared to say a rural area. This labour pool will not necessarily be specialist skills (in terms of localisation economies) but more generic business or soft skills. These are transferrable from one industry to another such as management and administration. These skills could be promoted by local colleges and universities. There are often local recruitment agencies. The ability to easily hire suitable labour when required by reducing the time taken to fill vacated or new positions lowers costs.

There are a wider range of real estate properties to choose from in a city. This means that it is easier for a firm to find an office or an industrial unit to meet its requirements. Competition between landlords should also lead to lower rents than in small settlements where landlords could have monopoly power. In turn, this should lead to reduced real estate costs for business.

Location in a city also permits the sharing of public amenities such as universities, transportation services and other elements of the common urban domain. A large city often has an airport nearby or is on a mainline rail link to facilitate business travel. Similarly, freight costs can be reduced by proximity to container terminals.

Reduced risks to business can also be seen as part of urbanisation economies. Uncertainty can be lowered by locating in an established centre of economic activity. New firms in particular benefit from urbanisation economies as the setup risks are reduced by the low input costs including real estate properties in rundown areas and an immediate market nearby. It is easier for the assets of business failures to be recycled. Similarly, real estate development is less risky in a city than in a rural area.

In general, the benefits of urbanisation economies are greater the larger the city as there is an enhanced scale of firms providing services and inputs. Greater urban concentration may also stimulate new services to be developed. Similarly, the more diverse the city the benefits are likely to be greater by providing a wider range or scope of services, thereby enabling a better matching between firms and suppliers. Larger cities also have a greater range of educational institutions such as universities and can justify and support excellent transport links with other cities.

To summarise, urbanisation economies are external to the firm but internal to the city. The economies revolve around the shared use of inputs with other firms. They are a function of the scope (diversity) of the urban concentration, which in turn is closely correlated with its size. The benefits are very broadly defined. They are particularly important to small firms.

Activity-complex economies

A final form of agglomeration benefit is activity-complex economies. These economies can arise when an industrial sector of a city can be viewed as one large producer with many individual firms being part of a sequential production process. Within this process, an individual firm has backward linkages in terms of inputs from other firms and forward linkages to customers who may be another set of firms or households.

The combining of the industrial process in a vertical chain of individual autonomous firms is an activity-complex economy. There may be a number of these chains in different sectors of an urban economy. There are two forms. First, the process outlined above with firms engaged in distinctive processes or stages of production. Second, individual firms supply different inputs to a firm for a final assembly.

In the case of manufacturing, for each constituent firm in the production chain, the proximity to other firms is a great advantage by reducing transport and inventory costs. The need for inventories is less, thereby permitting 'just-in-time' manufacturing in some cases through cooperation between firms. This cooperation enables cost savings through efficient information flows and the ability to coordinate activities between firms. In this way, input supply issues are minimised.

Activity-complex economies are also important in the finance sector. The central business district with its co-location of banking, financial institutions and associated business can be seen as an activity complex–localised economy. Face-to-face contact is an essential element in the business processes located there. They support backward and forward linkages to develop a range of final products including pensions, insurance and investment vehicles. The labyrinth of these processes may not fit the simplified manufacturing activity-complex economy set out above, but it is essentially the same phenomenon. The technology, media and telecommunications sector are probably another example of an activity-complex economy. In both cases, co-location and face-to-face interaction reduce the costs of products.

Commentary

Agglomeration economies arise in a number of different ways. The three classes of agglomeration cost economies highlight the various benefits of cities for firms, but they do overlap and it is very difficult in practice to separate them. It is not possible to identify with precision which agglomeration economies a particular firm benefits from. Even if some firms in the same industry co-locate together independently, say to be near raw materials, inevitably localisation and urbanisation economies will develop. The picture is even more complicated because firms may also benefit from the additional revenues generated by serving a large urban market.

Despite the ephemeral nature of agglomeration costs, it seems that the cost saving from agglomeration is high. This conclusion can be drawn from the fact that firms are prepared to pay a premium to operate in cities. This premium takes the form of higher wages for staff and higher rents for offices and factories. Given that the agglomeration benefits increase with urban size then so should these premiums. Indeed, the central office costs in global cities like Hong Kong, London, New York, Beijing and Tokyo can be the order of three times those in small provincial cities in the UK and the United States.

While this chapter has focussed on the benefits from spatial agglomeration, cities can grow to a point when there are diseconomies. The costs of production will begin to rise, caused, for instance, by traffic congestion, pollution, excess demand for skilled labour, etc. These issues are considered in Chapter 16.

Urban agglomeration benefits for households

The benefits to households of cities relate to both living and working. The scale of the labour market means that there are many work opportunities, hence choice about the type of employment. It reduces the search time for workers to find jobs and permits people to change employers easily. There is the potential to achieve work satisfaction, employment that matches a person's qualifications, or find a job that meets aspirations, for example with better pay. All this can be achieved without having to move home.

The range of employment opportunities in a city also mitigates the possibility of unemployment. It reduces employment risks by enabling a move to alternative employment if a worker is made redundant. A diversified employment structure in a city also reduces the industry-specific risk to the workforce so that a downturn in one industry still leaves opportunities in another sector of the urban economy.

Large spatial concentrations of people contribute to the quality of urban life by ensuring the viability of a wide range of shops, including specialist retailers, amenities and cultural opportunities. Cultural opportunities can range from theatres to attending sports matches and events. Consumers are able to find specialist goods and personal services. The larger an urban population then also the greater depth and range of services provided by the public sector. Examples include universities and hospitals.

Young people, in particular, are attracted to cities by the opportunity to study, the job opportunities and the social life offered. The large range of restaurants, coffee shops, bars and nightclubs that are viable in a city enables a nightlife buzz and the opportunity to meet new people and have a wide circle of friends. Further attractions include concert venues and cinemas.

Overall agglomeration economies make cities attractive to firms and households. However, while these economies induce spatial concentrations of both, the benefits are not the same for households and firms. There is the potential for residential and industrial location/agglomeration forces to be differentiated. In other words, in theory, the distinctive benefits of cities for households and firms could lead to different pulls on the spatial patterns of location activity.

Why cities grow?

So far, the discussion of spatial agglomeration has focussed on why cities exist rather than how they grow. Economic growth, in general, is driven by technical progress and productivity gains, which in turn depends on research and the application of innovations. In this section, aspects of agglomeration economies are shown to contribute to this process, hence the growth of cities. These can be referred to as dynamic externalities and relate to how productivity improvements transfer from one firm to another.

At an urban level, the close proximity of firms and people offers the basis for social interaction that also brings the potential for cross-fertilisation of ideas and innovations. Knowledge transfer in this way can happen with people in the same industry or from different walks of life. For example, academics at a university can informally talk and propose ideas to colleagues from the same or a different discipline. Parallel processes, if not so obvious and more diffuse, occur in the wider city.

Within cities, localisation economies, that have been earlier identified, include benefits from spin-offs in the form of technology transfer within individual industries. This innovation transfer occurs through staff transfer, imitation and even spying. In addition, ambitious individuals may be willing to set up risky creative spin-offs or as employees take positions in innovatory firms, knowing that failure would still leave them work opportunities in the adjacent large industrial cluster. An influential study by Porter argues that geographical clustering accentuates competition and promotes this technology transfer. The cooperation inherent in complex active economies also inherently promotes knowledge transfer.

An alternative view on the dynamics of knowledge transfer was formulated by Jacobs in her book, 'The Death and Life of Great American Cities'. She argues that most innovation comes from what can be called urbanisation dynamic externalities, namely the transfer of ideas between different industries. From this perspective, diversity and geographical proximity are most important in promoting new ideas and generating improved productivity.

Both Porter and Jacobs believed that local competition between firms speeds up innovation. There is limited empirical evidence on the relative strength of these theories of urban economic growth but what there is supports diversity over industrial clusters. In both cases, firms within urban areas should grow faster than those at remote locations.

The importance of these dynamic externalities depends on the state of a particular industry. In any given city, there are new and mature industries, some that are growing and others that are stable or declining, just as technological change can promote urban growth by the mechanisms set out above so it can instigate the opposite. Clusters of declining industries can accelerate urban stagnation. In any city, there are industries expanding while others are contracting (see Chapter 9). Urban growth requires employment in the 'new' to outstrip the 'old'.

Urban growth requires migrants into a city attracted by the rewards. In some cases, agricultural workers switch from the surrounding agricultural/farm work. In-migration flows can also come from other countries. Cities have traditionally been the first destination for immigrants into a new country. The attraction is often underpinned by the existence of ethnic neighbourhoods where migrants can find kinship and support.

The scale of urban development

The chapter has examined the reasons why cities exist and why they grow. In this section, the scale of urban economic growth is examined, primarily for simplicity and the availability of data by population size. A useful point in time to start is the industrial revolution with the de facto functional (re)birth of many of today's cities. A convenient place also to start is the UK as the first country to industrialise.

During the 1800s, the population of Great Britain rose from 10.5 m to 37 m in 1901. At the same time, there was a general movement of the population from rural areas into the cities. Manchester is emblematic of this new industrial age as a city effectively a product of industrialisation. It increased its population six-fold in the 60 years to 1831, when its population reached 142,000. Indeed, in the decade leading up to 1831, its population rose by 42%. Over the following 50 years, it gained another 200,000, so in 1881 the population stood at 341,000.

Manchester was just one of the British cities that expanded dramatically during this period. In fact, it was the fifth largest city in Britain at this time after London, Liverpool, Glasgow and Birmingham in that order. Setting aside London as the capital city of the UK, these provincial core cities – Birmingham, Glasgow, Liverpool and Manchester as well as Leeds – continued to grow in the coming decades through to the declaration of World War II in 1939. In the process, they extended their boundaries, engulfing surrounding settlements. Birmingham and Glasgow had populations of more than 1 million at this time.

World War II led to severe bombing damage to the cities and not surprisingly falls in their populations. However, the 1950s saw the core cities continuing to lose population, and this trend accelerated from the 1960s. The reasons are discussed in Chapter 10 but for now, it can be noted that the nature of urban agglomeration was changing, and it has continued to change. Since the millennium, these core cities (and others) have experienced a revival although with the exception of Birmingham (just over 1 m) their current populations are much lower than at their zenith. Manchester's population for example is just over half a million, two-thirds of what it was in 1940.

Urbanisation arrived later in the United States. It proceeded slowly in the first half of the 1800s but quickened after the American Civil War, 1861–1865, with rapid industrialisation. Chicago, for example, increased its population by more than 20-fold between 1860 and 1910 from 112,000 to 2.7 m. Following a similar pattern to the UK, American core cities also began to lose population from the 1950s. Chicago's population has thus fallen from just over 3.6 m in 1950 to around 2.7 m.

There is a close relationship between the rate of industrialisation and urbanisation. The rapid industrialisation in Asia (Japan earlier), parts of South America and Africa in the latter half of the 1900s has led dramatically to the growth of cities in these regions of the world. During the period from 1978 to 2014, China's level of urbanisation, for example, tripled from under 18% to almost 55% in 2014, and its urban population grew by 558 million.

Unlike cities in western economies, cities continue to grow rapidly in the developing world and their populations are in general much larger and denser than in Europe and North America. In broad terms, the younger the city the faster the recent growth. The largest of the world's cities now have populations in the tens of millions depending on how and where you draw the boundaries.

The term, 'mega-cities', has entered the language and is generally defined as a city's population exceeding 8 m (some definitions use 10 m). There are at least 36 of these mega-cities in the world compared to 11 in 1975. Most of these mega-cities are in Asia, 22 in total, compared to three in Europe (Paris, Moscow and Istanbul) and two (Los Angeles and New York) in North America. South America also has a major concentration of mega-cities with Sao Paulo, Mexico City, Buenos Aires, Rio de Janeiro, Lima and Bogota. The number of mega-cities is expected to rise. While most cities in developing countries are experiencing rapid urban growth, some cities in Japan, South Korea and Eastern Europe are losing population.

Nevertheless, more than half of the world's population lives in urban areas, up from 43% in 1990. Of the largest world economies, the figures are 92.1% for the UK, 87% United States, 84.4% Canada, 82% France, 80% Germany, 74.6% Italy and 73.7% Japan. Countries with 90% of their population in urban areas include Brazil, Chile, Puerto Rico, Uruguay and Venezuela in South America and Qatar and Kuwait in the Middle East. These figures belie the fact that urban agglomerations can be very different across the world with most people living not in very large cities, but settlements with a population of less than 500,000.

Summary

Agglomeration economies are key to the existence of cities. Spatial clustering of firms and households brings cost savings and higher revenues. These agglomeration economies apply to manufacturing and services. Agglomeration economies should be distinguished from internal economies of firms by locating all activities in one place. These internal economies on their own rarely lead to a town. The primary reasons for cities stem from external or agglomeration economies. These can be decomposed into three interrelated elements: localisation, urbanisation and activity-complex economies.

Localisation economies relate to the concentration of firms in the same industry. Such clustering of firms in the same business sector brings a range of financial benefits from cheaper specialist services through to a skilled workforce to draw on and contribute to productivity. The close proximity of firms from the same business sector is likely to lead to knowledge spillovers and improved efficiency.

Urbanisation economies are almost the reverse of localisation economies as they bring cost savings from the clustering of different types of firms. These economies arise partly from many firms sharing the services available in an urban area. This shared use of inputs enables the cost of these services to be lower than otherwise. These cost savings also extend the advantages of having a large generic labour pool, enabling easy matching of skills to jobs. There are similar savings from the wide range of real estate, permitting the finding of the right property to meet the needs of a firm. There are also urbanisation economies from public amenities, while cities can contribute to lower costs by having good interurban transport links. Overall urbanisation economies increase with city size as that brings greater diversity of services.

Activity-complex economies occur when effectively an industrial sector of a city is one large producer with many individual firms being part of a vertical production process. In manufacturing, this would be part of a just-in-time process where proximity reduces transport and inventory costs. In the services sectors, a central city office location is important to permit these complex localised economies to flourish through a face-to-face contact, supporting backward and forward linkages to develop a range of final services.

The agglomeration benefits to households of cities are seen through job opportunities and the range of services that contribute to urban life. The large number of potential employers means that there are a wide range of jobs available so that people can develop a career or maximise salaries without having to move home. Cities also bring with them access to a wide range of shops, public amenities and cultural opportunities.

The growth of cities is fundamentally propelled by technical progress and productivity gains. Agglomeration economies support this growth by encouraging knowledge spillovers. The social interaction generated by the close proximity of firms across all sectors of the urban economy leads to the cross-fertilisation of ideas and innovations. It is supported by workers moving from one employer to another, bringing with them technological expertise. These dynamic externalities are accelerated by the local competition between firms engendered by diversity and proximity. However, it is important to remember that in any city there will be growing and declining sectors so that urban growth requires the former to overcompensate for the latter.

Urbanisation has gone hand in hand with industrialisation. Britain was the first to industrialise in the 1800s and led the rapid expansion of cities in the form they are today. Other countries followed depending on the timing of industrialisation. For example, the United States followed in the latter half of that century. This initial surge in the growth of cities in advanced economies stalled or reversed from the 1950s. There has been a subsequent recovery in the populations of many cities in these countries but, in general, it has not been sufficient to return them to their former size. Instead, the nature of urban agglomeration in these countries has changed to a more diffuse pattern.

With industrialisation, in general, coming later to Asia, South America and Africa, urbanisation was also later. This urbanisation has seen the dramatic growth of very large cities, an expansion that continues. This phase of urbanisation has seen the emergence of mega-cities, many of which are at very high densities compared to those in western economies. The experiences of urbanisation clearly vary across the world in terms of scale, density and timing. Overall, though more than half the world's population now lives in urban areas.

Learning outcomes

Agglomeration economies provide the basis for the existence of cities.

There are different types of agglomeration economies for firms, namely localisation, urbanisation and activity-complex economies. All three represent different aspects of why the costs of firms tend to be cheaper in cities.

Households benefit from living in cities because of labour market opportunities, the services available and cultural amenities.

Cities grow through technological change and efficiency savings that are stimulated by proximity effects that are enhanced by competition between firms.

Urbanisation is linked to the timing of industrialisation but its nature has taken different forms around the world.

The nature of cities is evolving as the degree of urbanisation across the world increases.

Bibliography

Jacobs J (1961) *The Death and Life of Great American Cities: The Failure of Town Planning*, Penguin Books in association with Jonathan Cape, Harmondsworth.

Jones C (2013) *Office Markets and Public Policy*, Wiley-Blackwell, Chichester.

Parr J (2002) Agglomeration economies: Ambiguities and confusions, *Environment and Planning A*, 34, 717–731.

Porter M (1998) Clusters and the new economics of competition, *Harvard Business Review*, 76, November-December, 77–90.

4 Spatial structure of towns and cities

Objectives

Cities contain a myriad of land uses competing with each other, yet there is often a common internal spatial pattern of land uses across urban areas. City centres are historically the main locations for large shops and offices with subcentres in secondary locations. These features suggest that there are shared explanations for these phenomena.

The aim of this chapter is to understand the spatial structure of cities and the reasons for the pattern of different land uses. To achieve these tasks, it develops a series of economic models that aim to simplify the real world by making assumptions. The chapter is organised in the following way:

- Agricultural land rent determination
- Intra-urban patterns of land use and rents
- The hierarchical nature of services and retail centres, subcentres and associated catchment areas

Agricultural land rent determination

Before the chapter examines the urban pattern of land uses, there are useful insights to be gained from considering how early economists viewed the determination of agricultural rents. This section examines the contributions of Ricardo in 1817 and von Thünen in 1826 to land rent theory, and they provide essential building blocks for the next section.

Ricardo's theory

David Ricardo's writing in England in 1817 was interested in explaining why arable land rental values varied. To do this, he made a number of assumptions, namely that transportation costs to the market were minimal and could be ignored as there was a network of market towns close by to farms. He presumed labour, and capital could be used to help grow crops. His analysis focused on the role of the fertility of the land in this process.

If all land is of the same fertility, then Ricardo argued that no farmer would be prepared to pay a premium (or rent) for a particular plot. From the farmer's perspective, they are all the same. There is no advantage to be gained from farming one plot over another.

DOI: 10.1201/9781003027515-5

However, farmers are prepared to pay rent for land that is very fertile because they do not have to spend money on fertilising it. In other words, rent reflects the costs of growing or crop yields. In fact, the rent offered for a plot of land is related to its fertility and hence its profitability.

Given a market for land, farmers compete for plots by bidding against each other. In this bidding process, farmers frame what they are prepared to offer depending on the fertility of each plot. Through this competitive process, the rent of individual plots reflects the relative fertility. In this market, the rent of the most fertile land is the most expensive, but farmers will not pay more than the increased income it generates in extra output.

More generally, the rent of plots of arable land is linked to their fertility and the income the farmers can make from them. The price of the most fertile land is also defined as a premium on top of the price of poor-quality land. Actual prices paid depend on the degree of competition between farmers. Individual farmers want to pay as little as possible, but competition means that the prices will be bid up for the best land plots.

There is a limit to how much farmers are prepared to pay as they need to make a profit. If there is perfect competition, then there are lots of farmers bidding, and full information on the fertility of plots, so prices are bid up until each farmer makes 'normal' profits. Normal profits are generally defined as the minimum required to keep a producer in its line of production. In this instance, a normal profit would mean a farmer could receive income to ensure a reasonable return on any capital invested. If there is sufficient competition, then all farmers receive normal profits whether they rent low- or high-quality arable land. This is because the differential fertility of the land is reflected in the range of rents paid.

The rent of land in this model is determined by how much income is generated by the agricultural produce. The higher the price of say corn, the more the farmers are prepared to pay in rent. The competition between farmers means that higher arable food prices lead to higher rents. Rising food prices do not benefit the farmers but the landowners.

von Thünen model

Ricardo's theory is a useful starting point to look at how rents are determined. However, by assuming transport costs are constant, there is no spatial dimension, and this is crucial to understanding how towns work. The next piece of the jigsaw is to review the von Thünen model of 1826 that specifically looks at the role of location in the pattern of agricultural rents. His model of rent determination assumes fertility is constant and instead allows transportation costs for farmers to vary.

von Thünen's thinking relates to Germany at that time. His model has a market town on a featureless plain that is the (only) marketplace for agricultural produce in the surrounding area. Farmers can therefore only sell their produce at this town. The cost of transporting any produce from the farm to the marketplace is assumed to increase with distance from it.

At the time, the transport was horse-drawn so the main cost is farmers' travel time. The spatial pattern of market towns is therefore important. Historically, in England, the distance between such towns ensured that one could always be reached by no more than a day's travel there and back. However, from a farmer's perspective, there are strong financial benefits from being close to the market rather than spending a whole day travelling each week.

As all agricultural produce is sold in the market town, the price achieved by the farmer is not dependent on the location of the farm. The same price prevails for all agricultural producers who are competing for the same customers. Put another way, revenue is constant, irrespective of the distance from the market town, so viability is dependent on location.

It is evident that the further a farm is from the market town, the higher the costs, and there is a point at which farming is not viable. This point is where revenue equals transportation costs and the farmer's normal profit.

In contrast, farms at locations close to the market town have low transportation costs and get paid the same price for their produce as farms further away. These farmers are prepared to pay a higher rent for the privilege. Just like Ricardo's theory, given the competition between farmers and full information, the spatial pattern of rents reflects the transportation costs associated with each location.

The rents for each plot are bid up through competition until any transportation cost savings are completely eroded by higher occupation costs, leaving normal profits. A competitive land market ensures that the profits from farming are the same for all farms, irrespective of location.

The presentation of von Thünen's ideas so far has assumed only one land use. But this assumption can be relaxed, so that different agricultural uses can be bid for land surrounding the market town, as shown in Figure 4.1.

Forms of agriculture with high transportation costs are likely to bid the highest for locations close to the market town. These include horticulture, dairy farming or producing timber. Further away from the market town, the land will be used for arable farming and or cattle rearing and other agricultural uses with relatively low-produce transportation costs. By relaxing the assumptions in this way, a concentric model of rural land use is developed.

Ricardo and von Thünen have similar mechanisms that determine rents. One sees agricultural rents as a function of fertility and the other transportation costs and distance from the market town. In both cases, there is a competitive bidding process based on how much profit they can make at a given location. Farmers look at how much profit they will make for each plot to determine the rent they are prepared to bid.

Breaking this down, they are taking into account the revenue generated minus the costs at each location. This can be expressed another way as the surplus that can be made by each farm plot, and which is available to pay in rent. This is an important element of decision-making on rents that is applied further throughout the book. von Thünen's model also provides the basic idea for the explanation of intra-urban patterns of land use set out in the next section.

Intra-urban patterns of land use and rents

A useful reference point to think about understanding patterns of urban land use is to note that the centre of a city usually has a (central) business district and the primary shopping centre. The centre also represents the point of highest rents in the city. Following on from these observations, this section considers a model of the wider spatial pattern of land uses over a whole urban area encompassing housing, industrial,

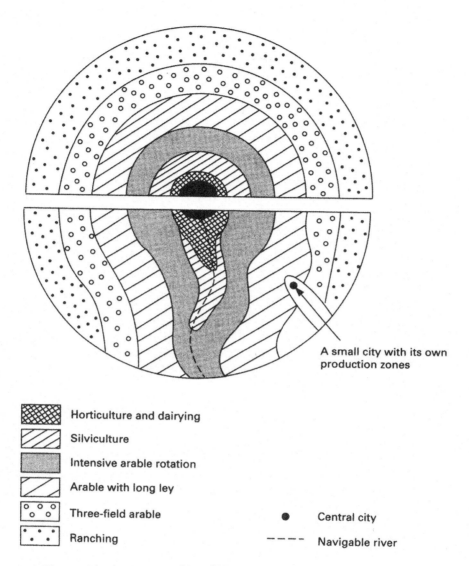

A small city with its own production zones

Horticulture and dairying

Silviculture

Intensive arable rotation

Arable with long ley

Three-field arable

Ranching

● Central city

– – – – Navigable river

Figure 4.1 The von Thünen Pattern of Land Use.
Source: Hall P (ed) (1966) *Von Thünen's 'Isolated State': An English Edition*, Pergamon, Oxford.

warehousing, etc. It examines why different uses are located and where they are. A key is the rent that each use is prepared to pay at different locations. The model derives a declining rent gradient for a city from its apex in the centre.

The model first published by Alonso in 1964 focuses on the rental value of the land. It is based on the view that the value of a plot of land is determined essentially by competition between potential occupiers for its use. It assumes that the owners of the land always supply it to the use that is prepared to bid the most. Before looking at how the land market works in the model, the assumptions are now spelt out.

This Alonso model creates a hypothetical city which lies on a featureless plain where the city centre is assumed to be the central marketplace. The key features of this hypothetical city are as follows:

- City centre is central marketplace
- Land is allocated to the highest bidder
- Featureless plain
- City centre is the point of greatest accessibility and the accessibility of individual locations are defined simply by distance from the centre
- Perfect information
- Free market in land
- Firms maximise profits
- Prices of goods other than transport and land are constant

Looking at these assumptions in more detail, the role of accessibility is crucial to the model. The centre of a town or city is generally the point of greatest accessibility as the road and public transport networks converge on it. All urban areas have such a point that at least some land users will be prepared to pay a premium for, for example to maximise the accessibility of potential customers.

As the city is set on a featureless plain, it does not matter to land users whether they are located to the east, north, south or west of the city centre. It also means that the analysis in the model enables the location to be defined simply by distance from the city centre. The prices of goods other than transport and land are constant so that the model can focus only on the spatial variation in land values with respect to location.

The other assumptions relate to the operation of the land market. Land rent is determined in the here and now. There is no past, so leases that set rent at some point in the past do not exist. There is a free market in land with perfect information. Firms are presumed to maximise their profits in choosing a location.

Land users calculate the rent they are prepared to pay for land at a particular location based on the financial benefits that it offers. Alonso's model uses the concept of a bid rent curve to demonstrate not only how individual firms decide on the rent they will offer for each location in the city, but also how their decisions relate to the wider urban land market.

Bid rent curves

Location within a city can be important to the profitability of a firm. Logically, the rent a firm is prepared to pay is maximised at the location that generates the highest profits. Profits are the difference between revenue and costs. It is the case that the pattern of costs and revenue will vary across the city.

Alonso took the view that revenue is maximised at the city centre and then fell away, while costs are minimised at the centre and rise with distance from the city centre. However, individual firms and land use sectors would have different spatial patterns of revenue and costs.

The role of location is most evident for the revenue of retailers. Variations in shop revenue are dependent on the accessibility of locations for customers. For many shops, the city centre represents the traditional point of maximum revenue. It is the most accessible for customers where passing footfall is highest.

A retailer is prepared to offer a high rent for a central location, based on the profits it can make, but a low rent at a suburban location with less footfall. This example shows that the rent bid at any given location is dependent on profits, defined as revenue minus costs. To be more precise, the surplus, the difference between revenue and costs, generated at any one location is available for paying rent.

In the case of other land uses, the relationship between location and revenue and cost is not so overt. For offices, the point of maximum profits is not necessarily the most accessible point in the city. Profitability is not so clearly related to location, but a prestige address in the city centre may be important for generating business. For industrial properties revenue is not likely to be influenced by location, and instead, the costs associated with locations could be more important. Households decide the rent they are prepared to pay based on the level of satisfaction and commuting costs from the centre of the city.

Based on the underlying surplus principle, a firm (setting aside households for the moment) can calculate what is the highest rent it is prepared to bid at every location. If all these different rents are mapped with distance from the city centre, then this is a 'rent bid' curve. There is a different bid rent curve for a particular level of profit. So, for each firm, there is a family of bid rent curves. These are graphically illustrated, assuming just three for simplicity, as br1, br2 and br3 in Figure 4.2.

The question is then which bid rent curve does a business want to be on? The answer is that the firm wants to be on the lowest possible bid rent curve because the lower the rent it pays the higher its overall profit. The bid rent curves can now identify where a firm should locate and what rent it should pay.

A firm is assumed to be a 'price taker' which means that they see the spatial pattern of rents in the city and choose an optimum location based on that. In Figure 4.2, this optimum location is where the firm's lowest bid rent curve is at a tangent to the rent gradient of the city. This choice of location also determines the rent the firm pays.

Bid rent curves, and hence rents, within Alonso's model are highest at the centre and decline with distance as shown in Figure 4.2. As firms in any one industry are similar,

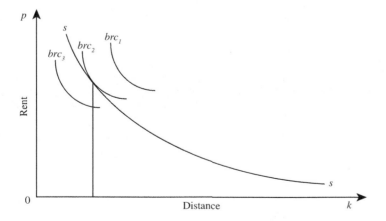

Figure 4.2 The Interaction of Bid Rent Curves with the Urban Rent Gradient.
Source: Jones (2013).

it is possible to think in terms of an average bid rent curve for an individual business sector. These different sectors may have distinct slopes to their bid rent gradients.

A set of average bid rent curves can be visualised, one for shops, another for offices and so on. For example, retailers on average have steep bid rent curve gradients reflecting the importance of central locations which bring passing trade. This potential trade falls quickly away from the city centre as shoppers are reluctant to walk to less accessible locations.

Office occupiers have a shallower average bid rent curve. Central locations bring prestige addresses, opportunities to conduct face-to-face business and access to a larger potential workforce. However, profitability does not rely on passing trade and so less central locations can still be valuable.

The profitability of industrial businesses is even less dependent on a central location, and so, they are not prepared to pay a high rent to locate there. Their average bid rent curve is likely to be the flattest of these three land uses. While households are assumed to prefer a central location to reduce commuting costs, they do not have strong desires, and their average bid rent curve is likely to be shallow compared with commercial and industrial land users.

Spatial equilibrium

The analysis so far has examined the locational decisions of individual firms, households and types of land users who make decisions assuming that the rent structure or gradient of the city is given or fixed. In reality, the urban rent gradient is actually the outcome of all the land users competing with each other. The rent gradient can be seen as the spatial equilibrium outcome achieved through the interaction between individual bid rent curves within the land market.

To consider how this happens, four average bid rent curves are visualised for shops, offices, industrial uses and residential purposes. Each distinct land use is associated with a different average bid rent curve as shown in Figure 4.3. Competitive bidding for land means that rents are bid up until each firm in each use can pay no more and stay in business (i.e. making only normal profits).

The resultant rent structure of land moving out from the centre to the urban periphery from these four land uses competing is given by the curve *abcd* in Figure 4.3. Land Use A can be seen as retailing which bids the highest rents at the centre, while Land Use B, offices, outbids other land uses just beyond the central core and so on. Similarly, C is industrial and D residential for whom accessibility is not so crucial.

It is important to note that the competitive bidding not only determines the rent gradient but, as Figure 4.3 demonstrates, also determines the pattern of urban land use. There is a series of concentric rings with retail at the centre because it has outbid other uses for this location, following by surrounding bands of offices, industrial and residential uses.

Figure 4.3 assumes rents fall to zero at the edge of the city, but this is a simplification. Land beyond the periphery is used for agriculture. It means that the lowest urban rent must be higher than what farmers are prepared to pay at the city's extremity. A further simplification of Figure 4.3 is that it only has four land uses. Different types of shops could have been included, and there could be a differentiation between office users, and other land uses such as hotels. The inclusion of more land uses would have given a more realistic picture but reduced the power of the message.

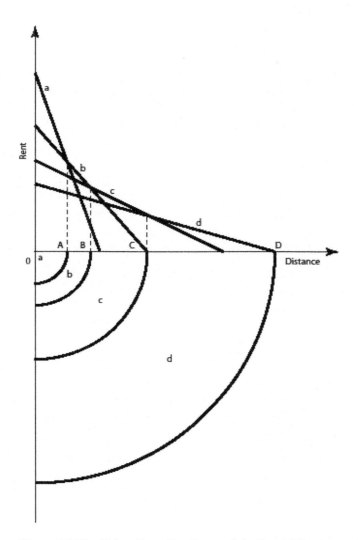

Figure 4.3 The Urban Rent Gradient and the Spatial Structure of Land Uses.
Source: Jones (2013).

It is also important to remember that it is not a land use that is competing with other land uses in the market but individual firms and households. Within the concentric band, aA, identified in Figure 4.3 where retailing outbids all other uses, there is competition between shops bidding up the rent of the most accessible locations. At the micro-level, this competition is seen within a local high or main shopping street by a negative rent gradient away from the most accessible point.

Alonso's model and the real world

The model has many assumptions both about how a real estate market works and the nature of a hypothetical city. In the real estate market, it assumes competition between land uses in towns is sufficient that any surplus (revenue minus costs and an allowance for profit) generated is fully translated into rent, as in the agricultural models earlier. In reality, there will be incomplete information, although property professionals will advise clients on what rent to bid. It is possible that not all the surplus will be bid away in the real world because of insufficient competition and information.

The operation of a free market, as in the model, leads to the use and rent of a plot of land being simultaneously determined. The highest bidder decides what to use the land for. This occurs because there is no planning to moderate the market. The role of planning and its impact is considered in Chapter 6.

Alonso's model at one level is a theory of how the urban land market works but its output includes the spatial structure of a city. The usefulness of a model and the insights it offers can be tested by its outcomes as well as the relevance of its assumptions. In terms of the predicted outcomes of Alonso's model, it is striking that despite the host of assumptions, it does present a picture of a traditional town with a concentric ring pattern of urban land uses. Rents too are highest at the centre and broadly decline with distance from the centre reflecting the importance of accessibility in urban areas.

Alonso set out his model at the beginning of the 1960s, and later in Chapter 10, its relevance for today is revisited. The one element of a traditional city missing from the outcomes is the lack of subcentres, and this is considered in the next section.

Hierarchical nature of services and retail centres and subcentres

Subcentres are an integral part of cities. They stem from the role of cities as service centres that can be traced back to their original functions as local markets. A typical city has many shopping centres within its bounds. There is a hierarchical pattern to these shopping centres. But subcentres are not just about shopping but services more widely, such as leisure amenities. These observations afford the base for the underlying economics presented in this section.

Providing services is a central function of a town or city. These services are provided not just to the immediate urban area but also to the surrounding places. From this perspective, a city can be seen as a central place for services. These services can be visualised within a hierarchical framework starting with basic or low-order goods and services such as newsagents and hairdressers. In contrast, specialised services (e.g. fashion shops, universities and hospitals) are higher-order services.

Christaller in 1933 published his central place theory that links centres to the services they offer in the hierarchy. In the system of central places, levels in the hierarchy have distinct functions aligned to each centre. The centres at the lowest level in the hierarchy offer a narrow range of low-order services. The next level of the centre would encompass an additional level of services and so on up to the top-level centre. This centre would provide the widest range and choice of goods and services including those of the highest-order services, but also all the lower-order services.

Beneath this pyramid of service centres, there is an economic explanation linked to household behaviour and the economics of the business that is easiest to understand by reference to retailing. The model is developed on a theoretical base that presumes a uniform plain on which transport costs are equal in all directions. On this plain, there is an evenly distributed population with equal incomes. In this world, sellers enjoy equivalent costs and free entry into the market and behave as profit maximisers.

On top of this structure, there are also a series of underlying behavioural assumptions. These include households shop at the nearest retail centre where the desired good is available, and shopping is undertaken via single purpose trips. The range or distance households are prepared to travel to buy particular types of goods vary.

Longer shopping trips are accepted for 'higher-order' goods that people buy less often. Necessities such as weekly food shopping trips would be undertaken locally involving only a short distance.

At the same time, retailers who are profit maximisers have spatial market thresholds required to make their business viable. To be precise, a threshold is the minimum demand for a commodity that would prompt a shopkeeper to operate its business. Expensive and infrequently purchases (comparison goods) require a higher threshold or catchment area than inexpensive or everyday (convenience) purchases.

Higher-order goods that people buy less often require the large market areas served by the top tier of the retail hierarchy. There are a high number of sellers of convenience purchases locally, while comparison goods are available only at the top or towards the apex of the pyramid.

Catchment areas

By combining the concept of range and threshold, spheres of influence or hinterland are defined. Comparison goods are only supplied from central places serving large populations, while convenience goods are sold locally in neighbourhoods. The different sized market areas can be linked into the hierarchy of centres by reference to a system of overlapping hexagons. Figure 4.4 demonstrates such a system with centres at three levels or orders.

The use of hexagons as market areas is just a convenient device to ensure all locations belong to a market area. It also enables the size of market areas/hexagons to reflect the centre's standing. Centre A, as a level 3 centre, is at the centre of the largest hexagon defined as the thickest line. The levels 1 and 2 hexagons/market areas are embedded within this large hexagon.

The application of hexagons is also useful to illustrate how a hierarchy of market areas could coexist. Their use is just one of a series of restrictive and unrealistic assumptions yet just like Alonso's model the outcomes are close to reality. Traditionally, cities have been characterised as having an internal hierarchy that incorporates the central place concept of a core retail area offering comparison shops. There are also a series of suburban high streets selling primarily convenience goods with neighbourhood centres offering a more limited range of goods and choices.

In such a hierarchy, it is presumed that say households visit corner shops for daily needs, neighbourhood parades for specific shops, district/town high streets for weekly shops and regional centres to purchase comparison goods such as clothing. This hierarchy has four levels, but it is possible to identify systems with more or less levels.

So far, the central place hierarchy has been applied just to retailing. But the theory applies more generally to services. As offices are used to provide services, so the theory can also be used to explain the existence of office subcentres. Within this perspective at the top of say a sub-regional hierarchy, the city centre offers business services to large companies. Small towns/suburban and neighbourhood centres provide business services to small firms and personal services. These local personal services include accountants, solicitors, financial advisors and estate agents.

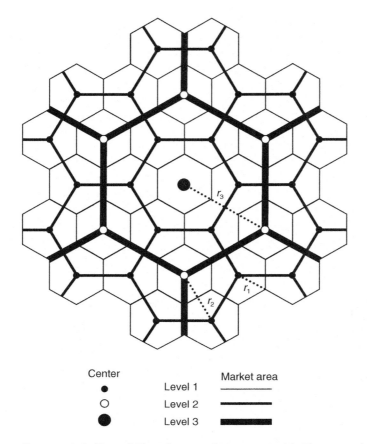

Figure 4.4 A Central Place System of Hexagons with Three Levels of Centres.

Criticisms and alternatives

There are many strict assumptions required to construct a hexagonal central place system. The real world is non-uniform, and purchasing power and the spatial distribution of the population are unevenly distributed. Nevertheless, it is arguably possible to adapt the central place principles. Instead of regular hexagons, market areas become irregular polygons.

Their areas are larger and centres more widely spaced where there is low population density or high purchasing power. Conversely, market areas are more compact and centres closer together where population and purchasing power are densely concentrated, most notably in urban areas.

The assumption of central place theory that everyone goes to the nearest shopping centre where the desired specific good is sold is much less easy to embrace and adapt. Further, shoppers undertake multi-purchase shopping involving goods of different orders.

Another school of thought, spatial interaction theory, proposes a more nuanced view that shoppers trade off the distance to individual centres and their relative attractiveness. In the world of the spatial interaction model, catchment areas of shopping centres are overlapping. Shoppers are prepared to travel past their nearest shopping centre to one that they view as more attractive.

The significance and nature of attractiveness vary with shopping purpose. Large shopping centres attract custom by the variety of different stores offering a broad choice of merchandise giving *economies of scope*. There is therefore scope for comparison between alternative products.

As no two consumers will have identical tastes and preferences, it enables retailers selling similar products (with different prices and qualities) to coexist. But attractiveness encompasses also the physical form and accessibility, including in advanced economies the availability of car parking.

The importance of economies of scope and the role of choice are dependent on the purpose of the shopping trip. There is a simple dichotomy between 'comparison' and everyday 'convenience' goods in this regard. Shopping for the former usually requires a degree of choice and visiting a centre of some size or agglomeration.

Convenience shopping is more concerned with proximity to home. A household's choice of shopping destination reflects the trip purpose, and while it generates a hierarchical or central place system, it is not necessarily defined by the rigidities of Christaller's formulation.

Summary

The goal of the chapter has been to understand the spatial structure of towns and cities drawing on simplifying theories or models. It began with a focus on the values of farming land as a building block to help ultimately understand cities.

Theories of agricultural rent determination were developed by Ricardo and von Thünen at the beginning of the 1800s. Ricardo's theory focuses on the role of fertility assuming that the other characteristics of land are constant. He demonstrates how market dynamics lead to farmers' rents reflecting the relative differences in fertility.

von Thünen assumes fertility is constant and instead lets transportation costs vary with distance from a market town. Applying the same competitive bidding process based on farmers' profits at a given location brings a negative rent gradient from the market town.

These models treat rent as a 'surplus'. It is the amount available to pay the rent on each farm after taking into account revenues and costs, plus acceptable or 'normal' profits. This surplus mechanism is also the basis by which urban land users determine the rent they are prepared to pay.

Understanding of urban spatial structure is given by a model developed by Alonso that is based on a hypothetical city which includes a featureless plain and a free market in land. Accessibility is key to determining land values in the city, and this is maximised at the centre.

Alonso saw revenue as generally maximised at the city centre and then fell away, while costs are lowest at the centre and increase with distance away from the centre. Given this pattern of revenue and costs, bid rent curves, and hence rents, within Alonso's model, are highest at the centre and decline with distance.

However, the need for and role of accessibility will vary by land use sectors, and hence so do the precise spatial patterns of revenue and costs. As a result, different sectors have distinct slopes to their bid rent gradients. Retailers on average have steep bid rent curves as they depend on passing trade that is highest at accessible central locations.

The profitability of offices is less contingent on the most accessible location, but a status address in the city centre could still lead to a preparedness to pay a high rent. In general, shops and offices outbid industrial and residential uses for the central area. These land uses have flatter bid rent curves.

The model assumes that firms and households choose their locations by reference to how much they can afford to pay in rent and the interface with the urban rent gradient that is observed. Specifically, their optimum location is achieved when the lowest bid rent curve meets the rent gradient.

However, the urban rent gradient arises from the collective market forces of all the land uses' rent bids. The outcome of this competitive bidding is a spatial equilibrium that establishes not only a rent gradient but also a pattern of urban land uses comprising a series of concentric rings.

Alonso's model incorporates extensive assumptions. While these assumptions can be queried, it does portray the basic land use structure and a typical rent pattern of a traditional urban area. There are inevitable limitations as the presumption of a free market excludes the role of planning.

An important aspect of a traditional city that is absent from the outcomes is the occurrence of subcentres. The nature of the city has also changed significantly in recent decades creating a more complex urban system, and this is considered in Part 2 of the book.

Cities can be viewed as comprising a series of subcentres with associated catchment areas. This aspect of cities is captured in central place theory with its hierarchical pattern of subcentres offering different levels of services.

In the theory developed by Christaller, there is a system of service centres with distinct functions associated with each centre in the hierarchy. His theory is constructed on a foundation of simplifying assumptions including a uniform plain that has an evenly distributed population with equal incomes.

In addition, there are a series of assumptions about how people shop. Households shop at the nearest retail centre they can, and shopping is undertaken in single purpose trips. People can travel long distances for higher-order goods that are bought less often. Necessities are purchased locally.

Retailers require a threshold level of demand to be viable that varies by type of good. Higher-order goods that people buy less often require the large market areas served by the top tier of the retail hierarchy. Christaller presents a model that combines these facets of shopping and retailing within a system of overlapping hexagons.

Just like Alonso's model, Christaller's model encompasses many strict assumptions necessary to produce a hexagonal central place system. The use of hexagons is only a device to present the ideas, and it is possible to develop a more generalised set of irregular polygons that accounts for variable population density and incomes.

The assumption of central place theory that customers shop at the nearest centre that has the good they demand undermines the application of the model today. It is now accepted that shoppers choose centres by balancing the distance to travel and their attractiveness.

The importance of attractiveness varies depending on the shopping trip purpose. Shopping for comparison goods commonly necessitates a measure of choice implying a trip to a large centre. Convenience shopping is generally associated with a short trip from home.

The differences between these two types of shopping trips in terms of distance and choice of destination inevitably generate at least a limited hierarchical or central place system that is not demarcated by the inflexibility within Christaller's model.

The two urban models considered in this chapter are incomplete in themselves. Alonso's model provides an insight into urban spatial structure based on competition between land uses. The concentric rings of land use that are generated by the model provide a powerful picture of a town, but it is incomplete.

It needs to be overlain with an internal central place system giving a hierarchical set of retail/services (sub)centres. The existence of these commercial centres implies that the rents would be bid up at these locations. The result is an urban rent surface rather like a mountain range with a dominant peak and many subsidiary summits.

Learning outcomes

Models of agricultural rent determination are the basis for an urban land use model.

Rent payable is generated as the surplus between revenue and cost, plus an allowance for profit is taken into account. In a competitive market, rents could be bid up until a tenant only makes just enough (normal) profits to stay in business.

Alonso's model of urban land use focuses on accessibility as the key to determining land values in the city, and this is maximised at the centre. Revenues are maximised at the centre and costs minimised. The model generates a land rent gradient that is highest at the centre and declines with distance.

The profitability of different land use sectors varies by location in the city and in particular the degree of accessibility to the centre. Some sectors value accessibility more than others and bid a higher rent for a central location. In general, shops and offices outbid industrial and residential uses for the central area.

Competitive bidding by land uses ultimately creates a concentric ring pattern of urban land use.

Overlain on this pattern are a system of commercial service subcentres. An explanation of the existence of these subcentres was first given by the central place theory developed by Christaller.

Central place theory derives a system of service centres with distinct functions associated with each centre in a hierarchy. Each centre has an associated catchment area that is defined in terms of overlapping hexagons.

It is possible to obtain a hierarchical or central place system of shopping centres based on the nature of shopping trips without reference to Christaller's model.

Bibliography

Alonso W (1964) *Location and Land Use*, Harvard University Press, Cambridge.
Jones C (2013) *Office Markets and Public Policy*, Wiley-Blackwell, Chichester.

5 Spatial structure of the housing market

Objectives

The focus of this chapter is on the spatial dimensions of local housing markets. It develops a theoretical basis for the internal structure of housing markets in cities. The initial section outlines an economic model that considers the key trade-offs that households make when they are making decisions about where to live in a city. The model is primarily based on that published by Muth in 1969. It provides an explanation of the spatial pattern of house prices and where (and why) different income level households live within a city.

In the following section, the role of neighbourhood differences in the housing market is considered by relaxing the assumptions of the model. In the final part of the chapter, the analysis is extended to view the concepts of travel to work and urban housing markets within a wider regional context. It considers how they can be defined in both a theoretical sense and in a practical way.

The structure of the chapter is as follows:

- Access-space model of the urban housing market
- Spatial patterns of house prices and incomes within cities
- Neighbourhood submarkets
- Urban functional areas

Access-space model of the urban housing market

There are two elements to the access-space model. First, a simplified hypothetical city is set out, and this provides a base for considering household decisions about where to live in it. The hypothetical city is based on the following assumptions:

- the town or city occupies a featureless plain, so any topographical features that might distort key relationships are ignored,
- employment is concentrated in the city centre, the central business district,
- travel costs are the same in every direction and are directly proportional to the distance travelled, and
- the prices of goods and services and taxes, including property tax rates, are the same wherever a household is located in the urban area.

The model, therefore, follows Alonso in Chapter 4 with a featureless plain and with travel costs the same in every direction. Taken together, these two assumptions mean

DOI: 10.1201/9781003027515-6

that a household is indifferent about which direction it lives from the centre, so a location can be simply defined by the distance from the centre. Travel costs are directly proportional to the distance travelled and households are assumed to make a fixed number of work trips a week. The cost of travel also includes the time taken and it is assumed to be related to income. Richer households are assumed to value their commuting time more than the poor.

Employment is concentrated in the centre of the city so that most households commute. A minority of people work in the suburban areas providing services such as local shops. However, as there are so few of them, their location choice decisions do not influence the price structure of the urban housing market. That is dominated by the choices and decisions of the majority, those who commute to the centre.

The model now adds some simplifying assumptions about household behaviour. In the real world, households choose between an infinite number of goods subject to an income constraint. However, in the simplified world of this model, households have only three choices on what to spend their income. These are housing, travel to the city centre or expenditure on all other goods. To make choices even simpler, housing is taken to be of uniform quality so the decision about housing consumption is simply about quantity.

In choosing a location, it is assumed that people do not like commuting. However, households do not locate simply to minimise commuting costs but also in relation to how much housing they wish to consume and the price they wish to pay. To understand what this means for an urban housing market price structure, consider how an individual household achieves locational equilibrium. The price of all other goods other than housing is constant across the city so they do not affect the decision. Just the balance between consumption of travel and housing determines the location and households need to be content with their choice.

A household weighs up the benefits of each location in terms of the cost of the housing per square metre they require and travel costs. For an equilibrium to occur, house price per square metre falls with the distance from the centre while travel costs rise. A household's equilibrium location is when savings in housing costs (for the quantity they desire) from moving slightly further out just balance extra travel costs. In this decision-making process, a household faces a trade-off between greater accessibility/lower commuting costs and higher housing costs at an inner-city location or high travel costs and lower housing costs at the periphery. This is why it is called the access-space model.

Spatial patterns of house prices and incomes within cities

Housing costs per square metre fall with the increased distance from the city centre while travel costs rise. It can be shown that for the equilibrium to be stable this negative house price gradient falls at a slower rate with distance from the centre and so takes the shape of a negative exponential curve as shown in Figure 5.1. Looking in more depth into this relationship, it can also be seen that how much households are prepared to pay for housing at any location depends on travel costs. The house price gradient is therefore linked to the level of travel costs. The lower the travel costs the flatter the house price gradient. Under a set of further assumptions, it can be shown that there is a negative exponential density function from the city centre.

The basic model, therefore, produces negative house price and density gradients outward from the centre of the city. In practice, this does not happen in every city,

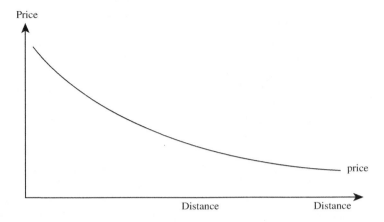

Figure 5.1 A Negative Exponential House Price Gradient.

simply because the central business district and central shopping centre outbid the use of land for housing at the centre as explained in Chapter 4. There is therefore a tendency toward the equivalent of a volcanic crater to these gradients in the real world with the central city having a residential density of zero.

Delving into the detail of the model provides insights into the spatial distribution of incomes in cities. Consider what happens if a household that is in equilibrium receives a significant increase in income, say because a member receives a pay rise. This will disrupt the household's equilibrium location and imply a move. Not only will the pay rise enable an increase in consumption but also increase the value of its travel time. Where is the new equilibrium location?

In deciding where to move, there are countervailing forces. On the one hand, the increased value of travel time would suggest a move nearer to the centre. On the other hand, increased income enables a household to purchase more housing. If there is increased demand for more housing, it would imply a move outwards, where the price per square metre of housing is lower. The model does not give us the complete answer on its own. However, in many advanced economies, housing is a luxury good, meaning that people spend an increasing proportion of their income on it as their incomes increase. For this reason, the desire for greater housing consumption outweighs the pull to the centre by an increased value of travel time, and households generally move further out with an income increase.

Another way of looking at this analysis is that two identical households except one with a higher income will have different equilibrium locations. The richer will live further out than the poorer and consume more housing. This can be generalised to the following observation: low-income households consume a small amount of housing at a high unit cost in inner high-density locations while high-income households consume a large amount of housing at a low unit cost per square metre in suburban areas.

These conclusions are based on the assumptions of a perfect market with no planning and represent a long-run equilibrium. This spatial structure of the urban market, in particular, is dependent on a uniform plain, the importance of city centre employment and a strong positive relationship between income and housing. In countries where real incomes are low and housing is an essential good for most of the population, the spatial structure may be very different. In these circumstances, the poor, many of whom are unemployed or self-employed and few working at the centre of the city, could be located at the periphery where housing is cheapest. This could take the form of informal settlements lacking public amenities.

Neighbourhood submarkets

The model has a series of significant assumptions designed to focus on understanding the spatial structure of an urban housing market and the important role of commuting. If the assumptions are relaxed, then more behavioural aspects can be introduced, both in terms of adding constraints and preferences. Financial constraints could include bank lending limiting or shaping households' choices in terms of location and quantity of housing purchased (see Chapter 7). Introducing different types of houses and an uneven geomorphological plain, in particular, can revolutionise household location decisions. If households have preferences for certain areas or neighbourhoods or house types, then this creates the possibility of supply constraints and premiums for particular areas. Households, for example, may wish to locate within the catchment area of a desirable school for their children and be prepared to pay a premium to do so.

Some households may also have a preference for city-centre living based on the social amenities and house types available. Certainly, there is a longstanding prevalence of affluent professional people locating in central city locations, such as luxury apartments in Manhattan, New York and town houses in the new town of Edinburgh and in European cities. There is also an element of preference for exclusivity in these location decisions.

Neighbourhood location decisions can be viewed as the result of preference and constraint. In fact, the distinction can be blurred. The concentrations of ethnic minorities in specific neighbourhoods of cities can be seen as resulting from the desire to be near similar people with specialist shops and religious centres nearby. But it also may result from discrimination, for example in restrictive bank lending to certain ethnic groups, so that such minorities find it easier to locate in these low-value areas. More generally, it can be seen as a special case of the poor constrained to cheaper neighbourhoods. The poor do not choose to live there but they cannot afford to live elsewhere. Constraint dominates choice. These spatial concentrations of poverty reflect the distribution of incomes in a city.

These neighbourhood differences are important in understanding the spatial pattern of house prices. They can be viewed as submarkets with either a negative or positive house price differential (premium) relative to the overall city gradient. Notwithstanding these divergences, statistical house price studies, that standardise for housing and neighbourhood characteristics, consistently find a significant negative distance decay effect from central urban locations. The implication is that the essential city-wide dynamic of the access-space model still holds under somewhat less restrictive conditions. The journey to work is therefore the key force in shaping local spatial housing markets.

Urban functional areas

The access-space model presumes a dominant city or town centre that represents the key point of accessibility and the major locus of urban employment. The urban housing and labour markets are implicitly the same. However, in many countries, the current pattern of settlements and commuting does not conform to these assumptions. There is not very often an urban system comprising a series of independent towns with separate commuting patterns.

In the real world, it is not as straightforward as the model. Administrative boundaries of cities do not equate to functional areas. A good way to visualise the cities in the spatial economy is as a web of overlapping markets or functional areas. These functional areas can be defined by spatial market 'flows', whether it be workers travelling to work or households moving home.

The spatial extent of commuting, defined by the longest distance commuters to the centre, can be seen as the functional boundary of a city or metropolitan area. At the same time, it represents the urban labour market because buyers (firms in the city) and sellers (commuters) of labour are interacting within that boundary to establish wage rates (prices). Very often this metropolitan boundary is also referred to as the travel to work area.

Following the access-space model, the urban housing market is constrained by travel to the work area and all buyers and sellers of housing live within these limits. However, the discussion above about neighbourhood submarkets suggests that there are subsystems of migration within cities. In general, households moving home go only short distances to adjust their housing consumption. Within these neighbourhood submarkets, there is a degree of migration self-containment so that the house price differentials discussed above arise from the interaction of localised demand and supply.

Not everyone will commute to a job in the city they live in. Similarly, not all house buyers move within an urban area. Indeed, some people will migrate between cities to take up a new job and buy a house. The city is not an island or a closed urban system. The functional areas of urban housing and labour markets therefore cannot be defined in practical terms by 100% closure rates or containment. Boundaries between spatial markets or functional areas are fuzzy.

A more appropriate commuting closure level is, say, 75%, namely three-quarters of households in an urban labour market commute within it. The actual number is to a degree arbitrary, but studies have found that the self-containment level is of this order. The higher the rate applied the larger the definition of the functional area. Following the logic of the access-space model, housing market areas are embedded within an urban labour market and empirical evidence suggests a migration closure rate of 50 to 60%.

Summary

This chapter considers the internal spatial structure of an urban housing market. It centres on the exposition of a model by Muth from 1969. Like Chapter 4, the starting point of the model is a hypothetical city on a featureless plain with the city centre as the point of maximum accessibility. Employment is concentrated in the central business district, so most households choose where to live by reference to their travel costs to the centre.

The economic model focusses on the key trade-offs that households have when they are making decisions about where to live in a city. To do so, it makes a range of simplifying assumptions about household choices. In fact, households have only three consumption choices – housing, travel to the city centre or expenditure on all other goods. The decisions facing households are made even easier by assuming all housing is of uniform quality, so the decision about housing consumption is simply about quantity.

A household chooses a location by trading off greater accessibility/lower commuting costs and higher housing costs at an inner-city location versus high travel costs and lower housing costs towards the periphery. The housing price per square metre falls with the distance from the centre while travel costs rise. Households choose their optimum location by balancing the benefits of each location to them in terms of the cost of the housing per square metre they require and travel costs.

Housing costs per square metre can be shown to follow a negative exponential curve from the city centre with the rate of decrease declining with increased distance. In developed countries where housing is a luxury, then the model explains that low-income households consume a small amount of housing at a high unit cost in inner high-density locations. At the same time, high-income households consume a large amount of housing at a low unit cost per square metre in suburban areas.

Relaxation of the assumptions to make the model more realistic enables an explanation of neighbourhood housing differentials. Neighbourhood preferences can explain why certain areas are popular and hence households pay a premium to live there. But, neighbourhoods with relatively low-average prices can be the result of income, poverty or ethnic discrimination constraints. Although there are neighbourhood influences on the spatial pattern of urban house prices, the overall framework is set by the trade-off between accessibility and cheaper housing at the periphery.

The access-space model defines the extent of a city as both the urban housing and labour markets, namely by how far people are prepared to commute and hence buy housing. This definition highlights that a city is a functional area but also suggests that they are physical 'islands', spatially distinct from other settlements. While administrative areas of cities have stark boundaries in practice, cities normally have blurred functional mergers into other settlements. Cities as functional areas often blend into physical contiguous neighbouring urban areas.

The spatial definition of functional city areas can be defined by reference to market flows. In this case, the focus is on commuting and migration flows. Given the impossibility of total urban self-containment in terms of commuting and migration, the choice of the level of closure is not based on theoretical considerations. The appropriate closure level for a functional area has been determined by empirical experimentation. For urban labour market areas, a daily commuting closure rate of 75% has been found to be broadly appropriate. Housing market areas based on periodic migration patterns have been found to have an approximate closure rate of 50 to 60%.

To conclude, the chapter has applied the access-space model to offer insights into the spatial structure of urban housing markets. By definition, models are simplifications and key assumptions include that workers are focussed at central locations, that there is no planning, and that the outcomes represent a long-run equilibrium state. Cities are not set in stone and employment is less focussed in inner urban areas now in many cities compared with the 1960s. The changing nature of cities and their impact on the housing market are reviewed in Part 2 of the book. The role and influence of planning is the subject of the next chapter.

Learning outcomes

A simplified model of the urban housing market that centres on the role of commuting can explain its spatial structure.

Housing price per square metre falls with distance from the city centre.

A household chooses an optimum location by trading off commuting costs against lower housing costs towards the periphery.

In developed countries where housing is a luxury, then low-income households consume a small amount of housing at a high unit cost in inner-city high-density locations. In contrast, high-income households consume a large amount of housing at a low unit cost per square metre in suburban areas.

In low-income countries, housing is an essential good and the poor could be located at the periphery where housing is cheapest.

Relaxation of the assumptions to make the model more realistic enables an explanation of neighbourhood housing differentials. Preferences for a particular neighbourhood can result in households paying a premium to live there. On the other hand, neighbourhoods with relatively low prices are likely to be the result of poverty/discrimination constraints.

The incomes distribution and the level of transport costs both affect the spatial house price structure of a city.

While there are neighbourhood variations within the spatial pattern of urban house prices, the overall framework is set by the trade-off between accessibility and cheaper housing at the periphery.

The functional definition of a city is primarily determined by commuting flows or more precisely how far people are prepared to commute. However, while the access-space model implicitly assumes a city is a physical island, the actual boundary of its functional area is fuzzy.

The appropriate closure level for a functional urban area has been determined by empirical experimentation. For urban labour market areas, a daily commuting closure rate of 75% has been found to be broadly appropriate. Housing market areas based on periodic migration patterns have been found to have an approximate closure rate of 50 to 60%.

The access-space model, as set out, has a number of simplifying assumptions that can be questioned in relation to urban forms in many cities today.

Bibliography

Jones C and Watkins C (2009) *Housing Markets and Planning Policy*, Wiley-Blackwell, Chichester, 2009.

Jones C, Coombes M, Dunse N, Watkins D and Wymer C (2012) Tiered housing markets and their relationship to labour market areas, *Urban Studies*, 49, 12, 2633–2650.

Muth R (1969) *Cities and Housing*, University of Chicago Press, Chicago.

6 Planning and the land market

Objectives

Land use planning is now accepted as an essential feature of a modern urban economy. However, it takes many different forms across the globe. It also generates much debate about its impact. As an intervention into land-use markets, there are inevitable questions about its effectiveness, efficiency and fairness. There is also often a heated debate in individual countries about the balance between market forces and planning regulation. In addition, while the concept of planning usually has a broad consensus, planning policies are often much more politically contentious.

In this chapter, there is a distinction between the generic goals of planning that is its starting point and specific planning policies. It also considers formally the relationship between planning and land use market forces, including its impact on land values. The final section assesses the extent to which policymakers can and should extract increases in land values brought about by planning decisions.

The structure is as follows:

- Functions of planning
- Real estate market implications
- Distributional implications
- Land value capture.

Functions of planning

There are alternative and overlapping justifications for planning. There is an economic argument for planning that focuses on what is called 'market failure'. The argument centres around the inefficiency of the property market given that it is very imperfect. These imperfections stem partly from the fact that unlike in the Alonso and Muth models (in the last two chapters) people do not have complete information about the complex market. Constraints may also dominate choice. There are a number of elements to this market failure, including:

- Externalities where costs of one firm are imposed on others without payment
- Failure to supply 'public goods'
- 'Unacceptable' market outcomes.

DOI: 10.1201/9781003027515-7

Externalities

The definitive example of an externality in the planning context is the impact of a factory with a smokey chimney, that is polluting the surrounding neighbourhood. The factory is effectively using a cheap production process that expels fumes out over land users nearby. It could have found ways to deal with this pollution internally but that would inevitably cost more. As a result, the factory does not pay the full cost of the production process. Instead, the pollution costs are in effect paid for by the nearby land users.

Another way to consider this is to distinguish between 'social' and private costs. In this example of the smokey factory, the production method is cheapest for the firm but it only pays its (private) costs, not the wider 'social costs'. If the social costs are taken into account, the production would be much lower or an alternative production process applied. Planning can insist on a production solution that addresses the social costs.

Other examples of externalities include the building of high skyscrapers that block a neighbouring building's light or view, new developments creating excess traffic noise or congestion, nightclubs in a residential area and houses in a natural beauty spot. Examples are endless but the problems they create are the most acute in urban areas because buildings and different land uses are closer together.

Failure to supply public goods

Left to its own devices, the free market is unlikely to provide the public infrastructure that a city needs to operate. This social infrastructure can be seen as a 'public good' because it cannot be easily charged for, as there are 'free riders', people can consume the product without paying for it. A true public good is publicly produced and consumed, in other words produced by the state and consumed by everyone.

This issue of the provision of public goods can be explored through the example of an urban park. This is arguably a public good as it is usually provided by a local authority and consumed by the public. However, in theory, it could be charged for as it is privately consumed by those who visit it. Indeed, there are examples around the world of access charges to parks, for example the Olympic Park in Beijing. But, these are exceptional and in practice, there are few privately produced parks reflecting the difficulties in applying an admission charge. It also undermines their purpose.

At the same time, residents overlooking a park may benefit from an attractive view that increases the value of their homes. In theory, the cost of a new park could be paid for by property taxes on new homes built on its circumference and the rise in values of other houses in the vicinity. This self-financing principle was applied to justify the building of Central Park in New York in 1856. In fact, such arguments held sway in the United States through to the 1930s. However, the funding of a new park in a poor area of a city is unlikely through increased property values.

An alternative 'solution' would arise if housebuilders build a large common open space and fund it by recouping the development cost by building houses nearby. These houses would have to be sold for very high prices to recoup the costs. This might also require a spatial monopoly for the developer and require the cooperation/intervention of the planning authority. It would also not be a universal solution to the building of urban parks.

A wider view of planning and the provision of public goods can be seen from two further examples. The countryside can be seen as a public good to be enjoyed by everyone. The best examples are the establishment of national parks with additional planning powers to preserve the environment. In particular, planning can be used to prevent the countryside being 'privatised' by house building. Such policies include the creation of 'green belts' around cities. Clean air is also an important public good and this can be achieved through 'smoke-free' zones where only special fuel can be burned in home fires.

Unacceptable market outcomes

As the property market is imperfect, uncontrolled development may result in buildings that do not meet society's needs. It might, for example, supply more shops than are 'needed' in a particular location or a large office development that there is no demand for. Following this line of argument, should developers be allowed to build housing/ offices that are unlikely to be let/sold because of where they are/poor quality? An example would be a long-standing empty office block built in an unsuitable location because it is not accessible for public transport. Alternatively, the market may produce insufficient supply for the community, e.g. a shortage of houses, parking spaces, etc.

The problem with these arguments is that planning solutions may also not be optimum. There are many examples of unsuccessful developments that have been promoted by the planning system. Planning intervenes in the market to address market 'errors' by using a range of measures that are principally passive. By using zoning or refusing planning permissions or applying density constraints, the planning system can control development. There are few proactive policies. Whether a passive or a proactive approach is undertaken by planning there is a fundamental issue about how can the planners decide what is needed rather than the market. This requires specially devised research tools.

More generally, society makes a value judgement and requires minimum standards for various aspects of the built environment linked to safety and public health, etc. for example so that people are not permitted to live in insanitary conditions. Incomplete information in the property market may mean that consumers are not always the best judge of their own interest. Decisions may not be based on complete information, for example someone building a house on the planned route of a road. Without planning controls, buildings/houses may be built next to hazardous sites, such as an oil depot or a fireworks factory and on contaminated land. There is a debate in many countries about whether housing should be built on flood plains that are subject to flooding. These views are arguably paternalistic but accepted as valid by the community.

These functions of planning can clearly be linked to practice, certainly to decisions about whether individual developments should happen, zoning and building standards. The function of planning can also be seen more as organisational activity. The land use market is therefore regulated by the state, and the planning framework creates order to enable it to operate efficiently. As an example, the market may produce an 'inefficient' spatial pattern of land use – in the sense of the time, it takes to travel around a town. Left to its own devices, the market develops the 'most accessible' land along the arterial roads. This would mean 'ribbon' development and elongated rather than compact urban areas.

This discussion about functions of planning ignores the fact that planning is more than just a regulation of the market. Planning began life in order to guide urban growth in the public interest in terms of where new development should be and incorporated the idea of model towns or garden cities. Today, planning policies are determined by politicians and their agendas, and to a degree by the values of planners. Governments from different political persuasions or in different countries may have distinctive perspectives on planning policies. Such differences could be seen in, for example, attitudes towards zoning, local externalities, or to the location of land uses within an urban area.

As part of its organisational function, planning can set out a strategic vision for an area. This could take a number of forms but could include a framework for urban change that incorporates designated land use patterns. In some cases, such strategic planning could be focussed on the expansion of a city or urban regeneration (see Part 3). Such planning can be proactive rather than reactive to market forces, with financial incentives to developers.

The potential wide remit of planning activities can be grasped by reference to policies aimed at urban sustainable development (see Chapter 16). Campbell in 1996 first considered this issue by reference to the 'planners' triangle' with sustainable development at its centre, as shown in Figure 6.1. The triangle exemplifies the balance planners have to make between environmental protection, economic development, and towards the third goal of social equity. Environmental protection could be at the expense of urban economic growth. The planning system has to address the tensions generated amongst these three fundamental aims that represent the points of his triangle. Inevitably, there will be different priorities and policies followed under the umbrella strategy of sustainable development.

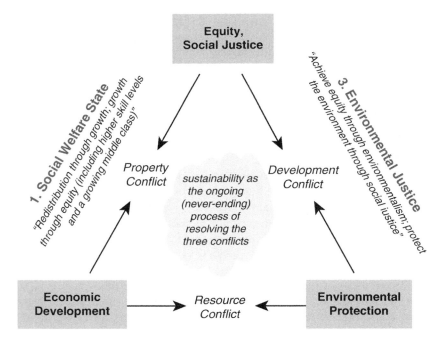

Figure 6.1 The Planners; Triangle.
Source: Campbell (2016).

Real estate market implications

Historically, planners have viewed themselves as physical or environmental planners and social engineers in terms of shaping neighbourhoods and cities or through creating new communities. The key problem for planning is that in a capitalist economy it is intervening in the urban land use market. Planners can be seen as shaping, regulating and stimulating the market. From this perspective, planners can embrace their role as market actors in order to manipulate the market to achieve social goals.

The planning system intervenes at many levels. At its most basic, it seeks to constrain the activities of individuals to control negative externalities. Planning intervention deliberately distorts the pattern of land values that one might expect to arise in an unfettered market. Yet, while there is abroad consensus that planning is an important and necessary element of a modern urban economy, there are arguably 'costs' as well as 'benefits'.

By definition, if planning intervenes to constrain the real estate market, then it distorts the equilibrium of supply and demand. It will also change the values of individual plots of land. Say if in order to promote economic development the planning system zones 'too much' land for industry, then this will lead to depressed industrial land values. If there are constraints on new city centre office developments brought about by strict rules on the conservation of existing buildings or the permitted floor area to plot size or building heights, then there are market consequences too. As a result of these planning constraints, demand is greater than supply and values of the existing offices increase over time while the low-density requirement could lead to an expansion of the area covered by the city. The result is distortions to the land rent gradient that is set out in Figure 4.3, although the fundamental spatial structure is unchanged.

Another example is designated green belts around cities that ensure easy access to the countryside for urban dwellers and also set physical limits to the urban area. The latter could be used to create a compact city (see Chapter 16). Such an urban containment policy acts as a boundary constraint on the market restricting the urban land supply. In terms of the housing market, it distorts the outcomes of the simple access-space model from Chapter 5. It is not simply a matter of creating a discontinuous price gradient as shown in Figure 6.2. Demand displaced from its 'natural' spatial equilibrium in the green belt is likely to seek close substitute locations. There is likely to be a piling up of demand on either side of the green belt as shown in Figure 6.3. Such an effect may be exaggerated if there are household preferences to live near the green belt.

More generally, the impact of planning, where it limits supply's response to rising demand, is the generation of higher land values that otherwise would have been the case. The market in turn will adjust the intensity of the use of urban land in the long term (see Chapter 13 that revisits this issue). In the short term, the increased price of land means that new development densities are higher than they would be without planning. Developers simply respond by using land more sparingly. For new housing, this means higher residential densities, in other words, flats rather than houses, or smaller houses or smaller gardens.

A further impact of planning is to slow development. Where development requires permission to build by definition, there is a lag that slows supply responses to demand. If new homes cannot be built without planning permission or rezoning,

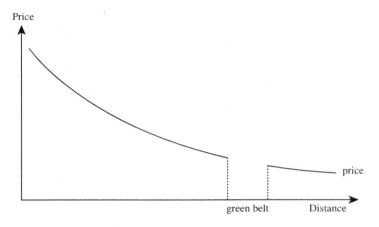

Figure 6.2 Discontinuous House Price Gradient Caused by Green Belt.
Source: Jones and Watkins (2009).

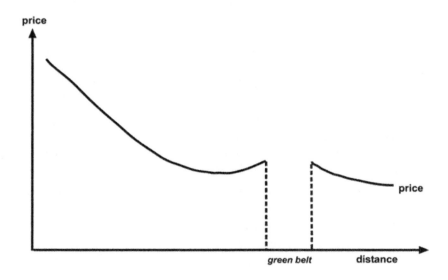

Figure 6.3 Distorted House Price Gradient Caused by Green Belt.
Source: Jones and Watkins (2009).

then supply necessarily lags demand pressures. Furthermore, positive planning responses to enable an expansion of new development may not be sufficient to meet
demand. There may be rational planning reasons for such decisions, for example
about social infrastructure constraints such as insufficient schools. Nevertheless,
it means new supply may be rationed and is slow to respond to demand, leading to
an increase in house prices. The time lag involved in planning, therefore, suggests
that the system tends to exacerbate price increases in periods of national and urban
economic growth.

Distributional implications

The land use market implications of planning in terms of increasing values are balanced by a broad consensus that planning of land use markets is essential to regulate market forces. There are benefits and costs from planning. The benefits in terms of addressing externalities and organisation of towns, etc. are set out at the beginning of the chapter. As noted above, costs take the form of increased land prices that, in turn, lead to higher urban densities, with less green space as it is built on and rising property values. On the one hand, the benefits can be seen as for the public good the costs are not uniformly distributed. In fact, the 'costs' in terms of higher land values can actually be a benefit to existing land and property owners.

Where there is a strong and expanding urban economy, then the impact of planning constraints has a greater impact on values and densities, especially on new housing. As population and incomes rise with a prosperous economy, the demand for housing increases, pushing up the prices and rents of the existing housing stock too. The negative impacts fall on lower-income groups and younger households. Households are rationed out of the owner-occupied housing market unable to afford the house prices or forced to live in a home too small or in a poorer condition than they want or are homeless.

Where planning is a strong restraining force then in particular land prices are much higher at the margins of these urban areas compared to agricultural land. This issue has been the subject of much research in the UK. A study in the late 1980s of the city of Reading exemplifies the impact. Reading was an expanding city at that time in the south-east of England to the west of London. Land in agricultural use near the city costs around £2,000 per acre compared with between £500,000 and £1 million once planning permission has been granted for housing. Thirty years on published government statistics report that in England as a whole (excluding London), agricultural land granted planning permission for residential use, would, on average, increase in value from £21,000 per hectare to £1.95 million per hectare.

These figures illustrate the impact of planning on land and property values. The discussion has primarily been in terms of the impact on the housing market but it could be extended to other land uses. The main beneficiaries of a highly constrained property market are existing property owners in these urban areas who make substantial capital gains. In one sense, they are paper gains for households as they will need to pay an equivalent sum to move to a similar house. But, they can move to say a cheaper rural area on retirement and realise their gains.

Land value capture

The impact of planning on land and property values raises important questions of equity or fairness. It can be argued that any rise in land values generated as a result of planning decisions is a windfall gain for the owner, whether it be through a change of zoning or via a planning permission. More generally increases in value can be seen primarily as socially generated, partly by planning and partly by urban growth increasing demand. The increase in value cannot solely be attributed to the efforts of the landowner.

Even before the advent of planning, there were forceful arguments that the growth of cities in the 1900s was creating increases in land values for owners who had done nothing to deserve the gain. Henry George put it in 1879,

> Our (land owner).... is now a millionaire. Like another Rip van Winkle he may
> have lain down and slept, still he is rich .... from the increase in population.

This led to calls for these land gains to be taxed. In fact, Henry George led a campaign called the 'Single Tax Movement' that advocated a tax on land value that would replace all other taxes and solve poverty. He stood for office on this manifesto for governor of New York in 1886 but just failed to get elected.

Rather than a tax on all land values, Henry George's ideas have been reformulated as arguments for taxing socially generated land values. In the UK, such a tax has been considered for more than a century, although policies designed to harness them were implemented only after World War II. Historically, the taxing of socially generated land values was referred to as the 'betterment' problem and more recently as 'planning gain' in the UK or now more generally as 'land value capture'. The arguments apply not just to urban growth and planning but also to increases in land values created by new transport infrastructure, particularly rail lines.

It is useful to consider the theory behind the taxation by taking the example when agricultural land is rezoned as housing. The uplift in land values arising from the planning decision to change the use of agricultural land into residential development could be argued to belong to the state as it was the state that conferred the right to develop. The tax is then on the difference between the agricultural land value and what it is worth for the building of housing.

This simple argument belies a number of underlying issues. First, what is the appropriate level of taxation? If all the increase in land value is socially generated, then there is a clear argument that the tax should be applied at 100% of the gain in value. However, there are cases where the increase in land value can be partly attributed to the landowner's efforts to make the land suitable for development by say improving access to the plot.

Second, unless the land is compulsorily purchased by the state, a landowner needs to be incentivised to bring forward the land for development. A farmer has to be persuaded to sell his land by a net of tax price that gives a value greater than its current agricultural use. A 100% tax would provide no incentive to give up his farming life. Together these issues imply that a simple 100% tax on increases in land values brought about by planning decisions is inappropriate. But the choice of percentage and mechanism has been the subject of much debate around the world.

International policy perspectives

Land value capture policies vary by whether a country has a public leasehold or a freehold land system. Land value capture is most simple to collect in public leasehold systems where all or most urban land is held on a leasehold basis from the state. Local authorities in China can receive the uplift in value from land identified for development by auctioning it for purchase by private developers. In Hong Kong, a lease modification premium is chargeable to the government for the acquisition of additional rights for land redevelopment. The premium is equivalent to the consequent enhancement in land value. The same is true in Singapore where public leaseholders may be required to pay a 'development charge' if they wish to increase the density of the land or change in its use.

In freehold systems, land value capture has been utilised by a number of countries to pay for the infrastructure associated with a development. Germany, for example, has a process that involves the purchase of land by local authorities prior to (re)development. The original landowners have their land returned to them on completion of the development scheme, but the authority retains the increase in value up to 30% for greenfield and up to 10% on brownfield land. The increase in value is therefore primarily to pay for the infrastructure costs but there is a potential for the community to receive a financial benefit.

An example of charging developers directly for infrastructure costs is the 'impact fees' in the United States that are one-off levies paid by property developers during the permit approval process. They are used for the provision of social infrastructure services, including roads, schools, parks and libraries, as well as water and sewers. They are not universally applied by local governments.

There are few countries in the world with freehold land systems that have sought land value capture directly by a proportional tax. The UK stands out as having a long history through a series of attempts linked to the granting of planning permission and dating back to 1948 as set out in Table 6.1. It is fair to say that all the attempts have their flaws and many were short-lived.

A number of issues arise from these attempts. The various approaches and mechanisms applied to taxing development gains demonstrate that there are considerable complexities to make what is a simple tax concept workable. Tax rates have varied as

Table 6.1 Evaluation of UK Land Value Capture Mechanisms

Scheme	Years	Aims	Design
Development charge	1948–52	Capture socially generated land values Lower land values through the threat of compulsory purchase to current use value	100% tax on increase in value from planning permission Paid by purchaser
Betterment Levy	1967–70	Capture socially generated land values Limit speculation	40% tax on the increase in value from planning permission Levied on seller
Development Gains Tax	1973–76	Capture capital gains from development Limit speculation	Tax levied on the sale of land and charged at an investor's marginal tax rate
Development Land Tax	1976–85	Capture socially generated land values to enable local authorities to pay for development at a low price	Land taxed at 80% of the difference between development value and current use
Planning Obligations	1990–	To pay for infrastructure required by development plus affordable housing	Site-specific negotiations between developers and local authorities Tax 'rate' is relatively low It has raised more than formal taxation regimes
Community Infrastructure Levy (only in England)	2010–	To help pay for general (not site-specific) infrastructure needs	National framework Standard charges set by local authorities in parallel with planning obligation

does who pays the tax to the government. Many of the initial attempts were short-lived because of a lack of consensus amongst UK political parties. Land value capture was introduced by a Labour Government only to be abolished by an incoming Conservative Government. As a result, landowners could defer development and paying any tax, by waiting for a change of government.

This analysis is a simplification as many of the early initiatives coincided with periods of low development activity. By the late 1970s, a political consensus did begin to emerge about land value capture if not about the tax rate to be applied. Eventually from 1990, there was a broad agreement over the use of planning agreements/obligations as a way of collecting some of the increase in value associated with planning. These agreements encompass obligations on the part of developers to provide the necessary infrastructure, as well as affordable housing units for the local community. Affordable housing obligations are expressed as a percentage of the housing to be built but are negotiated between developers and the local authority.

Planning obligations have been operative for over 30 years and so it can be argued that the approach has been the most successful in capturing land value uplift in the UK. However, there are many questions about the scheme, not least the role of negotiation and the rate of capture. Planning obligations have the lowest rate of capture of any of the schemes to date but it is unclear how much it is collecting (except the Community Infrastructure Levy that is collected in parallel). There is therefore an unanswered question as to whether the current obligation system is capturing a 'fair' proportion of land value uplift.

Who actually pays the tax?

Table 6.1 shows that the tax on land value gains at different times in the UK has been on the buyer and seller of the land. Most recently, it has been paid by the developer through contributions towards infrastructure and the provision of affordable housing. However, the purpose of land value capture collection by the state is a tax on landowners. In other words, landowners should not receive the whole increase in value generated by planning permission as it is socially generated and not the result of their actions.

The logic of how this tax works in a freehold system is universally dependent on developers knowing what their tax obligations are prior to land purchase. When the tax rate is set by central government and law, then the implication is straightforward. In the case of the obligations required of developers by a local authority, the decision is potentially more clouded. It requires the public/local authorities to set out a policy of exactly what is required of housing developers in terms of the provision of infrastructure and the scale of affordable housing.

Developers can then take the financial consequences into account when deciding how much to pay for the land. It is essentially a residual calculation by subtracting direct development and social infrastructure costs/tax from estimated revenue to give a land valuation. The developer should then take the obligations into account when deciding how much to pay for the land. As the policy applies to all potential developers, the cost implications are the same for all buyers. Competition between developers should lead to the land value being reduced by the cost of the obligations/tax.

The reality is more complicated, as illustrated by considering greenfield land. The landowner is not obliged to sell (unless compulsorily) and so needs to be incentivised.

A farmer may be happy to continue with his way of life and requires a significant premium on the agricultural value to sell. The farmer has market power in any negotiation if he/she does not have to sell. The developer is likely to therefore pay a lower price than the full value of the land for residential use but not the fully discounted price after the tax obligations.

The land is therefore sold above its market value for agriculture. However, the developer has to find a way of paying the rest of the tax obligations (not embedded in the land value) and still make at least a normal profit. The solution is to pass on the outstanding share of the obligation cost to the subsequent new homebuyers. This is possible because homebuyers expect to pay a premium for a new property and planning has constrained supply. Where this is not possible, then development may not happen as developer profits are reduced.

The analysis set out above also applies when developers purchase an option on land and then seek planning permission. Developers' financial calculations are essentially exactly the same. As such, the cost of any obligations required to obtain planning permission would result in a lower price being paid to the landowner in the event of the option to buy being taken up. The basic market dynamics also apply to 'inclusion zones' in the United States. In these zones, developers provide affordable housing as a percentage of the total, usually on a mandatory basis, albeit not formally linked to land value capture.

Overall, who pays the tax on land value increases from planning decisions is complex. Most of the impact of the tax should logically be transferred to the landowner rather than the developer. However, the balance of payment is a negotiation between the landowners and developers. Some of the tax may be passed on to the house buyer, depending on market conditions. There is also a possibility that land value capture reduces the incentive for land to be sold, and hence for development to take place.

Summary

Planning is an essential part of modern cities, regulating land use markets to ensure that the urban economy works efficiently and there is order to the inevitable and continuous process of development and redevelopment. In particular, it can address negative externalities and the provision of green space, moderate and control the scale and density of new development and avoid unacceptable low building standards.

Beyond these broad goals of planning, it is also driven by the values of national and local governments who set precise policies to follow. Much emphasis today has been placed on planning for urban sustainable development. Such policies have to triangulate environmental protection, economic development and social equity within a city. The interpretation and application of urban sustainable policies vary around the world and its challenges are considered in Chapter 16.

While planning traditionally has been seen in strict physical, social or environmental terms, it is fundamentally intervening and shaping land use markets. It 'distorts' market outcomes for the common good, but there are 'costs' as well as 'benefits'. The impact of planning is to limit and slow (re)development, and this brings higher land values that otherwise would have been the case. The market, in turn, will adjust the intensity of the use of urban land. In the housing market, this means higher residential densities, in the form of flats rather than houses, or smaller houses or smaller gardens.

Higher land values and higher house prices have a differential impact on elements of the urban population. Higher values benefit existing land and property owners by increasing their wealth. The negative impacts fall on lower-income groups and younger households who are not owner-occupiers of homes. Higher house prices create unaffordability with households being unable to purchase a home, or forced to live in overcrowding or in a poor-quality housing or homeless.

In a country like the UK where planning constraints can be severe in the most prosperous parts, then the difference between agricultural land values and reclassification for housing can be the order of 900%. This difference is a strong argument of taxing the uplift as it is socially generated by a combination of rising demand from population growth and planning decisions.

Taxing the uplift is relatively easy in countries with a public leasehold system where the state owns land available for development. It is more difficult in countries where there is a freehold system. There are questions about the appropriate rate of tax given that landowners' actions may contribute to the uplift in value and they need also to be incentivised to bring forward the land for development. There are different approaches around the world to such land value capture although many focus on using it to fund infrastructure associated with a particular development.

The UK, in particular, has a long history of attempting land value capture. One of the problems has been a lack of political consensus in favour of such a tax. It has meant that landowners could choose to defer development to avoid any tax and await a change of government. During the 1970s, a political consensus began to emerge and from 1990 the use of planning agreements/obligations was applied to collect some of the increase in value associated with planning decisions. Developers provide infrastructure and affordable housing as part of the agreement for planning permission. However, it is unclear what percentage of the uplift in value from planning is collected by this mechanism. The suspicion is that it is a low percentage.

The UK experience shows that there are many different approaches to taxing land value gains. If the developer nominally pays the tax, the logic is that it is passed on to the landowner by receiving a lower land value. For this to work, tax obligations need to be known prior to land purchase. A developer can then take the financial obligations into account when deciding how much to pay for the land. The difficulty in this reasoning is that a landowner is not obliged to sell at this price, and requires an incentive premium to sell. The result is that the tax is shared between the landowner, developers and the final customers, although the precise balance depends on market conditions.

Learning outcomes

The functions of planning are broadly to address market failures and provide better organisation of it.

Specific objectives include resolving negative externalities, ensuring green space in a city and avoiding unacceptable market outcomes. Such outcomes include buildings in the wrong place or constructed to an unacceptable low standard.

Distinct from the arguments for planning, specific policies vary depending on political beliefs.

While there is a general consensus in favour of planning, there are benefits and costs to planning.

Planning, as a rule, restricts real estate development, leading to higher real estate values and higher densities than otherwise.

Existing real estate owners benefit from the higher values through increasing wealth while the young, tenants and low incomes lose out.

The logic and principle of land value capture of socially generated land values are simple, but the practice of collection is complex in a freehold land system.

The tax on socially generated land values in practice is paid by a combination of landowners, developers and real estate purchasers.

Bibliography

Campbell S (2016) The planner's triangle revisited: Sustainability and the evolution of a planning ideal that can't stand still, *Journal of the American Planning Association*, 82, 4, 388–397.

Evans A W (2003) Shouting very loudly: Economics planning and policy, *Town Planning Review*, 74, 2, 195–212.

Harrison A J (1977) *Economics and Land Use Planning*, Croom Helm, London.

Jones C (2014) Land use planning policies and market forces: Utopian aspirations thwarted?, *Land Use Policy*, 38, 573–579.

Jones C and Stephens M (2020) Challenge of capturing socially generated land values: Principle versus practicality, *Town Planning Review*, 91, 6, 621–641.

Jones C and Watkins C (2009) *Housing Markets and Planning Policy*, Wiley-Blackwell, Chichester.

7 Urban housing markets

Objectives

This chapter examines the role of tenure structure and the **short-term** dynamics in the urban housing market. As housing is the dominant land use in a city, it is very important to an understanding of the workings of the urban economy. The chapter builds on Chapter 5 that considers the **long-term** spatial structure of the housing market, by focussing on the role of household mobility or migration within a city. Household movement is the key to understanding the housing market.

The chapter analyses how urban housing markets work drawing on evidence from different parts of the world and is ordered as follows:

- Housing tenure
- Choice and constraint, household movement in the housing market
- Affordability and affordable housing
- Financial constraints
- House price ripple effects
- Submarkets.

Housing tenure

It is useful to start with some definitions. The primary distinction in the housing market is between owner-occupation and renting. In addition, renting can be differentiated again between a tenancy from a private landlord and social housing. Owner-occupiers are households who own the legal title to the home in which they live. In a public leasehold system such as China or Singapore, households with very long leases are effectively owner-occupiers and can sell their legal interests if they desire.

The private rented sector comprises housing owned by landlords who let out their properties for profit to tenants. Tenancies are subject to leases that may prescribe a fixed time period. These tenancies are often regulated by laws that cover the extent of security of tenure for residents, housing standards and rents. The nature of these controls varies from one country to another, and in some there are none. Rents are generally subject to market forces but the rate of increase or the actual level can be controlled by the state.

Social housing is aimed at low-income households. It has taken many different forms with the original form being housing owned and subsidised by the state, for example in the UK. Instead of being directly provided by the state, such housing can be provided

DOI: 10.1201/9781003027515-8

by a public agency, a charity or a non-profit-making housing association. Tenants of this housing have to meet set criteria to be eligible.

In many countries such as the United States, there is no social housing but welfare housing on a very limited scale. These countries have a long-established private rented sector that represents a substantial proportion of the stock. In the United States, the state gives financial support to private landlords to house a limited number of low-income households. Welfare and social housing are sometimes referred to as affordable housing as discussed in Chapter 6. Forms of affordable housing are considered later in the chapter.

Tenure patterns in different countries are linked directly to housing policy, including any rent controls/regulation, the scale and allocation of social housing and welfare payments. The provision of housing by tenure is also dependent on the nature of the tax incentives and subsidies within a country. Some countries have a universal means-tested housing benefit/allowance available to all tenants. Nevertheless, many countries do not offer such a welfare safety net or social/affordable housing. At the same time, in many countries, property owners are able to claim significant tax relief on their expenditure, such as on their mortgage interest payments, and are not subject to capital gains taxation.

The dominant tenure around the world is homeownership/owner-occupation. There are very high levels of owner-occupation in Asia but numbers are unclear in Africa and South America. With the exception of Switzerland, in Europe, the majority tenure is owner-occupation. There are differences in the split between the proportions of social housing and private rented sectors. Social housing is important in many northern European countries although its role has been in decline through sales to sitting tenants. Where there is little or no social housing, then the private rented sector can be large, for example it accounts for more than a third of households in the United States.

It is important to remember that homeownership is considered distinctively in different cultures with implications for how housing markets work. For example, in southern European countries, households rarely move from their home which is often purchased with the help of the family. Often several generations live together under the same roof. In developing countries, many owner-occupiers adjust to their domestic needs by incrementally making improvements and adding extensions when funds permit so that a home is an evolving entity.

In the post-soviet world of eastern Europe, much of the public housing stock was privatised by giving individual units away to their tenants. The ownership of these properties is debt-free, but many are flats in multi-storey blocks and in poor physical condition. Homeownership in these circumstances comes with significant financial commitments to address repairs and is difficult to describe as an investment. In contrast, in many Anglo-Saxon countries, owner-occupiers are motivated by the investment attributes of homeownership. They seek to purchase at an early age and move to adjust their housing requirements through the family life cycle so there is a much more active housing market.

These examples illustrate that homeownership does not necessarily bestow benefits as an investment. For many in different parts of the world, it is primarily a home and simply a consumption good. There is also not necessarily a market in which homes are bought and sold. This is in direct contrast to the active housing market in most western economies, where households not only view homeownership as an investment but also as the principal source and store of private wealth.

Although the majority of households across the world are owner-occupiers, there is generally a higher percentage of private tenants in cities than in the rest of a country. A large number of American cities have more than half of their households as private tenants (based on administrative boundaries). Many African and South American cities are predominantly occupied by households who rent from private landlords although data are not readily available. Part of the reason is that the population of cities comprises a high percentage of young people who see the flexibility of renting. But many low-income households also live in cities and rent because they cannot afford to purchase.

Choice and constraint: household movement in the housing market

In Chapter 5, the analysis primarily considers urban housing markets by reference to the access-space model to explain the long-term equilibrium spatial structure of a city. The focus in this section is on the short term and moving home as a way of adjusting to a household's changing housing requirements. In countries where there is an active housing market, it is subject to potential constant change as households adapt to their circumstances of simply getting older, retirement or a change of job, or household composition.

At the centre of the workings of the housing market are the decisions households make about investment in and consumption of housing. These decisions, in turn, are based on the choices and constraints that households face. In terms of household movement, choices can be distinguished into two stages – to move or stay and where to move to. But this is a simplification, in fact, there may be a (prior) decision between moving and improving or extending.

In some developing countries, in particular, constraints on moving mean that the latter option is the only one available. Decisions are also influenced by attitudes towards housing as an investment that vary, with lower income groups tending to be less concerned with capital growth and seeing housing more as a consumption good.

The operation of the urban housing market is also set within the national macroeconomic context. As most households who wish to move to buy a home require a mortgage/loan, the owner-occupied market is significantly influenced by the availability and cost of finance from banks. The cost of finance, namely the interest rate charged to homebuyers, determines the monthly cost of repayments, hence what they can afford. Lower interest rates mean cheaper repayments and so households can afford to buy a more expensive and larger house.

The availability of finance is determined by two interconnected factors: the amount of finance available by banks to lend and by criteria used by them to award loans. Key lending criteria are the loan to house value ratio as that determines the deposit required and the loan to income ratio that determines the price ceiling that a household can bid for. The lower the deposit required the easier it is to buy a home.

The demand for owner-occupation is also a function of expected household incomes as this will influence what people feel they can spend on a house purchase, including paying back the mortgage loan. It means that in periods of economic stagnation real house prices tend to fall. Similarly, when an economy is booming and households see incomes rising, they are prepared to pay more for a house when they move. It translates into real house price increases. In fact, these three variables – interest rates, availability of finance and expected incomes – are the main short-term influences on the housing market and house prices.

These influences can be seen within the framework of a national house price cycle. The beginnings of a cycle and a rise in real prices can normally be partly attributed to a period of (initially) relatively low-interest rates, associated with the availability of mortgage finance and growth in real incomes.

Looking at the international house price boom around the millennium, it was preceded by a significant deregulation of mortgage finance in most of the industrialised world. It meant more flexible mortgage products that offered high loan-to-value ratios and longer terms were available. Together with falling interest rates and economic growth, these were the underpinnings of the international upswing in prices through the 2000s. The global financial crisis brought a drought in mortgage finance and a recession ending the house price boom, even though interest rates fell to almost zero.

Long-term influences on demand encompass demographic trends through the age structure of the population, the rate of household formation and migration. Land use planning constraints, as discussed in Chapter 6, influence the supply response to increased demand, hence house prices. This is a particular issue in the UK relative to other countries. However, in other countries with a weak planning system, there is the potential for overbuilding with consequences for the investment attributes of housing.

Another long-term influence on the housing market is the tenure structure. Owner-occupation and renting are substitutes, but traditionally in many western countries, households have rented while they save up a deposit to buy. Owner-occupation is generally preferred as a tenure because it provides more security of tenure, greater scope to adapt and an investment opportunity. Renting does have some advantages as it is accessible and offers greater flexibility to move that is attractive to young adults.

However, since the millennium the private rented sector has expanded in many western countries as tenants have found homeownership unaffordable and opportunities in social housing have declined. The role of the private rented sector historically has acted as a 'waiting room' for young adults as they wait to accumulate the capital to buy. However, the point at which young professional adults switch to owner-occupation is getting later and later in life. In fact, instead of a short stay in the sector, there has been a growth of long-term tenants.

Homeownership rates have fallen over recent decades amongst young adults in many countries as they are priced out of the owner-occupied market. There are a number of reasons. First, the boom in real house prices in the noughties led to the pricing out of young households from homeownership. Second, the global financial crisis then led to a mortgage famine as banks responded to their financial problems by requiring larger deposits. Third, the subsequent austerity policies in many developed countries have reduced real incomes, especially of young adults. To purchase a home, many have to draw on the resources of the (extended) family. Falling real incomes have also made it difficult to even afford to rent with the result that young adults are either staying at their parental home longer or moving back to it after a period of renting.

The unaffordability problem in western economies can also be attributed to a weak supply response to rising demand. Increasing demand is driven by a combination of the number of households seeking their own accommodation, and a rise in the urban population from inter-regional or international migrants. Part of this is simply the urban economy's ability to respond given the availability of land in a city. Part of the problem can be planning constraints as detailed in Chapter 6 and restrictions caused by bank lending considered later. In some countries, the risk aversion of speculative housebuilders has been blamed for an unwillingness to construct sufficient houses in order to maintain high profits.

In the developing world where there is rapid urbanisation, the impact of rising demand provides even greater pressures. The supply response of urban housing markets suffers not just from the scale of in-migration but also from a spectrum of impediments. In many cases, there are informal and complex property rights together with incomplete land registration. This creates difficulties for land ownership and the funding of new development. In addition, there are often unrealistic planning and construction standards for housing that can create further constraints on new development.

Affordability and affordable housing

The discussion above demonstrates that the causes of housing unaffordability are more than just the shortage of physical units of housing. It is partly also the culmination of macroeconomic forces such as high interest rates and falling exchange rates raising costs but also low real incomes and unemployment. It is partly the consequence of government policies towards social welfare, notably income support. It is partly the operation of the land market in terms of the role of planning and in some countries the weak land registration system.

At the household level, the issue of unaffordability at its crudest represents the importance of the income constraint on housing decisions. It has always been a crucial factor periodically in the housing market, but in many countries, it is now a long-term problem. As a result, housing affordability has become an important political and social issue. This section explores the concept in more detail and the challenges that it represents for public policy.

Defining affordability is not straightforward. In the previous section, affordability is primarily related to access to ownership but it can also be seen more specifically by reference to types of housing and different locations. Affordability, too, can be differentiated by specific groups of households, usually on the basis of income. In other words, affordability ranges from whether households can afford to buy a house to a more elaborate definition about access to different types of housing. Examples at the urban level include:

- Essential workers such as nurses or teachers unable to afford to live near their workplace.
- When young adults cannot afford to buy a home in the area they were brought up.

Affordability is a greater policy issue when it relates to more than low-income households. There are three main ways of measuring affordability. These are the ratio of housing price to income, the ratio of housing cost to income and the 'residual income' remaining after meeting housing costs. The first of these only applies to owner-occupation. The house price-to-income benchmark provides an indication of whether house prices are affordable in relation to incomes.

It suffers from being a relatively crude, if simple measure to calculate, because it takes no account of changing interest rates. The same ratio at two different points in time, when interest rates are 10% and 5%, implies the same level of affordability. However, taking into account the much lower mortgage repayments applicable, the latter is much more affordable.

The ratio of housing cost to income is a more direct indicator of affordability and often used by banks to make decisions about mortgage applications. It can be calculated by comparing mortgage repayments as a percentage of household income. Logically, for an individual house buyer, this initial percentage will fall over time as household income rises in line with inflation.

The cost-to-income relationship can also be applied to renting. A figure of 30% of income is often taken by governments as an acceptable upper ceiling for affordability. The higher the percentage of housing costs, the greater the housing cost burden becomes, and unaffordability is severe if this figure reaches 50.

The residual model is based on an assessment of whether the amount of disposable income remaining after paying housing costs, mortgage or rent still leaves the household with a sufficient level of income to live on. It can be argued that the residual income measurement is a more accurate measurement of affordability as the following example illustrates.

It is possible that a household with a high income may take out a large mortgage with monthly payments equivalent to say 50% of its income. On its own, this figure suggests severe unaffordability but because of the household's high income, there could be sufficient residual income to live comfortably.

The residual income measurement assesses whether, after paying housing costs, there is a sufficient level of remaining disposable income to provide basic needs, for example food, clothing, electricity, gas and other everyday basic costs. The residual approach, therefore, offers a more theoretically sound way of calculating affordability. However, it suffers from needing large data requirements and a degree of potential subjectivity in what are basic needs. There is no ideal measure.

Affordable housing

Irrespective of measurement, there is an unaffordability problem for policymakers if there is a gap between what a large number of households can afford and the market price or rent. The simple answer is the provision of affordable housing. The term originated in the United States in the 1980s but there is no precise definition and it takes various forms in individual countries. In some cases, affordable housing is aimed at the very lowest incomes, the unemployed and homeless. It can also be aimed at middle-income groups and key workers who are rationed out of homeownership.

Individual countries have approaches dependent on social and cultural norms. Here, the very different affordable housing practices of three countries are reviewed: the United States, UK and China. The United States is focussed on market-oriented solutions. Its main approach is therefore the 'Low Income Tax Credit' programme that has operated since 1986. The scheme accounts for 90% of affordable rented housing in the United States. It works by giving tax incentives/credits to investors on the condition that they provide a proportion of a development as affordable housing.

The total funding for these developments is limited and subject to an annual competition between investors organised by individual states. The criteria applied, therefore, vary by state but affordable housing in these developments is restricted to households whose income is say 50% or 60% of local average incomes.

Low-income tenants who rent this affordable housing are also charged a maximum rent based on a percentage of the average local income. The rest of the tenants in the development who are not on low income are charged market rents.

There is also a longstanding scheme by which means-tested choice vouchers are issued to help lower-income groups pay rent for private accommodation. Under the programme, households can use a voucher to lease a property and pay a proportion of the rent. It is designed so that most households pay 30% of their income, adjusted for the number of dependents and certain disability assistance and medical expenses. The means-tested assessment of tenant's resources includes capital in bank accounts. The rent charged must be a fair market/reasonable rent for the size and location of the property. Landlord's participation in the programme is voluntary.

In the UK, the definition of affordable housing is quite broad. As noted earlier, it can encompass social rented units provided by local authorities (known as council housing) and housing associations. It also includes 'intermediate'-rented housing with rents set at a maximum of 80% of the market value and low-cost homeownership including 'shared ownership'. Shared ownership involves part ownership by the household while renting the other part from a housing association.

From 1920 until 1980, local authorities were almost the sole provider of new social housing. From the 1980s, housing associations took over this mantle with a capital grant subsidy to build. Social housing in the UK is allocated on the basis of housing needs, which is a combination of household characteristics (excluding income) and the quality of existing housing. The criteria include overcrowding and poor housing.

From 1988, housing associations were able to access private finance in the form of mortgage loans to supplement the capital grants but only the grant counted as public spending. Rents went up to pay off the mortgage loans but were still at a sub-market level. These social housing rents have been affordable for the majority of low-income tenants because they also normally receive social security payments known as 'housing benefit' to cover the rent. The equivalent of housing benefits for low-income private tenants are housing allowances but the level of benefit does not necessarily cover the whole of the market rent. These payments are in the process of being replaced by universal credit.

Following the global financial crisis and public expenditure cuts, the government sought to achieve better value for money from its support of affordable housing. It decided to reduce the capital subsidy per housing unit to housing associations and thereby increase the rent. The new 'affordable housing' took the form of 'intermediate'-rented housing that as noted above charged rents at 80% of the market level.

In other words, a rent set at a level between social and market rents. However, most low-income households cannot pay this level of rent, especially as tenants on housing benefits were usually excluded. This intermediate-rented housing is best described as affordable housing for middle-income households.

Another affordable housing option targeted at middle-income groups is shared ownership. This scheme has been available for decades by housing associations and provides an opportunity for households to move in stages to homeownership. It permits a household to purchase a specific share of a property (between 25% and 75%) and to rent the remaining percentage, with a view to increasing their share of the property ownership over time.

The urban housing system in China has been transformed from primarily public provision to a free market in just over two decades. Historically, from the early 1950s, the majority of housing was provided by state owned industrial enterprises and public sector institutions for their employees. Since 1998, there have been major national housing reforms as there is a move towards a market economy. Amongst these reforms were the introduction of affordable housing for sale and the encouragement of commercial housing for the rich but the state is also providing housing for low-income households.

Initially, the government promoted the development of affordable housing in the form of owner-occupation for low- and middle-income households, known as 'economic and comfortable housing'. The land was provided to private developers for free who then built homes to designated standards. Development profits were constrained so that prices were attainable by low- to middle-income groups. There were also constraints on subsequent resale.

This programme failed to support sufficient low-income homebuyers or new young families. The state has turned to build public 'cheap rental' housing directly for low-income households. It is provided on a means-tested basis to qualified families for a rent determined as a proportion of the family's dispensable income. The rent subsidy is determined by the basic housing standard and the income level of the household.

These examples illustrate that there are alternative approaches to the provision of affordable housing in different countries. In some cases, affordable housing is directly provided by the state but in some cases, it is through a public–private partnership incorporating public subsidy. It is not primarily targeted at low-income households.

When affordable housing is aimed at middle incomes, it is a demonstration of the scale of unaffordability in a country. The examples are not exhaustive. It is also true that some countries have no affordable housing or a welfare system that gives income support. The absence of affordable housing is not a sign of a lack of unaffordability. In some countries, the provision of affordable housing for low-income households is not viewed as a social obligation by the state.

Financial constraints

One cause of unaffordability as noted above is financial constraints on the housing market that ultimately limit supply. These financial constraints can be seen through the lens of the crediting rationing criteria of banks. Earlier in the chapter, variations in lending are seen in terms of the loan-to-value and loan-to-income ratios a bank will offer a potential mortgagor. Variations in the terms available have been crucial influences on house prices over time, but other lending criteria are also important to the dynamics of urban housing markets.

A wider view of bank lending criteria incorporates assessments of the suitability of mortgage applications by reference to stability and surety of the household's income and the characteristics of their housing selected to purchase. The former relates to the ability of the household to pay the mortgage and creditworthiness, while the latter is associated with the attractiveness of the property as a surety for the mortgage loan. In particular, the bank needs to be sure of the future marketability of the home if it is sold in the future so any outstanding balance on the loan can be repaid.

These lending rules have been the subject of controversy in the past, especially in the United States. Banks in major US cities, certainly from the 1930s, have in the past discriminated against households of colour, denying them loans to buying a home in certain neighbourhoods. Mortgage lenders rejected loans for creditworthy borrowers based strictly on their race or where they lived.

As a result, geographical areas in cities were demarcated where loans would not be issued or only at a high risk of default. This became known as 'redlining' because of the red ink applied to identify the boundaries of these areas on maps. Such maps are not in general use today.

This issue is not confined to the United States. A number of studies in the 1970s reported on the existence of redlining in British cities. These areas were dominated by small properties in relatively poor repair in neighbourhoods with high proportions of privately rented housing.

A study of Glasgow at that time found that, while there were no redlining maps, the criteria used in the 1970s did explain the lack of lending in some inner-city areas. In other words, restrictive lending criteria based primarily on the characteristics of housing did translate into areas of cities that received little or no mortgage finance.

Over the next two decades, the housing markets in these types of neighbourhoods in British cities were absorbed into the normal lending realm of banks with the decline of the private rented sector in these areas, improvements in the stock and new buildings in these localities. A further factor has been the relaxation of lending criteria over time as signified by the evolution of 'subprime' lending.

Subprime lending originated in the United States. It uses a scientific approach to underwriting and pricing of mortgages based on credit scores and loan-to-value ratios, enabling lenders to shift down the risk spectrum. Households with weak credit ratings and histories can receive a subprime mortgage at a higher than usual interest rate to reflect the higher risk to the bank.

Many of these households are on low incomes living in inner-city areas. Households in the areas that were once redlined were now being offered subprime mortgages at a price. Unfortunately, mis-selling of these subprime lending products in the United States led to substantial mortgage defaults in the mid-2000s and eventually caused the global financial crisis. As a result, subprime lending has been heavily constrained, if not abandoned by banks.

The essential redlining problem has not gone away although spatial lending criteria may be less explicit. Banks are still concerned about the future marketability of properties in certain areas of cities. In Britain, for example, selling/buying a home in large former social housing estates where only a few properties have been sold to tenants is problematic. With minimal sales in these areas, there may be subsequent difficulty in selling and so banks are reluctant to lend. Similarly, it is said that banks in parts of the United States can be reticent in lending in black and Latino neighbourhoods.

Bank lending constraints are also a major problem for developing countries with extensive land subject to traditional customary ownership systems. Land acquired on this basis has no recorded legal title so owners have no formal security for a mortgage. It means that mortgage finance is often not available to most households because formal registration of land is a slow and expensive process. Low- and middle-income households are effectively blocked from accessing the formal housing market.

The result is barriers to new housing development as households have to finance building homes with their extended family resources usually on a gradual basis. One potential solution to this lack of funding is micro-housing finance that is tailored to this incremental construction process.

This embryonic approach has emerged in recent times as a way to meet the unsatisfied needs of the lowest socioeconomic groups. New institutions have been created that focus on providing small loans to support construction or home improvements. With the lack of collateral because of the absence of a legal title, documents such as tax receipts are often used as evidence to support the ability to pay back the loans.

Migration and spatial house price trends

The chapter now turns to household migration and its impact on house prices. It applies to cities where there are active housing markets in the sense that households periodically move home. Earlier in the chapter, the focus was on macroeconomic influences on the housing market, but the analysis now shifts to urban areas. In Chapter 5, it was noted that while an urban spatial housing market area is theoretically defined by the travel to work area in practice, the boundary is fuzzy.

Only about a migration containment defined by around 60% of housing moves represents a housing market area and there are subsystems of migration within cities that can be described as neighbourhood submarkets. Migration or household movement does not only determine the boundaries of housing market areas but given a fixed housing stock it is also the transmission mechanism that determines house prices in a local area.

The key message of this section is that owner-occupiers (or first-time purchasers) who move house and buy another are part of a market. In doing so and agreeing on a price for the next home, they are influencing prices in the market. The more a household is prepared to pay for a house the more it signals to the market as a whole that prices are rising.

The greater the number of households looking to buy a new home the greater the demand and the higher the prices. The reverse is true if the demand is falling because no one is moving or cannot afford to move. The same logic applies to the market for private rented housing, but the analysis here focusses on the owner-occupied sector for simplicity.

Given this migration transmission process, it is possible to explain why prices are different between areas and why the rate of change varies. It is useful as a starting point to remember that usually households move over only short distances unless they are changing employment too.

The result is that prices in a neighbourhood and a city are determined by the interaction of local demand and houses on the market from the existing stock of housing, plus those newly built. With this theoretical base, the chapter now considers two housing market phenomena – house price ripples and urban submarkets.

House price ripples

There are numerous examples around the world of regional or interurban house price ripples. Perhaps the most researched regional ripple is that in the UK that sees house price increases first in London and the South-East of the country and then transmitted first to nearby regions, such as the Midlands and the South West, then to the north of England and lastly to the peripheral regions. Similar ripples have been identified by researchers in the United States, China and Taiwan.

Why do these occur? The first issue is that regions of a country are administrative units and not functional housing market areas. It is more appropriate to explain the ripple phenomenon by reference to the operation and interaction of local housing market areas. There are two not mutually exclusive explanations:

1   The housing market ripple effect is caused by household migration across local housing market areas. This could occur for a range of reasons such as job-related moves or migration on retirement. Households moving from higher-priced

to lower-priced areas can afford to pay higher prices. The result is that high house prices in one housing market area are transmitted to another, and so on creating a ripple effect.

2   Variations in spatial house price trends are derived from the interaction of supply and demand within individual local housing market areas. This in turn is dependent on local supply constraints, household incomes, demographic trends and so on. Within this explanation, the transmission process does not occur via migration but by income.

Certain parts of the country are leading economies with higher incomes per capita. In the context of the UK, London and the South-East are say the first to see growth in the local economy as part of an upturn in the macroeconomy. Other parts of the country follow with time lags in turn leading to higher housing demand and house prices.

This ripple effect is also influenced by the variation in local supply constraints. Cities like London have a highly constrained land supply, so house prices are likely to rise quickly in response to increased incomes/demand.

Submarkets

The second explanation can also be seen as a way of explaining the existence of housing submarkets within cities. The existence of submarkets in cities can be seen where the same or equivalent houses (taking into account the distance from the city centre) have very different prices in different localities. They can also be seen in differential price changes in prices over time in different neighbourhoods.

The reasons for submarkets stem from many households moving only very short distances within a neighbourhood of a city. These migration patterns in turn reflect that many households when moving think only in terms of their local neighbourhood. As discussed in Chapter 5, this could be because of choices to be in a specific school catchment area or near friends and relatives or the local environment. It may also be because of constraints limiting their opportunities to a neighbourhood with cheap housing.

The result is a series of housing submarkets within a city in which localised supply and demand forces interact to determine price. These submarkets tend to be stable over time partly because any new housing built in these areas tends to be similar to existing housing. This stability occurs for a number of reasons – speculative housebuilders are risk-averse and concentrate on house types attractive to the existing population.

Where planning permission is required, authorities favour assuaging local residents' fears about new development, by supporting the construction of properties similar in character and design to the existing housing. New development, therefore, reinforces the submarket structure.

The stability of the housing markets is also reinforced by the high level of migration self-containment. Households' attachment to a particular neighbourhood, say in the west of the city, means that the existence of cheaper equivalent houses in the east does not necessarily lead to out movement. The result is that submarkets can have very distinctive house price levels and trends.

Summary

The purpose of this chapter has been to understand the workings of the housing market. Its starting point is the role of tenure. The predominant tenure in the world is owner-occupied housing. The remainder who rent are either tenants of for-profit private landlords or of public authorities/subsidised agencies. The latter are often called social-rented tenancies or affordable housing. Each tenure is taxed and subsidised in different ways in individual countries, and these fiscal measures frame how the housing market works and the attractiveness of tenures. In many countries, there is also a welfare safety net that at least partially supports housing costs.

Homeownership is not a uniform entity from one country to another, whereas in many western economies it is viewed as an investment in other countries it is simply a consumption good that is rarely traded. There are also differences in tenure patterns within countries.

Despite the high national proportions of owner-occupation in many cities, there are very high proportions of private tenants. In some cities, tenants are in the majority; the result of a combination of young people who want flexibility and poverty-constraining homeownership.

There are important macroeconomic influences on the urban owner-occupied housing market. In particular, expected future household incomes, interest rates and the availability of mortgage finance are the most important short-term drivers of the housing market and price cycles. Longer-term influences on house prices include demographic trends including migration, the role of planning and tenure structures.

There have also been changes to tenure patterns in many western countries over recent decades as housing market forces have made homeownership less affordable for young adults. Traditionally, the private rented sector acted as a 'waiting room' for young adults while they saved up a deposit to purchase their first home.

However, this process has got more difficult. There have been a number of reasons. First, young adults were priced out of the market during the boom of the noughties. Second, the immediate impact of the global financial crisis was to raise the deposits required by banks to buy. Third, the following decade brought falling real incomes for young adults.

The consequences are that homeownership rates have fallen over recent decades amongst young adults in many countries. The switch to owner-occupation is coming much later in life and there has been a growth of long-term tenants. In many cases, first-time purchasers need family financial support either directly or indirectly through staying at their parental home.

The unaffordability difficulties of would-be owner-occupiers is part of a wider problem of unaffordability that in part is the result of rising demand and an insufficient supply response. The inadequacy of the response is partly shortages of suitable land and (re)development constraints in cities. In the developing world, many urban areas also suffer from an informal customary land tenure system that creates difficulties for land ownership and the funding of new development.

Unaffordability is not a straightforward concept. It ranges from whether households can afford to buy a house to a wider perspective about access to different types of housing for specific groups of society. Economists have a number of ways of defining unaffordability although the starting point is affordability. These are the ratio of housing price to income, the ratio of housing cost to income and the 'residual income' remaining after meeting housing costs. The residual approach is arguably the most logical approach but it suffers from requiring large amounts of data and some subjectivity in what are the basic needs of households. There is no ideal measure.

Affordable housing, supported by the state, can be built to address a gap between what a large number of households can afford and the market price or rent. It can take a number of different forms, sometimes aimed at the very lowest incomes, other times targeted at middle-income groups and key workers who cannot access homeownership. Approaches vary by country as demonstrated by the examples of the United States, UK and China.

The main approach in the United States is to give tax incentives/credits to investors on condition that they provide a proportion of development as affordable housing. Low-income tenants who rent this affordable housing pay a maximum rent based on a percentage of the average local income. There is also an alternative scheme that issues means-tested choice vouchers that lower-income groups use to pay rent to private landlords.

In the UK, there are a range of affordable housing initiatives. In the rented sector, it ranges from council housing provided directly by local authorities and social housing available from housing associations to 'intermediate'-rented housing where tenants pay 80% of the market value.

There is also low-cost homeownership including 'shared ownership'. Shared ownership involves part ownership by the household while renting the other part from a housing association. A household purchases initially between 25% and 75% of a property, with a view to eventually buying the rest of it.

China's urban housing system has been radically changed away from the majority of housing being provided by state industrial companies for their employees. In the noughties, affordable housing took the form of owner-occupation for low- and middle-income households.

Private developers were given land for free in order to build affordable homes for these groups. Later, the government has turned to build public 'cheap rental' housing directly for low-income households, provided on a means-tested basis.

Bank lending criteria can be an important constraint on the working of urban housing markets. Banks assess households' creditworthiness and the surety of the housing they wish to buy when deciding on a loan. In particular, the bank takes into account the future marketability of housing. These lending processes can have significant implications for certain parts of cities.

In the past, neighbourhoods have been redlined, meaning no bank lending in these areas. Even if there are no red lines restrictive lending criteria based primarily on the characteristics of housing can still mean areas of cities where there is little or no regular mortgage finance. Such areas can get access to subprime mortgages on higher than usual interest rates to reflect the higher risk to the bank.

In developing countries, with customary ownership systems, bank lending constraints are more prevalent. Many households have no recorded legal title for their home so owners have therefore no formal security for a mortgage. Rather than using bank loans, households have to finance building homes with their extended family resources usually on a gradual basis. As a result, there are significant barriers to new urban development.

With the housing stock fixed, local migration, or household movement, is the transmission mechanism in the housing market that determines house prices. The larger the number of households seeking to buy a new home the greater the demand, leading to higher prices. Likewise, the greater the number seeking to rent in an area the higher the local rents.

Spatial house price ripples can be explained by reference to the operation and interaction of local housing market areas. There are two not mutually exclusive explanations. The first relates to migrants from higher- to lower-priced areas, pushing up prices in their destination locality. Such a process continues from one area to another and another.

The alternative explanation is that local house price ripples are caused not by migration between areas but by income ripples. Lagged increases in household incomes through a country ultimately lead to a house price change as rising income, with supply fixed, pushes up local house prices. In this way, a house price ripple is a reflection of the changing spatial economy.

Housing submarkets within cities that can be seen where the equivalent houses have very different prices in different localities. They can also be seen in differential price changes in prices over time in different neighbourhoods. Their existence can be traced to the very localised nature of household moves, often within the same neighbourhood.

The result is a series of relatively closed housing submarkets within which localised supply and demand forces interact to determine the price. These submarkets are stable because any new housing tends to be of a similar character and households are reluctant to move to other parts of a city even if the equivalent housing is cheaper.

Learning outcomes

The predominant tenure in the world is owner-occupied housing. The remainder who rent are either tenants of for-profit private landlords or of public authorities/subsidised agencies.

Each tenure is taxed and subsidised in different ways in individual countries, and these fiscal measures frame how the housing market works and the attractiveness of tenures.

Homeownership is not a uniform entity from one country to another.

Despite the high national proportions of owner-occupation, in many cities there are very high proportions of private tenants.

Expected future household incomes, interest rates and the availability of mortgage finance are the most important short-term drivers of the housing market and price cycles.

Longer-term influences on house prices include demographic trends, migration, the role of planning and tenure structures.

Housing market forces have made homeownership less affordable for young adults.

The unaffordability difficulties of would-be owner-occupiers are part of a wider problem of unaffordability that in part is the result of rising demand and an insufficient supply response.

In the developing world, many urban areas also suffer from an informal customary land tenure system that creates difficulties for land ownership and the funding of new development.

Economists have a number of ways of defining affordability. These are the ratio of housing-price-to-income, the ratio of housing-cost-to-income and the 'residual income' remaining after meeting housing costs.

There a number of different forms of affordable housing, sometimes aimed at the very lowest incomes, other times targeted at access to homeownership. Approaches vary by country as demonstrated by the examples of the United States, UK and China.

Bank lending criteria can be an important constraint on the working of urban housing markets, with potentially significant implications for certain parts of cities.

In developing countries, with a customary ownership system, bank lending constraints are very prevalent. Rather than using bank loans, households have to finance building homes with their extended family resources usually on a gradual basis.

With the housing stock fixed, local migration, or household movement, is the transmission mechanism in the housing market that determines house prices.

Spatial house price ripples can have two explanations. The first relates to migrants from higher- to lower-priced areas pushing up prices in the latter. The alternative explanation is that local house price ripples are caused by lagged increases in household incomes through a country so that house price change stems from rising local incomes.

Housing submarkets within cities occur because of the very localised nature of household moves, often within the same neighbourhood. The result is a series of relatively closed housing submarkets within which localised supply and demand forces interact to determine price.

Bibliography

Fears C, Wilson W and Barton C (2016) *What Is Affordable Housing?* House of Commons Briefing Paper No. 07747, House of Commons, London.

Jones C and Maclennan D (1987) Building societies and credit rationing: An empirical examination of redlining, *Urban Studies*, 24, 3, 205–216.

Jones C and Watkins C (2009) *Housing Markets and Planning Policy*, Wiley-Blackwell, Chichester.

Jones C, White M and Dunse N (eds) (2012) *Challenges of the Housing Economy: An International Perspective*, Wiley-Blackwell, Chichester.

8 Urban commercial real estate markets

Objectives

Factories, industrial units, offices, shops and warehouses provide the real estate infrastructure to support the manufacturing and service industries that comprise the urban economy. Each land use has different real estate characteristics and demand-specific to a singular industrial sector. The result is translated into separable markets with distinct internal dynamics. These land uses, therefore, represent individual real estate markets in a city in their own right, as well as competing with each other for locations as explained in Chapter 4.

Commercial real estate is not only occupied by industry but it is also bought as investment assets. In this chapter, the analysis therefore examines the determinants of occupation use and investment demand, together with the implications for rental and capital values. It further considers how rental and capital values influence development activity. The analysis also differentiates between the various sectors. The structure of the chapter is as follows:

- Tenure
- Occupation demand
- Investment
- Supply constraints
- Development activity
- Real estate cycles.

Tenure

Real estate is a complex good; in physical terms, it is a combination of land and buildings. Real estate is also unusual as the ownership and use or occupation can be separated. However, it is important to note that the 'goods' traded in the real estate market are not *physical units* but *legal rights* over land and buildings. There can be a range of legal rights applicable to one property, for example subtenants, but here a simple division between ownership and use is made.

While most occupiers of housing are owner-occupiers, the reverse is true in the commercial sector. In the commercial property sector, firms occupy premises primarily as tenants rather than owner-occupiers. For example, in the UK, it was estimated that just 3% of shopping centres and 14% of retail warehouses (power centers) were occupied by their owners in 2013. The reasons are that although tenants pay rent they

DOI: 10.1201/9781003027515-9

benefit from the specialist expertise of landlords who manage the property. Being a tenant also means that a business can use its capital more efficiently and effectively in its own area of expertise rather than having it tied up in real estate.

The legal rights involved in owning or renting vary across countries. Tenants' (and landlords/investors) rights are determined primarily by the nature of leases, security of tenure and regulation of rent (increases). The normal length of a lease varies between countries. In the UK, they typically range between five and fifteen years in length, with an average currently towards the bottom of this range. Larger space occupiers look for longer leases because of the potential relocation upheaval costs. Rent reviews are normally every five years.

Comparable lease lengths in other countries are shorter. In Germany, commercial property leases are usually for five to ten years, but leases up to 20 years are typical for supermarkets and department stores. In France, leases are commonly granted for a minimum of nine years with rent reviews every three years. Five- and ten-year leases are frequently applied in the United States. In Australia, a lease length for a large tenant would be eight to ten years. Commercial leases in China typically last three to five years. These lease terms provide the framework for occupation demand in the commercial real estate market.

Occupation demand

The demand for the use of property depends on the sector. For each sector, there are underlying similar economic trends that influence revenue and profitability of firms, hence the scale of take-up of space and rents applicable. The rent surplus model outlined in Chapter 4 applies generally to the rent a firm can afford to pay, namely how much is available from profits after paying other costs such as wages.

There are a range of types of shops from food through to fashion, and many shops also contain personal services, such as opticians, hairdressers, takeaways, etc. The demand for shops is therefore linked closely to retail spending and to a lesser extent consumer expenditure on personal services. Such expenditure is heavily dependent on the state of the macroeconomy and the level of real incomes.

Aside from tourists and day visitors, customers of shops come from the local catchment area. For some shops, this area is just the neighbourhood. Demand in small towns and cities is dependent on the prosperity of the local urban economy, including for example the scale of unemployment. In the case of large cities, the catchment area extends to the surrounding region.

Service industries are the predominant occupiers of offices, so demand is aligned with the profitability of this sector of the economy, although it is a quite disparate collection of firms. Many of these businesses do not necessarily provide business and personal services only to the city in which they are located. But, many will, such as accountants and lawyers. The demand for offices in a given city is therefore underpinned by the performance of the local economy.

Industrial property houses manufacturing firms much of whose output is for a national or even an international market so rents are not so influenced by local revenue generation. However, industrial units can be used for warehousing (some are built specifically for this purpose) and also for specialist retailers such as tyres/exhausts and builders' merchants, together with car servicing. These latter occupiers are dependent on local customers and the affluence or otherwise of the urban area.

To summarise, demand for commercial property is related
national and local economies and their impact on a particular
demand also relates to the current state of technology. How mu
quired will depend on the use of information communication
use of online purchases raises the demand for distribution war
less demand or shops. Manufacturing technology shapes the sp
industrial units.

Investment

Landlords or investors benefit from rental payments from tenants for the duration of a
lease potentially subject to periodic rent reviews to market levels. Rent levels are likely
to rise at review dates to keep pace with any inflation. Ultimately, at the end of the
lease, a landlord has ownership of a vacant property with a residual capital value that
is likely to have kept pace with inflation.

An investor can sell at any point and the value of a property depends on the current
rental income paid by the tenant but also the potential growth in income. An individual property's capital value is also partly dependent on the lease structure and terms.
Real estate can be viewed as a real asset. Its value is maintained in real terms by rental
values that keep pace with inflation every time they are reviewed.

Much of the commercial real estate in cities and towns is owned by financial institutions or property companies, called real estate investment trusts (REITs). To be more
precise, financial institutions are 'non-bank financial institutions'. There are three
types, namely general insurance companies, life assurance companies and pension
funds. Financial institutions invest in real estate and other investments to generate
future income for the benefit of policyholders and pensioners.

Investing in real estate by financial institutions is set within their portfolios of different types of investment assets. The other main types of assets that are held by financial institutions are company shares and government bonds. Real estate offers distinct
characteristics compared with these alternative investments so that it contributes to
the diversification of a portfolio. The role of real estate as an investment varies by type
of institution.

The business of general insurance companies relates to insuring against fire, motor,
accident, theft, travel, etc. It is an uncertain business and they need to hold a high percentage of liquid assets that can be drawn on quickly if necessary to pay any claims.
There is a slow and expensive process to sell real estate so that it is not suitable as a
general asset for insurance companies. Real estate is therefore only a minority asset
and an insurance company invests in it as a way of growing its long-term core assets in
support of its business growth.

Unlike insurance companies, the life assurance business relates to certain events.
Life assurance companies sell endowment policies and life insurance. These are long-
term contracts with specific maturity dates. There is little need for liquidity but there is
a need for assets that give capital growth to meet the promises to policyholders.

Pension funds have a similar business model to life assurance companies. Members
of the pension fund pay usually a set part of their income into it each week or month.
The basic idea is that workers pay into a pension scheme each month in order to build
up a pension on retirement. In this way, pension funds know their precise liabilities
and can design investments to match them.

esting in real estate as a real asset fits into a strategy to ensure long-term capital
wth. Real estate tends to be only a small percentage of the total assets of these
institutions. While real estate is a minority investment holding, it is still an important
component of institutional portfolios. Real estate investment strategies and decisions
are a dominant influence on urban real estate markets.

REITs are also important investors in urban real estate and exist in most parts of
the world. These property companies are quoted on a country's stock market and represent an alternative indirect way of investing in real estate. Often, they specialise in
one sector of the real estate sector, such as shopping centres, warehousing or offices.

In general, financial institutions and REITs invest in similar types of properties,
generally the best quality and the largest, known as the 'prime'. Prime is a nebulous
concept best defined as properties that are seen as having good rental prospects with
low risk, by reference to quality, size and location. Many 'prime' offices are let to public
agencies/authorities or large companies who are unlikely to default on rent payments.

A prime location would be one where there is always excess demand, either because
of accessibility or prestige. If a tenant vacates a shop or office, there would be another
in line to replace it. If a particular retail or office sector declines in profitability, there
would be firms in other sectors waiting in the wings to occupy their premises.

In contrast to prime, secondary properties are generally the converse, namely
smaller, older in poorer condition and in less soughtafter locations. In retailing, these
locations could be on the edge of city centres or suburban or neighbourhood centres.
These secondary properties are generally owned by small real estate companies and
are likely to be more prone to vacancies and to be empty for longer. The rental values
of these secondary properties are lower than prime ones, reflecting the relative profitability of their use. Capital values of secondary properties not only take into account
the lower rents but also the greater risk to the investor.

Supply constraints

Rental values are in theory determined by the interaction of local demand and supply. This interaction is not between two 'stock' variables but market flows. Although
supply can refer to the stock of buildings, real estate supply is the flow of properties
onto the market. Similarly, the demand flow is the potential tenants seeking premises
on the market.

The nature of these flows varies by real estate sector. The demand for shops in the
city centre is potential occupiers looking for very accessible locations with customers
passing by. Maximising the passing footfall of customers is crucial to success. There
is only a limited number of such locations so supply is relatively fixed at central locations, certainly in the short term.

In the long term, it is also difficult to respond to say rising demand for shops, simply
because of the specific location constraints that effectively require new development to
be adjacent to the established shopping area. Any increase in demand therefore simply
leads to a rise in demand for the existing shops, pushing up rents and capital values.

Potential occupiers of city centre offices, in general, have less spatial specific requirements than shops although often they still wish to locate in prestigious streets.
There is also the option for the supply of offices to be expanded by constructing higher
buildings in the longer term in response to rising demand. However, the supply of new
offices is constrained by land and planning constraints including the conservation of

existing buildings. As with shops to meet growing demand, there can be substantial land assembly problems. These involve the negotiation of sales from existing owners and permission for the demolition of existing buildings.

The demand for industrial buildings has traditionally not been so location-specific (but see Chapter 12). There are less supply constraints on new development and the simplicity of industrial sheds enables them to be built quickly. Demolition and replacement are also a more straightforward option.

In general, the more difficult it is to increase supply the more significant demand is in determining rental values. New supply also takes some time to develop if it happens at all. There may be limits to how much new supply can be produced because of the built-up nature of cities. In some locations, supply is not able to meet the growing demand, leading to long-run rent premiums for prestigious localities.

Development activity

Development occurs to meet the growing or changing real estate demands of a city. Commercial real estate property companies can build on a speculative basis with a view to finding a long-term investor to purchase and occupiers to let to. Ideally, the property would be pre-let or pre-sold before or at an early stage during the development period. Often it involves the redevelopment or refurbishment of buildings. Some commercial property companies develop normally only with a view to adding to their own investment portfolios.

Private development is dependent on the profit motive and is also risky, especially on a speculative basis. The potential viability of an individual development project is assessed by reference to existing land values and market trends. The bigger picture sees development as part of a wider circle of activity given in Figure 8.1. In this circle, the commercial real estate market is underpinned by occupational demand.

Without demand, there is no market. If demand rises, then rents tend to increase constrained by supply, and this, in turn, will lead to higher capital values. Investors pay more for the higher rental income. Higher capital values result in development

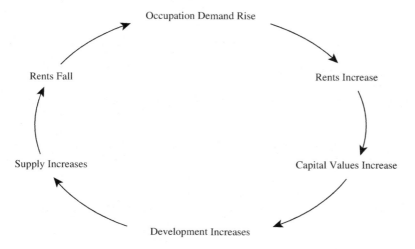

Figure 8.1 Commercial Real Estate Market Circle.

becoming more profitable. Increased development leads ultimately to increased supply and a dampening down of rents, and the circle is complete (to start again).

This real estate circle is influenced by urban and national economic forces. In terms of the impact of macroeconomy, interest rates are a good example of their influence on the real estate market. For instance, a cut in interest rates reduces the cost of borrowing and increases consumer spending in shops. Increased profitability of shops tends to lead to rising rents and capital values.

Similarly, a cut in borrowing costs indirectly means that investors can afford to pay more for assets and capital values are bid up. Lower borrowing costs from banks also mean that construction costs are cheaper, supporting increased development activity. Lower interest rates therefore can bring a significant boost to development profitability by influencing all three real estate sectors. A wider view of the macroeconomy's influence on the circle is considered in the next section.

In understanding this real estate circle and the impact of development, it is important to note that the role of new supply on the market may be limited. Supply flows onto the market comprise two elements:

A   Existing property being offered for a relet,
B   New property being let for the first time.

The percentage breakdown will vary between different areas and over time. New supply will generally be a lot less than relets.

The commercial real estate circle does not represent instantaneous adjustments. Supply of new property to let at any point in time depends on decisions taken some time in the past, maybe years in the past. As noted above, new development has to be seen as profitable but it is also dependent on a number of facilitating factors. It is the product of a number of sequential decisions by private and public decision makers, including meeting planning and building regulations.

Private decisions could include ensuring there is a bank loan available to pay for the construction, the choice of the building contractor and discussion with potential tenants and subsequent purchasers. The physical construction of large developments such as multi-storey office blocks or shopping centres can take years. Together all these processes can take a considerable time.

In addition, development requires the acquisition of land. Available land in a city is constrained as discussed in the previous section. While the total supply of land is fixed, the supply of land for a particular land use is not fixed in the long term. Land can be transferred between uses through redevelopment, thereby increasing/decreasing land supply of different uses.

For this to happen, it needs to be profitable and this is dependent on the differential capital values between land-use sectors. If land values for offices, for example, are higher than land values for industrial properties, it may be viable to change use.

Land use patterns therefore respond to market forces through differential land prices. However, viability is a function of profitability of development that is not just about differences in land prices but also the cost of land acquisition, demolition and construction. Development viability also has to take account of expectations about future real estate demand trends and capital values. They are dependent on the positive or negative prospects for the individual urban and national economies.

To summarise, development activity is part of the real estate circle with its inter-action between the occupation and investment sectors. There are considerable devel-opment time lags in the commercial real estate market. The result is that the decision to develop reflects market conditions applicable some time ago, and they are unlikely to be the same when the development project is delivered. The decision to develop therefore needs to embrace expectations about the future, especially with regard to the prospects for urban and national economies. The next section considers this issue within a framework of real estate cycles.

Real estate cycles

There are many examples of real estate development cycles in history around the world. Notable office development booms occurred in the second half of the 1980s in cities including London and New York. The 1980s boom in London added nearly 30% to the office stock of the central area. Figure 8.2 illustrates a series of office development cycles for the city of Calgary in Canada over the period of 1971 to 2017. There is some variation in annual building completions in the city with the most significant booms in the late 1970s and 2000s followed by rapid declines. The Calgary example also shows only a minor upturn in the late 1980s, demonstrating that local factors are important to explaining development cycles.

To understand why these cycles happen, a simplified stylised office development cycle is now outlined in a series of indicative stages.

Stage 1 The starting point is a strong upturn in business bringing rising demand for of-fices. It coincides with shortages in the available supply of offices after a period of

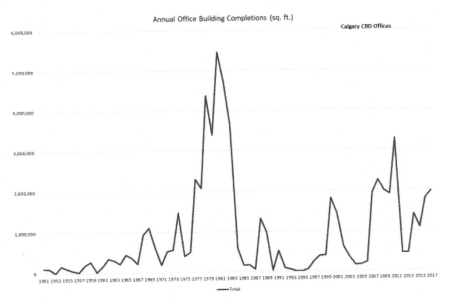

Figure 8.2 Annual Office Space (sq. ft) Completed in Calgary City Centre 1971–2017.
Source: Brooke et al. (2018)

low development activity in the previous business cycle. With demand increasing and supply fixed, rents and capital values begin to rise. Although capital values are rising, developers are unsure if these increases are sustainable and hold back responding.

Stage 2 There is a continued strengthening of demand while restricted supply continues. It leads to a sharp rise in rents and capital values. The capital values have now risen sufficiently to make development clearly profitable. It leads to a wave of new speculative development after the initial lag.

Stage 3 Given the time lags in the construction of offices, little new space has come on the market. Continued rising demand pushes rents and capital values further up. It stimulates more speculative development and a major building boom takes off.

Stage 4 The upsurge in business activity subsides just as the majority of office developments are completed and come available to be let on the market. There is now excess supply on the market with a high number of vacancies, leading to rents and capital values falling.

This stylised cycle is explained in terms of the changing balance of demand and supply and the role of development activity within a city. However, there are wider national economic influences. First, in terms of the national factors, the business upturn is often part of a wider growth spurt in the macroeconomy supported by low-interest rates and readily available bank loans. Such an upturn in economic growth feeds into increased demand for offices while low-interest rates and bank lending helps to boost development activity.

In Stage 4, the fallback in letting activity is likely to be part of a wider macroeconomic downturn. There may also be a rise in interest rates to combat an overheating economy that aggravates the problems of developers with empty properties. These property companies have to pay the interest on loans they took out to build the offices.

The companies are holding unlet or underlet buildings with no or insufficient income to pay (increased) interest payments. The outcome in Stage 4 is not just depressed rent and capital values but potentially widespread bankruptcies of property companies.

There are also regional and local economic factors that influence a cycle. First, the demand for office space is related to the structure and prosperity over time of the local economy. A city's economy may be dominated by one industry and that could have a significant influence on the demand for offices.

For example, Calgary, the example in Figure 8.2, has an economy focussed on the oil industry and its profitability over time is a major contributor to its cycles. Property time lags are also a core factor in causing real estate development cycles but their significance can differ between cities. In general, the larger the city the greater their role of constraints simply because of the lack of space.

Overall real estate development cycles are predominantly caused by a combination of a cyclical economic activity and real estate time lags. The result is that real estate development cycles have a larger amplitude than national economic cycles. Development is very profitable in the upturn of the cycle but the downturn brings severe losses. Development is therefore a precarious business and property companies often go bankrupt in the downturn of the cycle.

So far, the section has been concerned with real estate development cycles. However, the last global real estate cycle in the 2000s, that saw dramatic rises in capital values around the world, did not involve a substantial rise in development activity. There was also limited rental growth. Instead, there was a significant increase in investment only.

The rise in investment was simply the product of speculation. It was facilitated by low-interest rates and the ready willingness of banks to lend to purchase commercial real estate with high loan-to-value ratios. With many investors seeking to buy and no increase in the commercial real estate stock, the result was a severe growth in capital values. The bubble burst with the global financial crisis in 2007/2008 that saw the withdrawal of bank finance and then capital values collapsed.

Summary

Commercial real estate is a complex combination of legal rights. Most commercial occupiers are tenants so that they can apply their capital more efficiently and effectively to grow their businesses. Tenants have the right to occupy properties on a lease that normally runs for a period of years. The length of these leases varies from country to country but is typically around ten years.

Occupation demand for commercial property is derived from the demand for the goods or services provided in each sector. For shops, demand is aligned to retail spending and to a lesser extent consumer expenditure on personal services.

Service industries primarily occupy offices, so demand is broadly linked to this sector of the economy. Industrial property occupiers include manufacturing firms and distributors, as well as specialist retailers serving the local community. Influences on demand for such properties are varied.

Commercial real estate is seen as an attractive investment to financial institutions and large investors because it maintains its real value. Rental payments keep pace with inflation when rents are periodically reviewed. The result is that much of the high valued commercial real estate in cities and towns is owned by financial institutions or REITs. REITs are important investors in many parts of the world and often specialise in particular real estate sectors.

Financial institutions invest in real estate and other investments to generate future income for the benefit of policyholders and pensioners. Real estate is a minor asset class within their portfolios that are mainly comprised of company shares and government bonds. The influence of rent reviews means that real estate acts as a diversifier within their portfolios.

Financial institutions and REITs invest in prime real estate properties. These properties are chosen for good rental prospects with low risk, by reference to their quality, large size and location. A prime location would be one where there is always excess demand from occupiers, either because of accessibility or prestige. Many 'prime' properties are also regarded as low risk because they are let to large national companies or the public sector who are unlikely to default on rent payments.

In contrast to prime, secondary properties are generally the converse, namely smaller, older in poorer condition and in less soughtafter locations. The rental values of these secondary properties are lower, reflecting the relative profitability of their use. Capital values of these properties are relatively low taking into account their low rents and the greater risk of vacancy.

Rental values are determined by the interaction of local real estate demand and supply flows. Potential shop tenants seek out locations in the city centre that offer substantial passing footfall of customers. The problem is that such locations are in limited supply in central locations. A rise in demand pushes up rents and capital values.

Potential occupiers of city centre offices are more pragmatic in terms of location compared to shops. The supply of offices is also more adaptable to demand with the option to build upwards. Nevertheless, office supply is heavily constrained by land and planning constraints.

The industrial real estate sector is the least restrained sector given the flexibility of demand in terms of location and the ease of redevelopment. Overall cities suffer from significant real estate development constraints that are likely to lead to long-run rent premiums for prestigious localities.

Development that does occur is often undertaken on a speculative basis. It can be seen as part of a circle of commercial real estate market activity. The foundation of the market is occupational demand. The rental income generated from tenants provides the base for investment.

When rents rise, it stimulates investment and higher capital values, making development more profitable and eventually boosting supply. This commercial real estate circle is framed by urban and national economic forces that have a substantial impact on rental and capital market values.

The processes of adjustment in this commercial real estate circle can take considerable time because (re)development may require years to complete. The land has to be assembled/acquired, planning decisions may be required, demolition may be necessary and then there is construction on the new building.

There are considerable development time lags in the commercial real estate market. The decision to develop is therefore based on the state of the market in terms of the balance of supply and demand that may be very different from that applicable when a property is completed.

The development time lag is a contributor to real estate cycles. A stylised office development cycle can be identified as a series of stages. The cycle begins with rising demand and the slow response of supply exacerbates a rise in rental values.

Given the development time lag by the time new supply comes to the market, it is possible that there is a macroeconomic downturn and demand has fallen away. There is now excess supply and empty properties, and rent and capital values decline. The result is that property companies find themselves in financial distress.

While the national economy has a strong influence, there are urban factors that shape these real estate cycles. The economic performance of local industries is a key factor in determining demand over time. On the supply side, the development time lags can vary with the characteristics of the city, especially if it is large, historic and is subject to strong conservation policies. The existence of supply constraints means that real estate development cycles have a larger amplitude than national economic cycles.

Not all real estate cycles can be explained by development time lags. The global real estate cycle of the 2000s involved only limited development activity. It was driven by speculation with investors supported by readily available cheap loans. When the global financial crisis arrived, funding was removed and capital values collapsed.

Learning outcomes

Most commercial real estate occupiers are tenants so that they can apply their capital more efficiently and effectively to grow their businesses.

Tenants take on a lease that normally runs for a period of years, normally around ten years.

Occupation demand for commercial property is derived from the demand for the goods or services provided in each sector.

Commercial real estate is seen as an attractive investment to financial institutions and large investors because it maintains its real value.

Much of the high-valued commercial real estate in cities and towns is owned by financial institutions or REITs.

Financial institutions and REITs invest in prime real estate properties. A prime location would be one where there is always excess demand from occupiers, either because of accessibility or prestige.

Secondary properties are smaller, older in poorer condition and in less soughtafter locations. These are owned mainly by small property companies.

The demand for city centre shops is for locations that offer substantial passing footfall of customers, but there are very limited opportunities to increase supply. Any rise in demand pushes up rents and capital values.

The demand for offices in central urban areas has a degree of locational flexibility. While the supply of offices can be responsive to an increase in demand, it is heavily constrained by land and planning constraints.

Industrial units are less restrictive in terms of desirable locations, and new supply is easier to construct.

Development is usually undertaken on a speculative basis.

There is a circle of commercial real estate market activity with occupation demand at its heart. The rents generated by occupation demand are the basis for real estate as an investment. The level of capital values determines the profitability of development.

Urban and national economic forces have a substantial impact on rental and capital market values, hence on development activity.

Development can take years to complete as land has to be assembled/acquired, planning decisions may be required, demolition may be necessary and then there is construction on the new building.

Development time lags mean that the balance of supply and demand in a local market on completion of the project can be very different from when it was initiated.

The development time lag is a contributor to real estate cycles.

Development time lags can vary with the characteristics of the city, especially if it is large, historic and is subject to strong conservation policies.

The existence of supply constraints means that real estate development cycles have a larger amplitude than national economic cycles.

Not all real estate cycles can be explained by development time lags. The global real estate cycle of the 2000s was driven by speculation supported by readily available cheap loans.

Bibliography

Barras R (1994) Property and the economic cycle: Building cycles revisited, *Journal of Property Research*, 11, 183–197.

Brooke T, Jones C and Dunse N (2018) Commercial Property Cycles and Sub-market Emergence in Selected Canadian Cities, Paper presented to the annual conference of the European Real Estate Society, University of Reading, Reading.

Fraser W D (1993) *Principles of Property Investment and Pricing* (Second Edition), Macmillan, London.

Part II
Spatial change and public policy

9 Growth, decline and revival of cities

Objectives

While cities are products of the industrial revolution, their subsequent urban form has not been static. Cities have been modified, even transformed or reinvented with waves of technological change. These underlying changes encompass not just industrial production technologies and new products but also transport developments. The modifications to urban form have incorporated both a revised composition of industries and spatial structure.

This chapter is a platform for Part 2 of the book. It deciphers the different processes that comprise the underpinnings of urban change and the associated outcomes. These are:

- Industrialisation
- Deindustrialisation
- Decongestion
- Manufacturing decentralisation
- Suburbanisation
- Growth of services
- Inner-city donut
- Re-urbanisation and revival.

Industrialisation

Chapter 3 mapped out the history of urbanisation and demonstrated that it coincides closely with industrialisation. It notes that the UK was the first country to industrialise in the early 1800s based on inventions from the 1700s. It was followed quickly by Belgium and France. Other countries such as Germany and the United States industrialised in the late 1800s. Industrialisation in Asia, South America and Africa followed later, predominantly in the latter half of the 1900s, and brought with it rapid urbanisation to these continents.

The industrial revolution brought new machine-based industries, including iron manufacture, chemicals and textiles through to mechanised cotton spinning. It was stimulated by the use of water and the invention of steam power. Many of these developments were interrelated, such as the use of chemical bleaching and dyes in the textile industries. The steam pump enabled the expansion of coal mining. The combination of iron and steam enabled the creation of railways. The invention of a chemical process

DOI: 10.1201/9781003027515-11

to make cement, and hence concrete, supported the construction of the new social infrastructure associated with cities.

A further spurt to industrialisation and urbanisation occurred after 1870 with a range of innovations encompassing the manufacture of steel, greater mass production and the use of electricity. This is sometimes called the second industrial revolution and led to petroleum manufacture and car production in the twentieth century. Overall, this period of industrialisation was confined to Europe and the United States and driven by technological change.

This century of innovation provided the economic base for cities and the stimulus for their growth. The spatial adoption of industrial innovation also maps out the pattern of urbanisation around the world. The latter part of this period also saw the rise of large conglomerates benefitting from economies of scale. These giant industrial companies spread their tentacles around the world becoming multinational.

Deindustrialisation

Deindustrialisation is the decline of traditional nineteenth century industries in developed countries in the face of international competition and the globalisation of markets. In one sense, it is brought about by other parts of the world catching up with the industrialisation of North America and Europe. It is also a combination of the expansion of global trade and the ability of particularly Asian countries to undercut on price given high labour costs in western economies.

There are different dimensions to this process. There is an absolute decline in manufacturing employment in developed economies. It can be viewed as a severe structural change to the national economy through a lack of international competitiveness. It creates a fundamental problem of adjustment for the labour market as many of the workers have skills only applicable to their industry. The outcome is the creation of severe unemployment.

Deindustrialisation also has wider consequences for the macroeconomy. It leads to a decline in manufacturing industry's share of national employment. It changes the national balance of trade as manufacturing exports decline while the country now has to import the goods it once made. There are long-term consequences for finding replacement economic activity and the value of the currency.

Deindustrialisation began in earnest in the 1970s but can be traced back to the 1960s. It caused a shake-out of unprofitable firms that is demonstrated vividly by the statistics for Britain. By the mid-1970s, UK manufacturing employed 1 m less than ten years earlier, a decline of 15%. In the next ten years, it accelerated. The peak of the decline in manufacturing occurred in the early 1980s, partly as a result of a global recession. In the UK, between 1966, the peak year of manufacturing employment and 1983, the sector lost 3.14 m workers. This number is equivalent to 37% of the number in 1966. A long-term consequence beyond the collapse is that surviving firms suffer a drastic decline in profitability.

Deindustrialisation derives from changing international competitiveness, and so it is not directly an 'urban' process. However, cities that had a high concentration of the affected traditional industries suffered a severe economic decline, much more significant than the national picture. However, this structural explanation for the decline will vary widely between cities, depending on the nature of their traditional industries.

The force of deindustralisation was seen most at the city level. The industries that were the basis of their genesis and economic growth were those that now felt the brunt of global competition. Traditional manufacturing industries in these cities by the 1970s had outdated capital equipment and premises. The result was uncompetitive labour-intensive processes, in comparison with their international competitors. It is difficult to isolate the precise impact of deindustrialization on urban economies because decentralisation of industry was also occurring at the same time (see next section). Two extreme but different examples in the UK are Glasgow and Liverpool.

Glasgow is the centre of the Clydeside conurbation. A conurbation comprises a number of cities and towns that have fused over time to form a continuous urban area. Clydeside was once the world's foremost shipbuilding centre, building some of the largest passenger liners up to the 1970s. At its peak, there were 30 to 40 shipyards at any one time. However, beginning in the 1950s, it began to lose business to Japan and other Asian countries. By the mid-1960s, shipbuilding was becoming increasingly unviable. This culminated in the closure of major shipyards and the nationalisation of others. By the millennium, there were few private shipyards remaining and they mainly build warships for the UK Government or car ferries for the Scottish Government.

Liverpool is a long-established port in north-west England at the heart of a conurbation known as Merseyside (after the river on which it stands). A range of port-related industries had developed including sugar refining, flour mills, shipbuilding, chemicals and textiles. A study by Lloyd found that in the period 1966 to 1975, the inner part of the conurbation lost over 18,000 manufacturing jobs. These losses amounted to a fall of 24% from the 1966 total of 78,000. A breakdown of the losses showed a 26% fall in food and drink and a 25% fall in shipbuilding. There were bigger losses in chemicals and textiles of 50% while the furniture industry lost 40% of its jobs. Within the manufacturing sector, electrical engineering fared best with a loss of 18% of employment. Nevertheless, there were severe losses across the whole spectrum of manufacturing industries.

These job losses over a ten-year period were predominantly the direct result of large plants closing down rather than shrinkages. There was some modest decentralisation (see next section) but for the most part, the city saw the closure and abandonment of large manufacturing facilities, such as sugar refining plants, flour mills, etc. More than half of the losses were accounted for by 19 large plants.

The 1966 to 1975 time period is only a snapshot of a much longer process. As noted above, 1975 was at the beginning of peak deindustrialisation in the UK so the figures are an underestimate of the employment implications for Merseyside. In fact, the main shipyard on Merseyside struggled on through to 1993, primarily making warships and submarines for the UK Government, after which part of it turned to ship repairing.

Manufacturing decentralisation

Unlike deindustrialisation, decentralisation is clearly an urban economic process. It is partly existing firms moving out from the urban core. It is also partly new firms opening in suburban locations or in small towns rather than the traditional inner-city areas. The process can therefore be described as the redistribution of employment although clearly some localities lose jobs while others gain.

There can also be a regional dimension to this process of change. Manufacturing industry moving from high-cost regions to low wage regions with new plants opening at decentralized locations. Such a process is bound up with the changing industrial structure and the growth of multi-plant firms. It is possible to wind down an inefficient plant in a central part of one city and open a new one. This process therefore can be seen as a way of introducing new more efficient work practices to the company. Alternatively, it is a means to invest in technological change in the production process or the introduction of new products.

Decentralisation can be traced back to the 1950s. Certainly, in the UK, all industries, including services, were decentralizing between 1961 and 1966. It was stimulated by the use of production lines that required more land and the eventual death of the multi-storey factory. Shifts in demand for development in these areas are supported by lower initial land prices, easier land assembly and a better physical and social environment than many alternative inner urban sites. The globalisation of industry and markets has reinforced these trends. The establishment of large multi-national manufacturing companies has led to an increase in the size of an industrial plant. These plants require large land sites that are generally only available at peripheral locations.

At these locations, there were also better roads making the urban periphery more attractive as it has easy access to the national motorway networks (see Chapter 11 for more detail). There were also comparatively greater numbers of manufacturing industrial plant 'births' in decentralized locations as well as greater 'deaths' in inner locations. In the Merseyside study above, while there were 18,000 manufacturing job losses in the inner area, there was a net growth of 25,000 jobs in the outer part of the conurbation. Much of this growth in employment was from new firm formation.

A good example of the impact of decentralisation is the experience of the American city of Detroit. The primary reason for the growth of Detroit from 1900 onwards was the expansion of car manufacturing. Ford opened a large plant in 1927 employing 90,000 workers. The rapid expansion of the car industry led to inward migration. It provided a stimulus to population growth in the city whose population peaked in 1950. It became known as "Motor City".

During this period, the car industry in the United States coalesced into three large car manufacturers, namely Chrysler, Ford and General Motors. Between 1945 and 1957, these three companies built 25 new manufacturing plants in the metropolitan area of Detroit, but none in the city itself. The remaining plants in the city were eventually closed down. By the end of the 1950s, then car manufacture in the city had disappeared, partly to surrounding areas.

Interestingly, the car industry also began subsequently to suffer from global competition and then technological change in the 1970s. Robotics and automation also replaced assembly line jobs with machinery. Employment in the industry shrank substantially. The risks of the conurbation's reliance on a single industry were exposed with consequences for the wider economy and the housing market. In cities, like Detroit, entire neighbourhoods were decimated as high-paying manufacturing jobs in the car industry disappeared.

Decongestion

Decongestion is sometimes distinguished from decentralisation of manufacturing industries. Decongestion is a subprocess of decentralisation because it occurs strictly within the urban setting. It is a form of extended industrial suburbanisation that involves an intra-urban move as opposed to an interurban move. It is, however, a consequence of the same forces that cause decentralisation, but linked particularly to an interaction with the transport system. Improvements in transport technology and investment in the transport network have encouraged the process.

Suburbanisation

Suburbanisation is a parallel process to manufacturing decentralisation. The population moving out from the urban core is a longstanding process that precedes industrial decentralisation. Hoyt describes a suburbanisation process based on Chicago in the 1930s. It saw the rich moving out to modern peripheral housing so that high-income residential areas gradually move outwards from the urban core. There is a process of succession whereby lower status groups continually move out from the inner-city areas into the properties vacated by the higher-income groups.

The significance of suburbanisation and its precise timing vary around the world and by definition it follows urbanisation. In fact, it is a natural part of urban growth. With Britain the first country to urbanise in the 1800s, it was also the first to suburbanise as cities grew. It happened as cities moved from walking to work to commuting supported by the arrival of trams and railways in the latter half of that century.

However, the rate of suburbanisation is difficult to quantify. In some cases, it is masked by cities growing and annexing surrounding areas so suburbanisation is not evident in the formal statistics of population size. A useful way to look at the significance of suburbanisation is by looking at the balance between population change in the urban core and its suburbs as shown in Table 9.1 that is adaption of one by van den Berg and Klassen.

In Britain, the six out of seven core cities probably reached their peak population around 1940, although Leeds continued to grow until 1961. While most large urban core populations were in decline, the surrounding conurbation continued to experience population growth between 1951 and 1961. Overall, the conurbation populations continued to grow, and so in terms of the classification, they were now experiencing absolute suburbanisation. After 1961, the expansion of the populations in the suburban rings of London, Liverpool, Manchester and Newcastle was outweighed by the losses in their urban cores.

These conurbations were now in the disurbanisation stage. Meanwhile, the urban cores of Birmingham and Leeds were now in decline. Through to the 1980s, all large British conurbations continue to suffer disurbanisation, with Glasgow, Sheffield and Newcastle losing population in the suburban ring as well as the core. Smaller cities,

Table 9.1 The Balance of Population Change in Urban Cores and Suburban Rings over Time

	Urban Core	*Suburban Ring*	*Aggregate Growth*
Fast urban growth	++	+	Positive
Relative suburbanisation	+	++	Positive
Absolute suburbanisation	−	++	Positive
Relative disurbanisation	−−	+	Negative
Absolute disurbanisation	−−	−	Negative

+ population growth, ++ fast population growth.
− population decline, −− fast population decline.

such as Bristol and Nottingham, were still in the suburbanisation stage but had not experienced severe deindustrialisation.

Cities in other countries followed different time paths. During the 1980s, most large French cities were by then in a similar state of disurbanisation. Paris, Lyon and Lille were experiencing modest population decline in the suburban rings. The position in the United States was somewhat different. Eight large cities with over 1 million population lost core population in the 1970s, but for most of these cities the decline was short-lived. Cities such as New York, Philadelphia, Detroit, Cleveland, St. Louis, Pittsburgh, Milwaukee and Buffalo continued to lose population after 1980. With the exceptions of New York and Philadelphia that have experienced recent modest growth, all these other core cities in this list are still losing population in 2020, albeit some marginally.

At the same time, the population in the suburban rings were and are increasing. In terms of the classification in Table 9.1, most cities in the United States are in relative suburbanisation. Those that continue to suffer population decline at the core can be characterised as absolute suburbanisation.

Detroit is an extreme example of a city where the urban core has declined relative to the suburban ring. Between 1970 and 2010, the city's population halved. The suburban ring having undergone substantive growth between 1950 and 1970 then also experienced modest decline thereafter. The result was that the core's population as a percentage of that of the total urban area shrank from three-fifths in 1950 to less than a fifth in 2010.

The explanations for these trends are partly related to the deindustrialisation and continuing manufacturing decentralisation trends discussed earlier in the chapter. The urban cores that suffered most from deindustrialisaton also experienced the most substantial population loss. Decentralisation may also have contributed to the suburbanisation of the population but this relationship is more complex, and suburban life is attractive in its own right. The causes of suburbanisation are discussed in more depth in Chapter 10. The relatively quick turnaround from population decline to growth of some American cities can be attributed to a shift in the urban economy to services.

Growth of services

Many western economies have shifted from being primarily manufacturing based to services since the 1970s. By the millennium, there is no advanced economy in which the service sector does not represent the largest share of employment. However, the rise of the service sector has also been seen across both high- and low-income countries. Some of the latter have seen a rise in services without the transition from manufacturing.

The dynamics and timing of the change to services had transformative impacts on cities. In Britain, for example, while manufacturing employment fell by one-third between 1981 and 1996, banking finance and insurance rose by 73% to almost match the numbers working in manufacturing. At the individual British city level, Table 9.2 shows growth in this sector of 30% to 40% over the 1980s, although Liverpool is an outlier. In the United States, employment growth in finance, insurance and real estate services between 1980 and 1990 was equally dramatic as Table 9.2 also demonstrates. Typically, growth in this sector in large US cities was more than 40%.

Table 9.2 Growth in Financial and
Business Services Employment
in Selected Large US and UK
Cities in the 1980s

City	%
Glasgow	37.7
Liverpool	10.6
Manchester	30.4
Birmingham	43.4
Newcastle	41.8
Chicago	44.3
Detroit	32.2
Houston	43.4
New York	42.8
Los Angeles	58.9
Philadelphia	58.3
San Francisco	38.8
Seattle	47.3

Source: Jones (2013).

The 1990s saw continued growth in financial and business employment. In the United States, the sector expanded at a slower rate by 13.5% over the decade. But other new business services industries were expanding fast over this period. The most important of these were employment agencies and computer and data processing, and there were a wide range of new business services growing. A study of UK business services employment growth also demonstrates the scale of continuing growth in the 1990s beyond financial services. It finds a rise of 57.4% in knowledge-intensive business services.

The expansion of services employment has not been confined to business services. The growth of other services has led to an increase in the demand for administrative offices for activities, such as retailing and wholesaling. The rise of the use of information communication technologies has had a profound effect on business operations and online sales. Call centres have become an essential element in the delivery of services and goods to customers. The public sector employment in most countries has also seen a long-term increase in its workforce: Higher education has dramatically expanded with universities primarily located in cities. There has also been a major expansion of leisure services.

This transformation from manufacturing to services also heralded a shift from employment in a factory to an office, and this led to a rise in office development and the obsolescence of many old industrial premises in western economies. It also meant that traditional locations of employment changed. There was a fundamental restructuring of the urban economy in terms of the nature and location of economic activity. It has been supported by the rise of information communication technology and the motor car (discussed in later chapters).

Inner-city donut

The consequences of this transition, together with suburbanisation, were physical and population decline in the inner-city areas. They were in particular decimated by deindustrialisation and decentralisation of employment. These inner-city areas can be broadly seen, although not exclusively, as the urban areas at the end of the 1900s. At the same time, the suburban rings are growing in terms of population and employment.

There are inevitably the pains of change from the restructuring of the urban economy. The statistics for Britain emphasise the scale of the problems. The decline of employment (mainly manufacturing) in the urban cores was 27% between 1961 and 1981 as well as the jobs that were lost in the traditional manufacturing so were the skills of the workforce made redundant. The workers who lost their jobs also found that their skills were not necessarily easily transferred to other industries.

The closure of manufacturing firms in the urban core led to a reduction in demand for labour in these industries. While new services eventually replaced them, the skills applicable to these industries were not those held by the ex-manufacturing workers. There was a labour market skills mismatch between what the new industries wanted from their workers and the skills of workers shed by manufacturing firms.

Many of the older workers found it difficult to commute to the opportunities offered by the decentralised industries as they were often in low-income occupations. Younger, economically active and skilled workers moved away to new job opportunities. The result was high inner-city unemployment. In 1981, it was 50% above the national average in Britain, whereas in the suburban rings it was equivalent to the national average.

This profile of the inner city can be related to the urban 'donut' term commonly used in the United States that emphasises the divide between core and suburbs. There is a stark contrast between the suburbs and the inner area, essentially a ring of thriving suburbs surrounding a decaying central area. The suburban ring is growing and residents are wealthy, educated and low crime. The inner area is dominated by the poor and ethnic minorities, population decline and subject to high crime levels.

As a result of the deindustrialisation and industrial decentralisation, there was a negative impact on the inner-city real estate market. The industrial decline resulted in a large amount of unused even abandoned land or obsolescent industrial units or warehouses, vacated by closures. In some cases, there were large swathes of former industrial land or docks that were often contaminated. Poor housing, too, lay empty and in need of redevelopment. This land is known as 'brownfield' as opposed to greenfield at the periphery of the city.

The scale of this legacy of decline is not easy to address. From a physical perspective, contaminated land needs to be treated and derelict buildings must be replaced by new developments with alternative uses. The scarred townscape is also the home to many people on low incomes in poor housing with weak employment prospects. It is a social, physical and economic urban renewal task that began in the 1980s and is taking decades.

Re-urbanisation and revival

A revival of the centres of cities can arguably be traced to the growth of services providing a new economic base. It brought with it highly paid employment. There was ultimately a reversal, or a slowdown, of population decline in parts of urban cores. In Britain, the seven largest provincial core cities that suffered deindustrialisation and population decline began to increase their populations around 2001. It has been characterised as 're-urbanisation'.

Re-urbanisation, however, cannot be attributed just to the rise of services although it was a necessary condition. It was contingent on demographic rather than urban trends. There has been a rise of young people living alone or in two-person households as child bearing has been delayed for professional couples. These trends have been partly driven by financial constraints and partly associated with a change of lifestyles. Some young professionals are attracted to city-centre living with the proximity of social and leisure facilities. The expansion of higher education has also meant greater numbers of young adults seeking accommodation in the city where they are studying. In large cities, these numbers have been swelled by international students.

The result is that cities have seen rising concentrations of young professionals, students and young parents in central areas. In many cities, it has been encouraged by planning policies that have sought to promote new high-density developments as a means to stem urban sprawl. The result is that there has been some rejuvenation of parts of the centre of the donut and the intermingling of young professional households with the traditional poorer and older inner-city dwellers.

This process of re-urbanisation is amending but not fundamentally changing the donut spatial structure of cities. It necessarily involves the redevelopment of vacant brownfield sites and the repopulation of parts of the older urban areas. However, the suburban ring is still flourishing and the underlying but parallel drivers of decentralisation are still very much in action. Indeed, suburbanisation and decentralisation have revolutionised the spatial pattern of commercial and industrial land use as discussed in subsequent chapters.

Summary

Industrialisation began in the early 1800s based on inventions of the previous century. The first country to industrialise was Britain followed by Belgium and northern France; the United States followed at the end of the 1800s. As part of the industrialisation process, there was a move toward large multi-plant companies, many of which became multi-national.

These countries that were the first to industrialise were also the first to feel the brunt of global manufacturing competition beginning in the 1950s. Traditional manufacturing industries were unable to compete with competition from Asia where labour costs were much lower, and the requisite industry was more highly mechanised. The influence of this competition grew through subsequent decades, reaching a crescendo in terms of its impact on employment in the early 1980s.

Deindustrialisation led to a collapse in manufacturing in western economies that had the most dramatic consequences at the city level. The industries that had been the key to many cities' success were now the leading causes of their economic demise. Core cities were also enduring a double whammy with the decentralisation of industry.

Examples of deindustrialisation include the loss of shipbuilding as exemplified by the experience of Clydeside. It was once the world's foremost shipbuilding centre, but by the mid-1970s the prospects of terminal decline were evident. The remnants of the industry today survive with government contracts for specialist ships. Merseyside, another port city, saw a parallel and dramatic downward spiral of manufacturing employment. In ten years alone from the mid-1960s, it lost a quarter of its manufacturing jobs. These job losses were the direct result of large plants closing down, with more than half of the losses accounted for by just 19 plants.

Manufacturing decentralisation like deindustrialization has substantially shaped the urban economy. There are a number of different ways this decentralisation occurs. It can be existing firms moving out from the urban core or new firms preferring suburban rings. In some cases, it results from multi-plant firms opening new plants at decentralised locations while closing old inefficient plants in an inner location of a different city. Alternatively, it can arise for the relative growth and decline of plants in inner areas compared to outer areas.

Decentralisation is a longstanding process that has been propelled by technological change and industrial production requiring greater land input. This requirement for large land sites precipitates an attractiveness for suburban ring locations. The periphery is also where there is easy access to the national motorway networks for distribution.

The decentralisation of the car industry in Detroit is an extreme example of the problems caused by this process. Once known as 'Motor City,' the large car plants that established the city eventually closed down as manufacturing plants opened in the suburban rings. The loss of employment ultimately had wider ramifications for the city's economy and housing market.

Residential suburbanisation with decentralisation are twin processes dominating the change in the spatial structure of cities, certainly since the 1960s. Suburbanisation is a necessary part of urban growth as a city expands. In the 1870s, it was supported by the advent of transport to enable people to commute longer distances. This relationship between suburbanisation and transport improvements has continued ever since as Chapter 10 explains.

The importance of suburbanisation can be seen by charting the balance between population change in an urban core and its suburbs within the framework of a conurbation. Typically, urban cores begin to lose population while suburban rings continue to grow. In some cases, while the suburban population is still growing, a rapid decline in the core can lead to an overall decline for the conurbation. This latter phenomenon is known as disurbanisation.

Major British industrial cities that had experienced severe deindustrialisation followed this cycle, reaching disurbanisation in the 1960s. Not all western cities reached a stage of disurbanisation but many have seen the balance of their population move from core to suburbs. However, the timing and scale of suburbanisation vary by country.

The reasons for suburbanisation are only partly related to deindustrialisation and the decentralisation of manufacturing. The primary motivation for suburban life is that it is attractive in its own right.

The latter quarter of the 1900s saw the transformation of many cities from a manufacturing-based economy to one that was predominantly service. In particular, there was a large growth in business and finance services from the 1980s. These were just part of a much wider growth in business services linked to information communication technology and what has been called the knowledge economy. A call centre becomes a central component of the business models of firms, selling goods and services. The rapid expansion of universities predominantly located in cities created a new and significant education-based sector to the urban economy.

The shift to services in western economies has also seen the office become the main form of workplace. It has left many factories and industrial units redundant. The growth of office employment has led to substantial development activity. It has also meant that traditional locations of employment have changed, as part of a wider re-structuring of the spatial economy.

The transition from manufacturing to services together with suburbanisation contributed to a physical and population decline in the inner-city areas. There was an increasing contrast between the inner-city decline and a prosperous suburban ring. Residents in the inner-city areas struggled with redundancies of workers caused by the demise of the traditional manufacturing industries.

The skills of these workers proved to be obsolescent, not wanted by the new service industries. What opportunities there were in low-income occupations, paying much lower than their previously skilled jobs. These jobs were often some distance from their homes. Inner-city unemployment rocketed and has remained high. Younger, economically active and skilled workers left if they could.

Commentators in the United States refer to this as the urban 'donut', a term, underlining the divide between decay in the core and the prosperous suburbs. In the inner areas, there are physical remnants of a past manufacturing industrial age, poor housing and a relatively deprived population. The resolution of the consequences of decline is an enormous challenge that began in the 1980s but is taking decades.

In recent decades, the regeneration of urban cores has been supported by the growth of services employment establishing a new economic base and seen a rise in population. This rise in population is characterised by re-urbanisation. It is primarily a consequence of an increase in young adults including student numbers living in central areas. It has brought rejuvenation to parts of the centre of the donut but the suburban ring is still flourishing, and the spatial structure of a city is fundamentally unchanged.

Learning outcomes

Industrialisation began in the early 1800s with Britain the first country to industrialise followed by Belgium, northern France and the United States.

Deindustrialisation led to a collapse in manufacturing in western economies from the 1960s through to the 1980s and had the most dramatic consequences at the city level.

Manufacturing decentralisation together with deindustrialization has substantially shaped the urban economy.

Decentralisation from the 1950s has been thrust forward by technological change and industrial production requiring greater land input. Urban peripheral locations also afford accessible locations to the national interurban road system.

Residential suburbanisation as well as decentralisation are dual processes dominating the change in the spatial structure of cities, yet are broadly independent of each other. Suburbanisation is the result of low-density living.

Suburbanisation has seen urban cores lose population while suburban rings continued to grow. Ultimately, if the urban core is rapidly losing residents, then it can lead to the overall decline in the population of the conurbation. This phenomenon is known as disurbanisation.

Disurbanisation has occurred mainly in cities that suffered severe deindustrialisation. Many other cities have seen suburbanisation significantly change the balance of their population from core to suburbs. The timing and scale of suburbanisation vary by country.

The revival of cities' fortunes has been led by the growth of services. There has been a makeover from a manufacturing-based urban economy to one that was predominantly services.

Much of the increase in business services centred on information communication technology and the knowledge economy. In addition, there was a rapid expansion of the higher education sector in cities.

The move to a service-based urban economy in western cities has reorganised the spatial pattern of employment. Suburbanisation has further influenced physical and population decline in the inner-city areas.

Inner-city poverty has become starkly juxtaposed with affluent suburban rings, giving rise to the term 'the urban donut.' Urban policies to address the spectrum of inner-city problems have been ongoing since the 1980s.

Many traditional cities have in recent decades seen a rise in the population referred to as re-urbanisation. It is mainly the result of a significant increase in young adults living in central areas. However, at the same time, suburban rings are still thriving.

Bibliography

van den Berg L, Drewett R, and Klaassen L H, Rossi A and Vijverberg C H J (1982) *Urban Europe: A Study of Growth and Decline*, Pergamon, Oxford.

Jones C (2013) *Office Markets and Public Policy*, Wiley-Blackwell, Chichester.

Juday J (2015) *The Changing Shape of American Cities*, University of Virginia, Charlottesville.

Lever W F (1993) Reurbanisation – The policy implications, *Urban Studies*, 30, 2, 267–284.

Re'rat P (2012) The new demographic growth of cities: The case of reurbanisation in Switzerland, *Urban Studies*, 49, 5, 1107–1125.

10 Explaining intra-urban economic change

Objectives

The nature of the urban system in western economies has seen substantial change over the last half-century or so, as detailed in Chapter 9. The change has encompassed a reversal of decline for many major cities around the turn of the millennium and the reshaping of urban spatial structure. The fundamental forces of this change that drove the urban economy were a combination of deindustrialisation initially and then the growth of services. But the story also involves changing spatial structure that is complex. It is partly the outcome of the interaction of the decentralisation of industry and partly the suburbanisation of the population.

This chapter examines the underlying reasons for these spatial changes through the numerous facets of the motor age and changing real incomes with the consequences for urban real estate markets. In particular, the chapter is structured as follows:

- Intra-/interurban transport costs
- Changing real incomes
- Changing commercial rent gradients – revisiting the Alonso model of Chapter 4
- Changing house price gradients – revisiting the Muth model of Chapter 5.
- Re-urbanisation and the Changing Donut

The chapter does not consider the role of information communication technology. While this influence is increasingly drivingurban change, analysis of it is delayed until Chapter 12.

Intra-/interurban transport costs

Over much of the twentieth century, the central business district (CBD) flourished as a direct consequence of agglomeration economies that were generated from the clustering of firms and the maximising of revenue at the most (or close to the) accessible point in the city. Such central city benefits applied not only to offices but also to shops and manufacturing industry. This is seen in Alonso's urban rent gradient declining from the centre.

Intra- and interurban transport systems have an important influence in determining accessibility relationships in the spatial economy. These in turn have implications for households and firms in terms of the choice of location. Both the nature of intra- and interurban travel has seen a dramatic change since the 1960s with the motor age.

DOI: 10.1201/9781003027515-12

In 1913, Henry Ford started a revolution with the mass production of petrol-driven cars and heralded the beginning of inexpensive motoring for the middle class. But there were few roads suitable for cars at that time, and it was still many decades before they became widely available to the public. The most common form of urban transport was the bus through the 1950s and 1960s. Significant growth in the use of cars in western economies began in the 1950s.

More than 40% of personal travel was by bus at the beginning of the 1950s in the UK. By the 1970s, most households owned a car, and their popularity and demand continued to grow. At this point, three-quarters of all passenger-travel by distance was by car. The proportion increased through the 1980s to around 85% and has stayed broadly constant since. By 2020, cars have become seen as a necessity with the vast majority, 83%, of households with more than one person owning at least two cars.

A more quantitative way of looking at this growth is the actual numbers of cars on the road as given in Table 10.1 for the UK. The number of cars increased by 2.5 times in the 1950s and doubled again in the 1960s. These two decades were the periods of the fastest relative growth but in terms of the absolute increase of cars, the peak decade is the 1980s just ahead of the 1960s. The rate of growth has slowed since 1990 but the number of cars registered in the UK is more than six times that in 1960. At the same time, the population has risen by only 13%.

The switch to car travel has been reflected in the demise of local bus travel. In the 1950s, most local travel was by bus. However, the number of these passenger trips in 2019 was less than 40% of that in 1960. The numbers began to fall away significantly in the 1960s with a 27% drop over that decade, with a follow-on decrease of 24% in the 1970s. The rate of decline has slowed since 1980 but continued the downward trend until 2008 when free bus travel was introduced for the elderly. Since then, the numbers have plateaued.

The United States also experienced a similar dramatic growth pattern in car usage, although it began earlier. Vehicle registrations increased approximately 2.5 times between 1960 and 1980. The rate of growth has slowed in recent decades, as in the UK, but the absolute number of cars on the road is now 4.5 times the number in 1960. This rate of growth compares with a population increase of 83%.

The rise of car ownership in western economies was a result of its falling real cost, but it was also facilitated by the expansion and improvement of urban road networks. New bypasses have been built, roads upgraded to dual carriageways or simply widened as well as urban motorways created in some cases right into the heart of cities.

Table 10.1 Number of Cars Registered in the UK 1950–2019

	Number '000s	Index 1960=100	Increase in Registrations
1950	1979	40	
1960	4900	100	2921
1970	9971	203	5071
1980	14660	299	4689
1990	19742	403	5082
2000	23196	473	3454
2010	27018	551	3822
2019	30165	616	3147

Source: Transport Statistics.

The interurban road system has in the main been replaced by motorways as they called in the UK, freeways and highways in the United States, autobahns in Germany, autoroutes in France or expressways in other countries.

These interurban national roads were built at different points in time. Germany's first autobahn opened in 1932, inspired by Italian expressways beginning in 1925. The United States' freeways programme was instigated in 1955. The French autoroute network system took shape initially in the 1960s. The Chinese motorway network was substantially built in the 1990s but extended westward in the 2000s.

In the UK, the motorway network has been incrementally constructed over decades beginning in 1959. The 1960s and 1970s were the principal decades of motorway growth with only relatively minor additions to the network subsequently. Most British cities are now connected to motorways with the latest important link being the completion of the motorway between Edinburgh and Glasgow in 2017.

At one level, the development of the motorway/trunk road system has had important implications for both interregional trade and for relative spatial development between towns and cities (see Chapter 11). In particular, the direct impact of the motorway network is a changed accessibility landscape within a country defined in relative and absolute terms. But there are also implications for accessibility relationships within towns via the changing freight distribution process.

The growth of motorways has brought a shift in the way freight is transported. Until motorways arrived, freight was transported long distances by rail. In 1953, the number of 'tonne kilometres' transported by rail in the UK was 37 billion but by 1995 the total had fallen to just 13 billion. This decline was despite the rise of global trade and rising freight levels. The railway market share of the freight travel market, road and rail combined in 1995, was only 8%. It has stayed at approximately this proportion ever since.

Much of the freight tonnage transported by train is bulk mineral materials such as coal but the railways also transport containers from ports to inland depots where they are transferred to lorries. These container depots (see Chapter 12) are located in proximity to the motorway network Overall, rail has found it difficult to compete with the flexibility of interurban road transport via the motorways, and the distribution of goods has been revolutionised.

The implications for the spatial economy of a city are in the location of freight depots. Historically, the rail freight terminals were to be found on the edge of city centres, sidings off the main train lines. Freight was transported to a city by rail to these terminals and then distributed by road to customers within the city and its hinterland. Alternatively, outgoing freight was brought to the central shipment/transportation terminal for onward delivery by rail. In the motor age, these depots moved to the periphery.

The consequences of these transport developments, namely the rise of the car and motorway networks, for urban real estate markets are considered later in the chapter. However, first, another important influence, changing real incomes over time, is discussed.

Changing real incomes

The role of household incomes is identified in previous chapters as a key factor shaping the urban housing market. But average household incomes in a city are not stationary over time. Western economies experienced a rapid growth in real incomes in the boom that followed World War II. The rate of growth slowed from the 1960s but one study finds that across the world the average person, after adjusting for inflation, is 4.4 times richer in 2019 than in 1950.

In the UK, real wage growth has averaged an annual positive rate of 2% since 1950, although there were periodic short-lived recessions when incomes fell. Like many western economies noted above, this growth has slowed over time. Real wage growth averaged 2.9% in the 1970s and 1980s, 1.5% in the 1990s and 1.2% in the 2000s. The cumulative effect was that average real incomes doubled between 1979 and just before the global financial crisis in 2007.

The global financial crisis brought first a recession and then fiscal austerity policies in many countries through the 2010s that cut back government borrowing and public expenditures. The consequences have been that average real incomes have stagnated. For many households, especially young adults, income has fallen in real terms. There has been a growth of part-time employment with many people having more than one job.

Each country has its own pattern of real income change over time, although all Western countries experienced a spurt from 1950 that levelled off in the 1960s. One further common feature is increasing income inequalities between the rich and the poor. In the United States, for example, there have been stagnant real incomes for middle-wage workers, while wages have declined for low-wage workers since 1980.

These changes in real incomes influence directly the demand for housing. As noted in previous chapters, as households' incomes increase, they spend a higher proportion on housing, and this rising demand also has consequences for where they want to locate. And the same is true in reverse when incomes fall. A later section examines the spatial outcomes for the housing market and the development of cities.

Changing commercial rent gradients

Intra-urban location theory originates with Alonso's seminal land-use model as outlined in Chapter 4. It may be useful to reread that chapter before continuing. This model implicitly assumes agglomeration economies in the city centre. To recap, it is based on a city located on a featureless plain where land use is allocated to the highest bidder in a competitive land market. In this notional city, CBD is the point of maximum accessibility where business revenue is at a maximum and costs (other than land costs) are minimised.

As the point of maximum accessibility, rental values are highest at the city centre. Differences in the optimum locations of industrial and commercial land uses relate directly to the responsiveness of revenue and costs to distance from the centre. These in turn determine the rent bid abilities of firms in particular land uses for any given location. In the original Alonso model, it is presumed that revenues fall and costs rise with distance from the CBD. In turn, this means that the bid rent curves of individual firms slope downwards from the centre. The land market is presumed to be competitive so that a land plot is allocated to a specific land use on the basis of the highest rent bid. In this way, a downward sloping rent gradient from the CBD is generated.

In the Alonso model set out in Chapter 4, there are distinct bid rent curves for different land uses. In this chapter, the model is extended by decomposing both retailing and offices into two types to illustrate the implications of the expansion of car usage.

City centre shops are distinguished from out-of-town retailing in the forms of super-markets, malls, individual big box units or retail parks. Similarly, offices can be differentiated into offices in city centre locations and office parks in decentralised locations. The former could be linked to firms in financial services while the latter could be used by a call centre or for administrative purposes.

The distinction between offices that require city centre locations and those that seek dispersed locations can be represented in different bid rent curve patterns as shown in Figure 10.1. These differing shapes to bid rent curves reflect the nature of the businesses. Financial services depend crucially on face-to-face contact to undertake their business and maximise revenues. These firms are prepared to bid high rents to locate in the city centre where the financial service industry is focussed and has steep bid rent curves.

Administrative activities have no such requirement and their revenues are not dependent on location within the city. Their bid rent curves are shaped primarily by costs. These firms require a substantial land area to provide car parking and accessibility via the road network for their mainly suburban-based workforces. These considerations lead to a relatively flat set of bid rent curves for administrative activities for whom a face-to-face contact with clients and subcontractors is not necessary.

Turning to retailing the dichotomy between types of shops and their location decisions is more clearly driven by the rise of car usage. Out-of-town locations for retailing are a response to car-borne shoppers looking for more convenient locations in terms of proximity to their homes and the availability of spaces to park their cars. To meet these demands, retail centres require more space than the traditional high (main) street offers and suburban locations. The bid rent curves for these out-of-town retailers are analogous to the administrative activities in Figure 10.1. However, the shape of these curves arises not just from the need to pay for a large amount of space but also from the revenues that can be achieved at these localities.

At the same time, central shops continue to operate catering for a different market (but see Chapter 12). The city centre is the focal point of public transport although car parking is generally restricted in comparison to out-of-town. The bid rent curves of

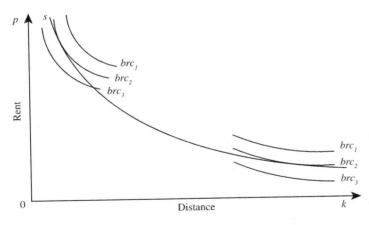

Figure 10.1 A Comparison of Bid Rent Curves for Financial Services and Administrative Offices.

retailers at this location have a steep slope, reflecting the high revenue generated and the importance of accessibility to customers walking from one shop to another.

These examples illustrate that Alonso's assumptions about the role of accessibility and the shape of bid rent curves is no longer so straightforward. He assumes that the revenue of firms is maximised at the city centre and then falls away with the distance from it. At the same time, transport costs are minimised at the city centre and then rise with distance to the periphery. In the original model, the further away from the city centre the location of the business is the higher the transport costs and the lower the revenue.

Bid rent curves and the rent gradient therefore slope negatively from the centre. It means that the rent paid by firms would have to be lower with distance from the centre in order to maintain the level of profit. Of the four examples above, financial services and city centre shops definitely follow this pattern. However, the locational factors of administrative offices and out-of-town shops encompass a preference for accessibility to suburban locations rather than the urban core. It suggests that the fundamentals of the Alonso model no longer universally hold or are blurred. Bid rent curves of some businesses do not necessarily have a negative slope from the city centre (although presented as such in Figure 10.1).

The Alonso assumptions are challenged even more with a review of the bid rent curves for industrial property. As discussed earlier, the establishment of a motorway network has meant that freight is no longer shipped to and from railway terminals near the city centre. It is therefore no longer necessary to be close to the centre for distribution purposes, indeed the city centre can be congested and a real disadvantage for certain industrial businesses. As discussed earlier, the replacement of freight distribution by rail with road has fundamentally changed the internal accessibility relationships within cities.

The most accessible locations for distribution are now usually motorway junctions on the edge of the urban area. In some cases, it can be airports. With labour costs broadly constant within cities, these peripheral urban locations are the point of cost minimisation rather than the city centre. Furthermore, as discussed in Chapter 9, the shift of industrial production to assembly lines requires large inputs of land more readily available at peripheral locations.

In addition, with industrial products now being sold to regional, national or even global markets, a firm's revenue will not vary according to the spatial location within a city or region. Overall then the most profitable location is often beside a motorway junction and firms have upward-sloping bid rent curves to these points, bidding up land prices accordingly.

To recap downward sloping bid rent curves away from the centre are crucial elements of the Alonso model but he was writing around 1960. Downward sloping bid rent curves and an urban land gradient made sense then, but the urban economy has since adapted to a motor age. The decentralisation of offices, shops and industry has led to suburban land prices being bid up. At the very least, these trends have flattened the intra-urban land price gradient as suburban localities have become popular for commercial land use. In addition, the importance of accessibility to the periphery/suburbs for some industries implies that they could have upward-sloping bid rent curves.

The implication is that the urban land gradient is no longer downward sloping from the centre to the edge of the city. The CBD is logically still the peak of the rent gradient where many offices and retailers are prepared to pay high rents. However, the motor

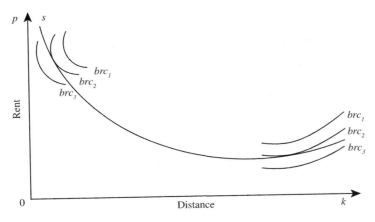

Figure 10.2 An Urban Rent Gradient Encompassing Downward and Upward Sloping Bid Rent Curves.

age has changed the nature of accessibility relationships in a city in favour of suburban locations. At the very least, the accessibility advantages of locations close to motorways have led to a rent premium for the immediate surrounding area.

Figure 10.2 illustrates a simplified urban rent gradient that encompasses twin peaks at the centre and the periphery. It includes upward and downward sloping bid rent curves that underpin this rent gradient. In reality, the rent gradient would be a more complex surface taking into account specific accessibility relationships created by the road network.

Changing house price gradient

In Chapter 5, the spatial structure of an urban housing market was explained by reference to the access-space model, based on a wide range of simplifying assumptions. It is useful to reread this chapter to remind yourself of its detail. The essentials for our discussion here are that households make decisions about where to live on a featureless plain, so that location is determined simply by distance from the centre. Households have no preferences for particular locations/neighbourhoods as the city is located on a uniform plain.

A household chooses its optimum location by trading off reduced commuting travel costs with proximity to the city centre, where most people work, and cheaper housing per square metre the further out from the centre. Each household chooses its equilibrium given how much housing it wishes to consume. In western countries, as income increases, a household spends more of its income on housing, so high-income groups tend to live in large housing in the suburbs, where unit housing costs are relatively low. In contrast, low-income households live in small accommodation in inner-city areas paying high unit costs. The collective decisions of households create a negative exponential house price (per square metre) gradient from the centre.

This model provides a framework for understanding the increasing suburbanisation of cities. The key underlying forces are falling long-term transport costs as car use has become universally cheaper and available and rising real incomes. Both of these trends are described earlier in the chapter. In addition, Chapter 9 chronicles the rise of decentralised employment so that commuting to the city centre is not necessary for many households. Such employment potentially leads to a reframing of household residential location decisions for some.

Rising real incomes means that households have more to spend on housing. In addition, as noted earlier, there is a high propensity in western cities to spend proportionally higher amounts of increases in income on housing. At the individual level, it has led to households living further from the city centre, consuming more housing at a lower unit cost. With incomes, in general, having risen over the twentieth century, households have generally moved outwards to consume more housing at a lower density. In this way, individual demand decisions have collectively contributed to and extended the suburbs.

The model can also be applied to assessing the role of falling transport or commuting costs with the increased uptake of cars, for example in the UK from the 1950s. In fact, falling travel costs is equivalent to a rise in real incomes for households, as a household has more money to spend on other goods. Just like rising incomes, then it means that households reappraise where they want to live, and as a consequence, there is a general movement outwards from the city centre.

Rising incomes and the falling cost of commuting therefore both work in the same direction in promoting the suburbs as a place to live. The result is that rising demand in suburban areas in the long term leads to higher house prices and a greater supply of new homes. Meanwhile, in inner-city areas, the movement of households outward leads to lower demand. As a result, house prices fall and there is excess (poorest) housing that may be abandoned or ultimately redeveloped into alternative uses. In terms of the model, these trends would be reflected in a flatter house price gradient from the city centre through time (although no studies have undertaken such a time series analysis).

The increasingly decentralised employment also has an impact on the spatial structure of the urban housing market. Workers at decentralised locations do not face a trade-off between commuting costs into the city centre and housing space. At the very least, these workers can live in suburban localities and have short commuting trips. Alternatively, these work locations can provide the opportunity, even encouragement, for people to live even farther from the city centres. Decentralisation of employment and suburbanisation are therefore mutually reinforcing.

Re-urbanisation and the changing donut

This increasingly flatter negative house price gradient with suburbanisation through the twentieth century does not necessarily apply to central locations. As Chapter 5 points out in some cities there is a long history of affluent professionals living in inner-most city locations often in luxury apartments, townhouses or converted warehouses. The reasons are at least in part a preference by fashionable elites to cluster together in these locations with the cultural and social amenities close by. These households are unaffected by the suburbanisation forces noted above.

Adjacent to the city centre are inner-city areas described as urban 'donut' in Chapter 9. This area is characterised by industrial and population decline resulted in a large amount of unused, even abandoned land, under-occupied housing or obsolescent industrial units and warehouses. Chapter 9 also notes that there has been a process of re-urbanisation that has eaten into the donut 'void' with new housing in city central residential areas. The households moving into these areas in city centres or their fringes are not professional elites but young adults many working in junior professional occupations or students.

The questions are why this has happened and how does it relate to the wider suburbanisation process? Inner-city living can be partly attributed to a lifestyle choice or preference for young adults similar to those of high-income households. However, this is a simplification as the discussion about gentrification in Chapter 17 illustrates. It can also be characterised in the UK as the result of recent government planning policy that has constrained suburban development and promoted the development of small city centre flats. However, there are also constraining market factors at work.

In recent decades, the long-term rise in real incomes in western economies as noted earlier has plateaued and reversed for some households. This reversal affected particularly households aged less than 40 years. The significance of income constraints has been seen through falling homeownership rates in young households (see Chapter 7). City centre rented flats offer the 'best' option for young adults priced out of owner-occupation, as well as catering for rising numbers of students and immigrants.

The cost of motoring has also stopped falling in real terms. In the UK, between 1991 and 2011, the cost of running a car including its purchase rose by 4% in real terms. Perhaps, for this reason, from the late 1990s, the number of driving licence holders has dropped. It suggests that car ownership was falling amongst younger adults. Overall income and motoring cost constraints for young people have contributed to the growth of city centre living.

These are key reasons why at least for young adults the underlying forces are encouraging centralisation rather than suburbanisation. The pivotal point for this turning point certainly in British cities was between 1991/02 and 2011/12. The impact has seen a restructuring of the donut shape with a revival of the decaying city centre, but that is not to say the long-term suburbanisation process is being reversed. It is rather that segmentation of the housing market means that both centralisation and suburbanisation are happening simultaneously. Some of the childless households occupying an inner area flat in all probability will eventually move to the suburbs.

Summary

The second half of the twentieth century saw significant changes in intra- and inter-urban transport systems bringing substantial rearrangement of accessibility relationships in the spatial economy. The motor age also led to modal switches in personal and freight travel, from buses to cars and from rail to lorries. Significant growth in the use of cars in western economies began in the 1950s and the vast majority of journeys are now undertaken by car.

In the UK, the number of cars more than doubled in the 1950s and then again in the 1960s. In absolute terms, the fastest growth was the 1980s but the growth has slowed since. The cumulative impact is that there are six times more cars on the road than in 1960. The hegemony of the car in the United States began earlier than the UK but followed a similar pattern – rapid growth before levelling off.

The long-term spread of car ownership among households has been the consequence of falling real costs, and it has been supported by the evolution and improvement of urban and interurban road networks. National motorway systems have been built in different countries beginning in the 1930s in Germany. The growth of the UK motorway network was predominantly in the 1960s and 1970s, but outlying areas are still being connected to the system.

The motorway network has stimulated the growth of freight transported by road. The flexibility it provides has meant that rail freight has been confined principally to large mineral loads. This transference of freight to lorries has had consequences for the location of depots. Formerly, freight was transported to and from a city by rail with terminals on the edge of the centre. It was distributed only locally by road. With the dominance of roads, freight depots have moved to the urban periphery.

Another major influence on the urban economy from the second half of the twentieth century in western economies has been the rise of real incomes. The 1950s saw a rapid growth and although the rise subsequently slowed the cumulative impact has been substantial, the order of a fourfold increase. Even so, in the last three decades, real incomes have been growing at historic low levels and in fact, there has been stagnation since the global financial crisis in 2007.

Beneath these recent trends, there has been a redistribution of incomes and greater inequality. As the growth of real incomes has levelled off, young adults and low-income households have seen their real wages fall. There has been a growth of part-time employment with many people having more than one job. These changes in incomes have implications for the spatial structure of the housing market.

The motor age has had a transformative impact too on the spatial pattern of commercial land uses. The chapter uses Alonso's model of land use to revisit and rethink relationships underpinning urban spatial structure. It does so by examining the shape of bid rent curves for different types of firms. The examples demonstrate that some firms such as those in financial services and many shops still have a preference for the city centre. At this location, their revenues are maximised and hence they are prepared to pay high rents to locate there. In addition, as accessibility to the centre is the key to revenue generation, they have steep bid rent curves. These bid rent curves are aligned with the Alonso model that assumes downward sloping bid rent curves from the urban core.

The shapes of the bid rent curves are less clear for new property forms that have emerged such as office parks and out-of-town shopping centres. These uses require a substantial land area to provide car parking and accessibility via the road network either for their mainly suburban-based workforces or customers. For administrative activities located in office parks, revenues are not dependent on location within the city. Their bid rent curves are shaped primarily by the costs of the extensive land requirement and accessibility to the suburbs, not the centre. In relation to out-of-town retailing, the shape of these curves is also a function of their revenues that can be achieved at these localities. In both cases, the shape of the bid rent curves could be upward rather than downward from the city centre.

The position is clearer for industrial properties. With the markets for industrial products now often regional and beyond intraurban location has generally no impact on revenues, and hence bid rent curves. On the other hand, costs are minimised, and profits maximised, at the most accessible locations for distribution, namely motorway junctions that can be on the edge of the urban area. Industrial firms therefore have upward-sloping bid rent curves to these points, that in turn leads to land prices bid up close to motorway junctions.

The impact of these changes is that suburban land prices are bid up, and the intra-urban land price gradient has become at least flatter and not necessarily downward sloping through to the periphery. With the importance of accessibility to the periphery/suburbs, for some industries, there is a strong potential for upward sloping bid rent curves. At the very least, the accessibility advantages of locations close to motorways have led to a rent premium for the immediate surrounding area.

In Chapter 5, the spatial structure of an urban housing market was explained by reference to the access-space model, based on a wide range of simplifying assumptions. It is useful to reread this chapter to remind yourself of its detail. The essentials for our discussion here are that households make decisions about where to live on a featureless plain, so that location is determined simply by distance from the centre. Households have no preferences for particular locations/neighbourhoods as the city is located on a uniform plain.

The rise of the car and the falling cost of commuting have impacted the spatial structure of the housing market. Rising real incomes too have changed the opportunities for where households can and want to live within a city. The chapter applies Muth's access-space model to assess the consequences. In this model, households choose an optimum location by trading off commuting accessibility and lower housing costs further away from the centre.

People in western economies generally spend a higher proportion of their income on housing as it increases. As incomes for most households rose over the twentieth century, they have taken the opportunity to consume more housing at a lower unit price. At the same time, car commuting costs have been falling, similarly boosting the flight to the suburbs. The rise of suburban living has not only meant more housing at the periphery but also increased house prices. Suburbanisation is also reinforced by the decentralisation of employment as it enables commuting from further distances. Meanwhile, demand and prices have fallen for inner-city areas and the result is a flatter urban house price gradient.

In parallel to this suburbanisation, there has been a recent trend toward increased city centre living. There has always been the phenomenon of affluent professionals living in upmarket housing in some cities enjoying nearby cultural and social amenities. The recent re-urbanisation of inner-city areas has been by young professional adults impinging on the urban donut.

There are a number of potential reasons. It could be simply that young adults collectively are making a lifestyle choice, similar to that of the long-standing high-income earners. In the UK government planning policy has also encouraged the development of new city centre small flats. However, there are also a range of important constraining market factors in play linked to the falling real income of young adults over recent decades and the rise in the cost of motoring.

Re-urbanisation is regenerating the decaying donut and fringes and parts of the city centre. It is a parallel process to suburbanisation that continues unabated. Re-urbanisation can be seen as part of a revised life cycle pattern to housing careers, with households moving from an inner area flat to a home in suburbia as they move into child-rearing.

Learning outcomes

The motor age in western economies began in the 1950s and now most personal travel is by car and freight has swapped trains for lorries.

After car numbers rising rapidly, the growth in the UK slowed after 1990. The cumulative impact is that there are six times more cars on the road than in 1960.

The rise in car usage has been brought about by falling real costs and the expansion and upgrading of urban and interurban road networks.

This transference of freight to road from rail has led to fright terminals moving from the edge of a city centre to a depot adjacent to a motorway junction, typically near the urban periphery.

A major influence on the urban economy from the second half of the twentieth century in western economies has been the rise of real incomes.

In recent decades, the growth of real incomes began to stall in these countries with stagnation after the global financial crisis.

This stagnation in income growth has also led to greater inequality with young adults in particular suffering.

The motor age has changed accessibility relationships within a city.

Revisiting Alonso's model of urban land use in the current context shows that some firms still have a preference for the city centre where revenue is maximised. However, not all bid rent curves now slope down from the city centre.

Firms occupying new property forms in decentralised locations have bid rent curves that are shaped primarily by their extensive land requirement and accessibility to the suburbs, not the centre. The shape of their bid rent curves could be upward rather than downward from the city centre.

Industrial firms that are serving regional or national markets have upward-sloping bid rent curves near motorway junctions, that in turn leads to land prices being bid up at these locations.

The impact of the motor age has been a redrawing of the urban land gradient. The city centre still has the highest land values. But the gradient has become flatter and likely to rise in the suburbs and certainly close to motorway junctions.

Following Muth's access-space model, a household chooses its optimum residential location by trading off reduced commuting travel costs near to the city centre and cheaper housing per square metre in suburban locations.

The balance of this location decision has changed with rising real incomes and lower commuting transport costs since 1950.

Rising real incomes enable households to spend more on housing. Typically, households in western economies spend a high proportion of any increase in income on housing.

As incomes rise, the increased desire for housing leads to households living farther from the city centre, consuming more housing at a lower unit cost. Rising incomes have translated into greater suburbanisation.

Falling travel costs have similarly led, for broadly equivalent reasons, to a general movement outwards to the suburbs.

The rising demand in suburban areas fostered by falling commuting costs and rising incomes has led to higher house prices and new housebuilding. In contrast, reduced demand in inner areas means lower house prices, vacant stock and ultimately redevelopment. The result is a flatter house price gradient from the city centre.

Increasingly decentralised employment has encouraged people to live even farther from the city centres. Decentralisation of employment and suburbanisation are therefore mutually reinforcing.

The recent re-urbanisation of inner-city areas by young professional adults impinging on the urban donut has a number of potential explanations. It may represent a new fashionable lifestyle choice. Alternatively, it could represent a response to the falling real incomes of young adults.

Re-urbanisation and suburbanisation are not mutually exclusive. Re-urbanisation is likely to be part of a new form of housing career life cycle, with households eventually moving to suburbia as they move into child-rearing.

Bibliography

Buzar S, Ogden P, Hall R, Haase A, Kabisch S and Steinführer A (2007) Splintering urban populations: Emergent landscapes of reurbanisation in four European cities, *Urban Studies*, 44, 4, 651–677.

Jones C and Watkins C (2009) *Housing Markets and Planning Policy*, Wiley-Blackwell, Chichester.

Jones C (2013) *Office Markets and Public Policy*, Wiley-Blackwell, Chichester.

11　The changing urban system

Objectives

So far, the book has looked at urban change at the city level, but in this chapter, it is considered as part of a broader urban system perspective. In other words, the economies of individual cities are part of a (changing) national order. The sizes and functions of nearby cities are not necessarily independent as they are part of a national urban system. More widely, it can be seen that the distribution of the sizes of individual cities and towns in a country or even a continent is linked together. A fundamental factor in the changing urban system is the role of transport, both within and between cities.

The chapter examines these issues and is structured in the following way:

- Stages model of urban change
- Facets of urban dispersal – suburban gridlock, edge cities, city regions and archipelagos
- Individual influences on cities
- Regional growth poles.

Stages model of urban change

The core of the model examines how and why a national population is spread between cities and within cities. It does this formally by considering the distribution of populations *within* cities with the distribution of populations *between* cities. The starting point for this model is that towns and cities are part of a wider urban system and an individual city cannot be seen in isolation. The premise of the model is that there are general underlying relationships that shape the spatial pattern of a national urban system. It addresses questions about the reasons for the distribution of cities sizes in a country. The model is based on a stages model of urban change from Parr and Jones (1983).

The focus of the underlying reasons for change within the model is on how transport costs have evolved over time:

- between cities (inter-urban)
- within cities (intra-urban)

As a preamble, let us consider an example. If we take the case of an improvement in inter-urban transport, say the introduction of freight transport by rail. This transport

DOI: 10.1201/9781003027515-13

development favours economies of scale because it enables large-scale production for national markets that could not be reached before. It could mean that some large cities will benefit and grow even more. In terms of the national spatial pattern of population, this leads to an increase in inter-urban concentration in the urban system within a small number of cities.

This process of urban change could also impact the intra-urban pattern of population of cities. The growth of the large cities puts pressure on the spatial structures of these cities. The outcome will depend on the state of intra-urban transport. If it is not very efficient, then central densities will rise, because in effect, the increase in the population/workers will need to be accommodated within short distances of the employment locations. If it is efficient, then the increased population can be accommodated by suburbanisation.

These ideas are brought together in a stages model that here identifies four steps (the original paper has five).

Stage 1 Pre-urban or pre-modern stage

Low-quality inter-urban transport

In this stage, because of the time it takes to travel, and the difficulty/cost involved in transporting goods, there is little interaction between cities. There is therefore very modest inter-regional trade or trade between cities. Cities tend to be self-sufficient in the production of many goods, meeting the demands of the local market. As a consequence, production is at a very low level in a large number of locations within a country. There is very little urban concentration, so there are a large number of small cities.

Poor-quality intra-urban transport

Transport of goods and people within cities is also of a poor standard, but this is of little influence on the performance of the urban economy.

Stage 2 Urban specialisation stage

Dramatic improvement in the level of inter-urban transport

As set out in the example above in the preamble, a dramatic improvement in inter-urban freight transport increases the scope for large-scale production/urban concentration. As a result, certain cities benefit (the ones with good transport links) and experience very rapid economic growth with their populations increasing in parallel.

Intra-urban transport poor

The continuing weak transport system within cities acts as a constraint on the growth of cities that are benefitting from the improvements in transport between cities. It means that this growth in the selected cities has to be accommodated by increasing urban densities in areas immediately surrounding the city centres.

Stage 3 Urban transformation stage

Improvement in intra-urban transport

The main driver of change in this stage is the improvement of travel within cities. People can now travel further to work. This improvement removes the internal constraint on city growth existing in the previous stage. The result is that the urban system undergoes a really dramatic change and the largest cities expand substantially. There is a transformative impact on the urban system.

Inter-urban transport also improves

While the inter-urban transport of goods continues to improve in this stage, it is not a significant influence. Overall, while city growth accelerates, urban decentralisation has begun by permitting cities to spread out.

Stage 4 Urban dispersal stage

Improvements in inter-urban transport dominate

The origins of this stage lie in decentralisation begun in the previous stage, but it is now driven by improvements in inter-urban transport which radically change the accessibility relationships across the country. As a result, towns outside the old urban cores can now exploit new locational advantages created by better transport networks. As discussed in Chapter 10, this is seen in the development of a network of motorway/expressways as well as the upgrading of trunk roads between urban areas including dual carriageways and bypasses.

These improvements mean that the advantages of concentrations of production evident in previous stages become subordinate to decentralised locations. The focus of growth has moved to free-standing towns and small cities. These towns can now link into the national transport network but also have advantages for serving regional markets. These decentralised locations also have the advantages of space for industrial expansion and their lower residential density/housing stock/quality of life means that they can attract families and skilled labour.

This growth of towns and secondary cities has been characterised as the creation of urban archipelagos. They are so called as they represent islands of growth or hot spots with diffuse boundaries. These urban islands are often located in former rural areas or at the urban and rural interface.

The *inter-urban transport* improvements are now working to the detriment of the larger cities. At the same, the *intra-urban transport* infrastructure is no longer improving, even deteriorating, relative to the demand on it in large cities. The negative externalities that are created, particularly congestion and the polluted atmosphere, and ever longer commuting travel times, contribute to the pressures to decentralise. The concerns apply whether commuting by car or public transport.

Whereas the minimum average daily commuting time to the centre of a small city is at least 40 minutes at the other end of the spectrum, it can be more than double that time. In large South American cities such as Bogota, Rio de Janeiro and Sao Paulo, commuting can take more than 90 minutes on average. In the biggest North American and British cities, the average is on par if only marginally less. These statistics may not be entirely reliable but give the order of time involved in commuting to core cities. It can also be costly in terms of the percentage of monthly income.

Typical speeds in city centres are uniformly very slow, averaging 10mph or just over in most large cities. This is true for major cities in Europe, North America and South America. It is said that the travel times in Central London are now slower than in the eighteen hundreds during the era of horse-drawn carriages.

The original stages model was related to urban development in the UK by looking at the timings of transport improvements. This involved the introduction of canals, railways and road networks between cities and the evolution of the use of trams, suburban trains, buses and cars within cities. However, the model can be applied to different countries and relate to different types of transport innovation. In doing so, the precise timings and nature of the stages will vary depending on when transport changes occur. There could be less stages.

In the dispersal stage, the distribution of populations between cities becomes more even as small cities/towns grow while older centres stagnate or exhibit modest re-urbanisation. This can now be seen in developing countries as their primary/capital city becomes less dominant as second-level cities expand.

Facets of urban dispersal

The nature of the dispersed form of the urban system can be seen through a number of phenomena that are considered in this section. In a sense, they are a collection of un-related observations but they are illustrative of recent patterns of urban development. They also offer insights into how the spatial economy now works.

Suburban Gridlock

The traditional model of a city is of commuting into the central business district and inner industrial areas. As manufacturing and services have decentralised, commuting patterns have inevitably become more diffuse. This trend is supported by continuing suburbanisation. As a consequence, there was, for example, a decreasing share of com-muting trips to the cores of American cities during the 1980s.

The result is the emergence of working and living within the same suburb, or suburb-to-suburb commuting or travelling outwards to work at the urban periphery. By 1980, suburb-to-suburb commuting was already predominant in the United States. There was also a growth in core-to-suburbs commuting in the 1980s. The United States led these trends, but these patterns are common place in western economies. Commuting is no longer simply from the suburbs to the city centre but is more often circumferential from suburb to sub-centre.

The decentralised urban form and travel flows reflect the rise of car usage. How-ever, public transport networks are devised to be radial from the city centre reflecting historic travel patterns. Travel between two suburbs to get to work could involve two journeys by public transport into the centre and then out again. These public transport time and cost constraints have inevitably reinforced a shift to commuting by car.

The term suburban gridlock emerged in the United States in the 1980s. The stimula-tion of suburb to suburb car traffic flows led to congested circumferential commuting and created long traffic delays. The same phenomena occurred later in European countries as the capacity at peak times of, for example, city bypasses reached saturation.

Edge cities

Edge cities is another term that originated in the United States and was coined by Garreau. He was a journalist who wrote a book that was published in 1991 on his travels around America. In the book, Garreau notes that there has been the development of large-scale outer suburban centres that have the following features:

- Service-oriented economies
- Mixed use – offices/shops
- More jobs than bedrooms

He saw these 'new' urban forms resulting from high car ownership and the benefits of out-commuting from the centre against the primary flow. They also benefit from low land values and hence the ability to provide extensive car parking. Examples he noted included Jersey City and Midtown Atlanta. Garreau sets out a precise definition of an edge city as mixed-use cluster comprising:

- 5 million sq ft of office space
- 600,000 sq ft of retail space
- Large daytime population

These requirements appear arbitrary. This precise definition does not translate around the world, but similar mixed-use developments on the urban fringe are seen elsewhere in the world.

Garreau's edge cities had shopping malls as an essential component. In the UK, it has been argued that some out-of-town shopping centres are becoming edge cities as offices development is attracted and built nearby creating a new distinctive form of peripheral urban development. These centres involve the intensification of land use on the fringe of a city but are at a lower density than central urban locations. The underlying economic reasons for these are considered in more detail in Chapter 12.

City regions

The phenomenon of urban dispersal has changed and is transforming relationships in the urban system. The spatial pattern of economic activity is changing in terms of location of land uses and hence commuting patterns and residential location choices. In previous chapters, these have been considered individually, but here an overview of change is presented including the concept of the city region.

Traditionally, large urban areas, defined by the physical built-up areas, have been thought of as a city and its conurbation, or as a core city and suburbs. In the new dispersed urban world, it is no longer appropriate to equate the urban functional area (as in Chapter 5) by the physical built-up areas. The notion of a core/periphery model of a city is also being rethought, and in particular, spatial economic relationships are now seen as more complex.

Instead, a polycentric urban network comprising a series of interlinking centres with a range of functions can be visualised. As part of this process of dispersal, the intra-regional system of urban areas has become more integrated, particularly in terms of commuting linkages. Travel to work areas have become more open as commuters can travel further to work. Inter-urban commuting has become more common. Travel to work areas are therefore larger with implications for the spatial definition of housing market areas.

The prevalence of two worker households and the geographical spread of work locations have arguably made residential location decisions more complicated. One member of a household may work locally and the other commute over a long distance. The classic trade-off between accessibility to employment in the urban core and lower unit house prices is perhaps no more. However, migration still tends to be over short distances, and local. In some cases, it seems people choose to change employment (and associated location) by reference to where they live.

More widely, a new dispersed sub-regional urban system has emerged in which the key accessibility relationships have been transformed. Accessibility is no longer seen just in terms of travel to the centres of individual urban cores but also to regional centrality nodes. One impact of this change is a 'rationalisation' of the spatial retail hierarchy considered in Chapter 4.

As shoppers can now more easily travel longer distances by car, the pattern of shopping has changed. In this context, Chapter 10 discussed a change in accessibility within cities and the evolution of out-of-town shopping centres. However, improved inter-urban road networks can also lead to a revision in shopping habits at a regional level. In the UK, for example, comparison shoppers are now able to travel to larger centres with more choice. The result is that large centres have grown at the expense of small-town centres that have gone into decline.

The traditional idea of a market or rural town illustrates the transformation. It was traditionally a focal point of trade and services for a rural, primarily agricultural, hinterland. The reality today is that market towns provide public services such as health and schools, leisure activities and food purchases. However, increasing car ownership has led to a restructuring of the inter-urban retail hierarchy (as noted above) with comparison goods now predominantly purchased in core cities.

Reflecting today's dispersed spatial economy, the nature of urban forms needs to be seen in a sub-regional context. In particular, a city region can be conceptualised as a functional economic area based on core urban areas that draw people for work and services such as shopping, education, health, leisure and entertainment. This concept of the city region incorporates the possibility of more than one decentralised centre, such as towns or an airport. It therefore may encompass the nearby urban archipelago islands noted earlier. A city region can be seen as the pinnacle of a sub-regional urban system of towns. They are increasingly used as the basis for urban planning.

It is also possible to think in terms of a polycentric urban region (PUR) with linkages between settlements or centres giving rise to a functional system of areas. Unlike the city region, the PUR does not necessarily contain a dominant node. It could be a region having two or more separate cities, with no one centre dominant, in reasonable proximity and well-connected. These centres could have high levels of interaction but with specialisation between centres. PURs suggest an implied shift away from a central place hierarchy. The classic example is the Randstad region in the Netherlands.

It comprises Amsterdam, Rotterdam, The Hague, Utrecht as well as a number of smaller towns within an area of 90 km by 90 km.

Beyond city regions or PURs, some authors have identified mega-city regions seen as between 20 to 50 urban areas, forming a functional network or cluster around one or more dominant cities, and representing contiguous urban areas. They are similar to 'consolidated metropolitan statistical areas' in the United States. Examples of such mega-city regions include London and the South East of England, and Rhine-Ruhr and Rhine-Main in Germany.

Individual influences on localities

To the general changes to the urban system recorded so far in the chapter must also be added the influence of the individuality of cities. There are differential experiences for individual cities that can contribute to the fashioning of the wider urban networks. In addition, there are variations in urban systems between countries that reflect national circumstances. In this section, the potential reasons for these variations are considered, namely:

- **Population Structure**

  The socio-economic-demographic characteristics of a city's population can influence the pattern of urban activity. Important inter-related factors include age, household incomes and ethnic composition. There are a number of aspects to age that centre around the number and proportion of households who are economically active. Settlements that cater primarily or predominantly for the elderly are likely to have distinctive characteristics.

  Where household incomes are high, there is likely to be lower residential densities and decentralised urban forms are prevalent, which consume more land. Societies where extended families live together may be more likely to have strong traditional local communities, and households are less likely to consider decentralised locations.

- **Nature of Hinterland**

  Urban systems are likely to evolve differently dependent on the historic distances between cities. In the UK, provincial cities are often the order of 50 miles apart, but in many other countries, the distances between urban cores are much greater.

  The physical environment of surrounding areas can also be significant in influencing urban development patterns. Such features include deserts and mountain ranges that can constrain spatial economic activity and in some cases also stimulate industry to meet local demand. The proximity to the sea or river estuaries can offer economic opportunities but also physical limits.

- **Structure of Economic Activity**

  Local urban systems are linked to the nature of industrial activity. There is a close two-way relationship between an area's industrial specialism and its urban form. It extends to the wider network of surrounding towns. There are, for example, different local urban systems associated with say areas that are dominated by textiles or computer software industries. These differences are interwoven with the characteristics of the residential and commuting patterns of the local labour force in these industries.

• *Functions*
Most cities will be multi-functional, but with each function having a different order of importance to the immediate region or country. All cities have economies
comprising elements of manufacturing and services, but specific elements can be
of great significance in promoting their role in the national urban system. This is
most obvious where the city has a tradition as an important retail centre.

Cities may have a history as primarily an educational centre with say a
long-standing university. A university can be the sole basis for the growth of a
town, such as St Andrews in Scotland. The student population can provide the
basis of the local economy. Cities can act as administrative centres as national, regional or state capitals. In some cases, such a designation can be relatively new, for
example, Abuja, the capital in Nigeria since 1991. It provides a focus for long-term
economic growth as many businesses wish to be located nearby seats of power.

Religious centres, too, tend to offer a base for economic growth. Visitor numbers generate a demand for associated services such as accommodation, restaurants and shops. An example is Lourdes, a pilgrimage centre in France. There is
often an overlap between the location of universities, administrative capitals and
major religious sites creating heritage attractions for tourists. Together, these amplify the local economic benefits of each individual function.

• *Specialist linkages between centres*
Cities can be linked to surrounding towns through forward and backward industrial linkages in production as discussed in Chapter 3. An example is subcontract-
ing repairers in towns nearby to a large steel mill in a major urban centre. The
significance for the urban system lies in the economic performance of the core
firm. The closure of the steel mill would have implications for employment in the
wider sub-regional system.

These factors influence the position of a city and its surrounds within an urban system.
Many of them are long-standing but are not set in stone. It means that the performance
of an urban economy and its spatial structure is partially a consequence of its past.
However, while the strength of a local economy is partly dependent on its historic
path, it is also dependent on a city's ability to adapt. In some cases, for example, as
cities have shifted from manufacturing to services centres, the original reasons for
an individual city's emergence often no longer explain its existence. Failure to adapt
could see a city's role in an urban system be 'demoted', even terminal.

Regional growth poles

It is potentially possible to identify regional growth points within an urban system
partially linked to individual characteristics identified above. Harking back to the
stages model earlier in the chapter, core cities were the growth points in Stages 2 and
3, while free-standing towns were in Stage 4. Similarly, at a different spatial scale, edge
cities can also be seen as growth points. Individual, or parts of, cities as growth points
therefore relate to a specific period of development. They involve industrial concentration and are often induced by new transport infrastructure. Their existence is also an
inevitable component of a dynamic spatial economy.

These growth points are not necessarily the same as the concept of a 'growth pole'
as envisioned by Perroux in 1955. He saw a growth pole as based around the spatial

concentration of an expanding industry or group of industries. They provide the basis for innovation that induces further industrial and urban development nearby through backward and forward linkages. In addition, local growth is stimulated by rising employment and improving wages generated by the expanding industry. It leads to growth in other local industries including services such as retailing.

There is an argument that any urban centre above a certain size that demonstrates industrial growth, irrespective of the basis for the expansion, can be called a growth pole. This follows from the fact that the essential dynamics of growth outlined by Perroux are universal. His specific emphasis on the role of innovation is potentially a differentiating factor. However, it is part and parcel of technology-based growth.

Growth poles can therefore usefully describe all expanding localities of employment, with associated impacts on their zone of influence. Given also there is no definitive view on the size requirement for a growth pole, the distinction between whether expanding urban centres are growth points or poles is arguably spurious.

The concept of a growth pole was historically perceived in terms of being a manufacturing centre, centring on say a steelworks. But in western economies, urban areas are now predominantly services based. Today, growth poles can be visualised in these countries as clusters of high-tech or knowledge-based services feeding into cutting-edge manufacturing.

Revitalised urban core cities can be viewed as the epitome of these new reformulated growth poles. They encompass a concentration of expertise or human/intellectual capital, and knowledge infrastructure such as local universities. These industries promote entrepreneurship, creativity and innovation and lead the economic growth in the city region. As such, they generate spillover growth impacts on the neighbouring areas that comprise this functional area through commuting, additional services and industrial linkages.

These industries are not necessarily spatially focused within parts of the city but could be located in office or science parks. The precise location is not normally a key driver of performance or the subsequent impact on the wider urban economy. However, airports and major road network junctions can also generate spatial economic activity in the form of warehousing clustering. They are akin to growth poles, although without significant positive linkages for urban economic growth.

These growth poles discussed above can be described as 'natural' rather than planned. They are distinct from a growth pole established as part of a planning strategy to enhance the performance of a regional economic system. Following the ideas of Perroux from the 1960s on, there have been many attempts to apply such a strategy in both developed and developing countries. Such initiatives continue to occur.

Planned growth poles involve the deliberate focus of investment in a limited number of locations by providing enhanced transport routes and social infrastructure such as housing. In some cases, there are development incentives to private investors. These planned new towns could be designed to enable the deconcentration of regions by siphoning population and economic activities away from high-density cities.

In other cases, the new towns are located to boost the economies of depressed regions. More recently, growth poles have been utilised as tools to modify the urban system to improve national urban competitiveness. However, it is important to note that many have not achieved the promised outcomes and that the use of growth poles is no longer fashionable.

Summary

A stages model has been presented that addresses the evolving pattern of the urban system. It focuses on the relationship between populations *within* cities with the distribution of populations *between* cities. Within the model, the underlying forces for change are changing transport costs between and within cities through time.

In Stage 1, there is little interaction between cities because of poor inter-urban transport. Industrial production is low, just sufficient to meet local needs. The urban system comprises a large number of small cities. While intra-urban transport is also poor, its influence on the urban economy is unsubstantial.

Dramatic improvement in the level of inter-urban transport occurs in Stage 2. The cities with good transport links take advantage of economies of scale to sell to national markets. Economic growth in these cities takes off together with their populations. However, the poor transport system within cities acts as a constraint on growth, and urban densities rise.

The urban system is transformed in Stage 3 by improvements in intra-urban transport that permit people to travel further to work. Cities are no longer spatially constrained by extreme travel limits on their workforces and are able to decentralise from the centre. The increased space now available enables the local economies to substantially expand output, and in parallel, as a result their populations increase dramatically.

In Stage 4, the urban system is now characterised by dispersal. Decentralisation of the urban system now predominates as inter-urban transport improvements rearrange national accessibility relationships. Free-standing and small cities are now able not only to link into the national transport network but also to exploit their advantages of plentiful space and a low-density quality of life. The existence of these dispersed urban areas is referred to by some as urban archipelagos.

The large cities are now suffering from congestion and pollution. Commuting travel times are getting longer emphasising the benefits of a location in a small city. In the worst-case scenarios, commuting takes 90 minutes or more in some cities, and it can be very expensive. Typically, vehicle speeds are only around 10mph in most large cities, on a par with horse-drawn transport over a 100 years ago.

As part of the urban dispersal, the distribution of populations between cities becomes more even. New urban centres have emerged and grown, sometimes rapidly, while older centres often lost populations although subsequently experienced modest re-urbanisation. In developing countries, this rebalancing of the urban system is reflected in a primary/capital city, which becomes less dominant as second-level cities expand.

The dispersed urban system has a number of features. The decentralisation of industry and the suburbanisation of the populations have led to working and living within the same suburb, or suburb-to-suburb commuting or travelling outwards to work at the urban periphery. With public transport networks essentially radial within cities, suburb-to-suburb commuting is by car. The growth of these car traffic flows has led to the phenomenon of suburban gridlock, as the capacity of roads is overloaded at peak times.

An edge city is a new distinctive form of peripheral urban development that was first identified in the United States. These are large-scale outer suburban mixed office and shopping centres. They have extensive car parking to capture the benefits of out-commuting from the centre against the primary flow. The exact form of an edge city varies around the globe. In the UK, some out-of-town shopping centres are becoming edge cities by the development of adjacent offices.

Urban dispersal has involved not only new types of centre and the location of land uses but also the reformulation of the network of spatial economic linkages and flows. The traditional city represented as a core and suburbs has given way to a collection of interlinking urban centres. Accessibility is no longer represented by reference to the centres of individual urban areas but to regional centrality nodes.

Inter-urban commuting has become more prevalent and travel to work areas more extensive. While residential location decisions have become complex given the decentralised pattern of employment, housing market areas have remained localised. Car-borne shoppers have a choice between out-of-town shopping centres and traditional centres. Improved inter-urban road networks enable comparison shoppers to travel to larger centres with more choice rather their nearby small-town retail centres, which have suffered decline.

A city region can now be seen as the pinnacle of this dispersed interlinked sub-regional urban system of towns. It is a functional economic area based on core urban areas that draw people for work and services such as shopping, education, health, leisure and entertainment. The city region incorporates the possibility of more than one decentralised centre and encompasses the nearby urban archipelago islands. City regions as functional areas are the logical basis for strategic urban planning.

Besides city regions, there are the potential phenomena of polycentric urban regions that do not necessarily contain a dominant node. These regions have two or more separate cities that form a network of centres that combine interlinkages with specialism. Mega-city regions can also be identified as between 20 to 50 urban areas, forming a functional network around one or more dominant cities, and representing a contiguous area.

Within these wide networks or urban systems, there are individual city or country influences that can create some variations. A city's or a country's socio-economic-demographic population profile can be associated with distinct urban forms. Close-knit societies with extended families living together are likely to live in traditional compact local communities.

Other local factors influencing an urban system include the nature and topography of surrounding areas creating potential physical barriers to spatial activity. Local industrial specialisms can strongly impact on an urban network although the level of incomes associated with the industry. In particular, higher average incomes are likely to bring lower densities and a more decentralised urban system.

The (historic) functions of cities also play a part in defining their positions in the urban system, even refining it. These roles include acting as a major retail, university, administrative or religious centre. The urban system is also moulded by the specific forward and backward industrial linkages in production, often with small centres having a high dependency on the performance of a core city.

There are often long-standing reasons for the position of a city within an urban system. However, its continuing existence may be a function of its ability to adapt to changing economic circumstances. Cities that adapt quickly can be seen as regional growth points as a spatial economy evolves.

The concept of a 'growth pole' as proposed by Perroux in 1955 envisaged a spatial concentration of an expanding industry or group of industries generating technological innovation. These growth poles are then at the heart of industrial and urban change. In practice, all growing cities have these features so there is no real difference between growth points and growth poles.

Today's growth poles can be seen as clusters of knowledge-based services feeding into cutting-edge manufacturing. Many of these industries are located in urban core cities drawing on local expertise and universities. They represent an engine of growth for the surrounding city region. In addition, within a city region, there can be transport focal points, such as airports, that provide the basis for localised industrial clusters, particularly warehousing.

Planned growth poles were initiated around the world following Perroux's concept. Such planned new towns were sometimes located to promote regional deconcentration by attracting population and firms away from overcrowded cities. New towns have been at times established in depressed regions to stimulate change. It is fair to say that quite often their objectives were not realised and as an urban policy tool they have fallen out of favour.

Learning outcomes

Change in urban systems is partly a function of changing transport technologies between and within cities.

A stages model highlights how the interaction between developments in intra- and inter-urban transport can shape urban systems over time.

Urban systems have become more dispersed. Decentralisation of urban systems has been driven by inter-urban transport improvements, while travel within cities has become increasingly slow and congested.

Free-standing and small cities have benefitted from linking into a national transport network but also exploit their advantages of plentiful space and a low-density quality of life. These urban areas can be seen as part of an urban archipelago.

As urban systems have become more dispersed, the distribution of populations between cities has become more even. Secondary centres have grown in size reducing the dominance of major cities.

Decentralisation of industry and the suburbanisation of the populations have brought significant levels of circumferential commuting by car. It has led to suburban gridlock in many cities

Edge cities in the form of large-scale outer suburban mixed office and shopping centres have been established.

Urban dispersal encompasses a collection of interlinking urban centres with associated complex commuting flows often over long distances.

A city region is a functional economic area based on core urban areas that draw people for work and services such as shopping, education, health, leisure and entertainment. City regions as functional areas are the logical basis for strategic urban planning.

Polycentric urban regions can exist that do not necessarily contain a dominant node.

Mega-city regions are between 20 to 50 urban areas, forming a functional network around one or more dominant cities.

Urban systems are influenced by individual city or country influences. These influences include socio-economic-demographic population characteristics, physical topography, and historic functions and linkages between cities.

While there may be long-standing reasons for the position of cities within an urban system, they have usually had to adapt to economic circumstances.

Cities that adapt quickly can be seen as regional growth poles as a spatial economy evolves.

A growth pole is a spatial concentration of an expanding industry or group of industries generating technological innovation.

Clusters of knowledge-based services in major urban cores are engines of growth for the surrounding city regions.

Planned growth poles have been established across the globe to create new towns to adapt urban systems and promote economic growth. However, they have not always been successful.

Bibliography

Garreau J (1991) *Edge City: Life on the New Frontier,* Anchor Books, New York.

Hall P, Sands B and Streeter W (1993) *Managing the Suburban Commute: A Cross-National Comparison of Three Metropolitan Areas,* Institute of Urban and Regional Development, University of California, Berkeley.

Jones C (2017) Spatial economy and the geography of functional economic areas, *Environment and Planning B: Urban Analytics and City Science,* 44, 3, 486–503.

Parr J and Jones C (1983) City size distributions and urban density functions: Some interrelationships, *Journal of Regional Science,* 23, 3, 283–307.

Parr J (1999) Growth-pole strategies in regional economic planning: A retrospective view. Part 1, Origins and Advocacy, *Urban Studies,* 36, 7, 1195–1215.

12 Real estate impacts of urban and technological change

Objectives

The focus of Part 2 of the book to date has considered the nature of urban change to individual cities and the wider urban system derived from innovations in transport infrastructure. These changes are seen primarily in terms of the populations and functions of cities, in particular the growth of services, the spatial structure of rents and house prices, and density consequences.

This chapter charts the implications for the spatial pattern of specific land uses, notably offices, retailing and warehousing, and the emergence of new property forms. It also assesses the impact of information and communications technology (ICT) on cities. It asks questions about whether in an information age whether the fundamentals of cities still apply.

The chapter examines the evolution of retail, office and warehousing forms. In doing so, it looks at their changing location within and between cities. The structure of the chapter is as follows:

- Diversification of office centres
- New forms of retailing
- Changing retail hierarchy and the decline of high/main streets
- Evolution of warehousing
- Cities in the information age

Diversification of office centres

The revival of cities as services centres was mapped in Chapter 9 with the growth of services employment, in particular from the 1980s. Table 9.2 highlighted the growth of financial and business services employment in the core cities of Britain and the United States. However, as shown in Table 12.1, it is evident that this growth in UK services was not confined to core cities.

Taking a wide spatial perspective during the 1980s, there was actually faster business services growth in free-standing centres. There was also rapid growth in services in other decentralised locations including 'sub-dominant' satellite centres on the fringe of London such as Basingstoke and Guildford. This pattern of business services growth in part involves firms moving from traditional core cities, but business services employment was still concentrated in core cities. The higher percentage increases in the smaller centres reflect the lower original size of services employment.

DOI: 10.1201/9781003027515-14

Table 12.1 The Spatial Pattern of
Financial and Business Services
Employment Change in the
UK, 1981–1991

	%
Core cities	
Glasgow	37.7
Liverpool	10.6
Manchester	30.4
Birmingham	43.4
Newcastle	41.8
Free-standing centres	
Cambridge	67.8
St Albans	88.0
Reading	61.1
Aberdeen	58.0
Sub-dominant centres	
Basingstoke	75.7
Guildford	53.8
Solihull	86.0

It is interesting to note that this decentralisation is often not to suburban locations within a city region but to town centres further down the urban hierarchy, such as the sub-dominant centres. Decentralisation of offices is complex for while some offices are still attracted to urban centres others are no longer choosing traditional localities. Part of the reason has been that moves of office employment from larger to smaller centres are motivated to a degree by lower real estate costs. But the causes of the decentralisation are more complex, and so are the subsequent patterns of office subcentres.

Once upon a time, offices were primarily concentrated at the centre of a city. In some cases, there were separate streets of specialism that also represented prestige addresses. Lawyers congregated in one location and accountants in another and so on. It was essential for business to be located in the requisite street. The central business district was also closely defined, as exemplified by the historic square mile of the 'City of London' as an insurance and banking centre.

Today, there is more likely to be a spatial fragmentation of office locations that has to some extent revolutionised this picture. A new polycentric pattern to office locations within a city region that has developed can be characterised as

- Traditional *'downtown' or central business district* based on walking distances and served by radial public transport
- *Office centre* developed from the conversion of an old prestige residential quarter, often housing 'professional' firms such as lawyers
- A *central fringe centre* resulting from pressure of space in the traditional centres and often occurs via redevelopment of redundant industrial, public facilities or transport land
- *Edge city* often located on a main radial axis, and sometimes near the main airport
- An *office or business park* of suburban low-density developments

An individual city may only have a subset of these subcentres, but it almost certainly incorporates some of them. There are also many offices that are not located in a recognisable concentration, for example, above a suburban shop.

The classification identifies that city centre offices have become more diffuse with three types of central 'submarkets'. In many cities, a range of major developments have occurred on the fringe of the CBD but still sufficiently spatially distanced to be distinct entities or centres. In suburban areas, there are office or business parks set within a green environment and with extensive car parking for employees and visitors. There are also edge cities that often augment a business park environment with a shopping centre.

Widespread fragmentation began to occur in the 1980s in the UK and earlier in the United States and possibly other countries. The focus on the central core has broken down for a number of reasons, not least the growth of the role of services meant the expansion of the demand for offices. New ICT also transformed the nature of the office and provided the catalyst for a change to historic notions of ideal locations.

Decentralisation began with the flexibility provided by the 'motor age', but with the arrival of the 'information age', there has been a major step change. The ubiquitous adoption of new ICT into standard work practices has permitted the ready transfer of information and transformed business practices. It has also had implications for where and how we work, stimulating home working and hot-desking.

The restructuring of business also brought the separation of administrative 'back' offices from higher-order activities such as strategic decision-making, and the emergence of call centres. These activities can be undertaken at different sites. These 'new' demands meant that physical office requirements could not necessarily be accommodated in the existing stock. Some offices were out of date and needed to be replaced. The preferred new office accommodation was in the form of open-plan flexible space. It also led to a rethink on optimum locations.

In particular, ICT has meant businesses in some services sectors at least have been freed from traditional locational constraints such as dealing with customers or business partners face to face (F2F). Following broadly the nature of wider urban dispersal noted in Chapter 11, decentralised locations in the suburbs or in small free-standing centres also have the advantages of less congestion, space for expansion and cheap land. A potential consequence is that the freedom engendered by ICT and road networks could eventually lead to the decline in the dominance of the CBD.

Agglomeration economies revisited

The changing nature of the urban economy it can be argued raises questions about the significance of agglomeration economies. Agglomeration economies were explained in Chapter 3, but it is useful to have a reminder. Agglomeration economies can be decomposed into:

*   **Localisation economies** arising from the concentration of like firms or firms from the same broad economic sector
*   **Urbanisation economies** derived from cost savings from the concentration of economic activity through the mutual sharing of business services, public infrastructure and the availability of labour
*   **Complex activity economies** created by individual firms located close to each other being part of a sequential production process with each firm having backward and forward linkages

Careful perusal of these agglomeration economies reveals that firms require to be within an urban area but not necessarily within close proximity or at the city centre. Indeed, firms can benefit from many of these agglomeration economies at many locations within a city region. The availability of labour, for example, may be better at accessible dispersed locations be freed from central congestion constraints.

While decentralised locations within a city region can take advantage of urbanisation and localisation economies, only the CBD still offers complexity activity economies. Business and financial services (and headquarters functions) continue to locate in city centres for this reason. Firms in these industries require immediate spatial proximity as F2F contact is at the heart of their business.

Banking, financial institutions and associated business services co-locate at the CBD. It enables complex economic activity involving both backward and forward linkages with specialist firms incorporating information transfer. Together, they produce a range of final products encompassing pensions, insurance and investment vehicles. The technology, media and telecommunications sector is another example that congregates in the CBD to benefit from complex activity economies.

These business webs involving deal making and inter-firm personal relationships contribute to a central city 'buzz'. As a result, co-location and F2F contact at the CBD significantly enhance business returns and reduce costs. The specific cost advantages include:

• Knowledge transfer that can lead to creativity and technology spillovers
• Shared common inputs structures and clients
• Promoting trust and incentives in business relationships as meetings enable the detection of lying and commitment

The buzz is particularly important in these rapidly changing business environments and reduces costs.

The buzz of the CBD also extends to its attractiveness to workers. It encompasses the wide range of amenities including shops and cafes to visit at lunchtime or after work which are social agglomeration economies. Many workers are attracted to working in the CBD rather than the suburbs to benefit from these amenities. The inter-movement between firms, as a result of urbanisation economies, supports long-term personal friendships that can be fostered in work breaks.

The attraction of the city centre to some workers demonstrates potential limitations of dispersed office locations, even if they can benefit from urbanisation economies. It may be easier to attract certain types of employees to a core location. The solution is an edge city that combines shops and local amenities, such a childcare facilities, nearby with the advantages of the availability of space and cheap land.

The location of offices and types of submarkets can be usefully demarcated by two extremes. On the one hand, there are business sectors that require F2F contact and a well-educated highly paid workforce dealing with complex information. Such businesses tend to locate in city centres. In contrast, other business sectors can operate at dispersed locations processing standardised data using predominantly less qualified labour. The reality is more complex, and there is a spectrum of types of office users. The choice of location and type of office centre within the city region depend on the precise benefits from agglomeration economies for a particular business.

New forms of retailing

Shopping has been at the heart of the town or city centre for centuries first through (weekly) markets and then on high/main streets. A street of shops became the principal form of shopping centre from the 1800s, whether it was in the city/town centre or along an arterial road in the suburbs. There was also a hierarchy of these shopping centres as described in Chapter 4.

The arrival of the family car began to reshape the pattern of shopping and redraw catchment areas. People no longer simply walked to a local shopping centre or took public transport to shop in their town or city centre. A person did not have to shop at the nearest centre but could travel on to a more attractive one. The range a shopper could travel was now much greater.

At the same time, the urban population has been suburbanising necessarily on its own changing the spatial pattern of retail catchment areas, including their hierarchical structure. Together suburbanisation and new-found shopping freedom led to the decentralisation of shopping centres. It occurred initially in the United States, where the first suburban covered shopping mall opened in 1956.

In each country, decentralisation has ensued differently, in terms of both timing and retail formats. The chapter now focuses on how this decentralisation occurred in the UK. From the 1970s on, in the UK, the process was characterised by Schiller as three waves. It was first seen in supermarkets opening up outside town centres and was subsequently followed by new out-of-town retail forms from the 1980s. First, there were stand-alone retail warehouses (big box shops) and eventually out-of-town shopping malls.

Individual retail warehouses began to open in the first half of the 1980s and were sometimes converted units on industrial estates. By the end of the decade, these box units were being grouped together to form 'retail parks' with common car parking. These developments are known as 'power centers' in the USA. The units on the most recent retail parks increasingly look like shops and are often co-located with a grocery superstore.

Changing retail hierarchy and the decline of high/main streets

The early retail warehouses focused on selling bulky products including DIY, furniture, carpets and electrical appliances. Out-of-town locations suited these types of goods with the advantage of space and lower rents than the traditional high/main street, together with extensive car parking facilities. At first, these developments were seen as beneficial to central urban locations. It eased serious historic space constraints to expansion in high/main streets because of their surrounding urban infrastructure.

However, over time, the bulky goods stores that originally predominated in retail parks have been replaced by mainstream national 'high street' retailers. Retail parks now represent serious damaging threats to town centres. There is also a growing menace to urban retail cores from the development of out-of-town shopping malls.

The retail offering of city/town centres had also responded to customer demands for more convenience. New in-town centre shopping malls have been developed, often with more than one floor. The greater flexibility to consumers offered by this retail format, contributing to the decline of department stores that historically anchored high/main streets. Despite these developments of shopping malls and pedestrianisation, small-town centres have suffered from car parking constraints in comparison with out-of-town centres.

Hand in hand with these changes to shopping centres, there has been a worldwide long-term growth of multiple retailers, many of whom are now transnational companies. Many locally based retailers have been unable to compete with the rise and influence of 'brand power'. A larger specification for new retail units has been driven by these brands, contributing to the attractiveness of retail parks for these retailers. Many small shops at the periphery of major shopping centres have become obsolete and converted or being converted to alternative uses.

As a result of these trends, research finds that in the UK, nearly all the recent growth of retail space is out of town. The consequence is that in many cities, there is now more out-of-town retail space than 'in town'. However, the precise implications for the intra-urban retail hierarchy are uncertain. The hierarchical concept of tiers of centres within a city linked to ranges of goods sold, as set out in Chapter 4, is no longer clearly observed. The most significant discriminator between the city centre and the out-of-town centres in terms of retail offering is in the range of cafes and restaurants. The city centre generally has a higher proportion of space accounted for by cafes and restaurants so that leisure shopping is now most important in determining the top of the retail hierarchy.

The retail hierarchy is changing not just within cities but also between cities. As noted in Chapter 11, improved inter-urban road networks have driven a transformation in shopping habits at a regional level. The phenomenon of 'out shopping' has emerged. Instead of shopping in the local town for comparison goods, such as clothing, households travel to larger centres with more choice. The result is that large regional centres, in some cases modern out-of-town centres, have benefitted at the expense of small-town centres.

Small-town centres have also suffered from the competition from out-of-town centres as retailers have moved to the new retail locations on their periphery. The result is that most shopping centres in the second- and third-tier cities in the regional hierarchy have gone into decline. The inter-urban retail hierarchy has become much more compressed. Both traditional small-town and local suburban shopping centres have also suffered from the long-term rise of the supermarket. Supermarkets now account for nearly all grocery sales with a decline in high street specialist grocery shops, such as greengrocers or butchers, over many decades.

Suburban high streets have shifted from comprising primarily of small specialist shops, many selling different types of food, to a focus on personal services such as hairdressers, dentists, leisure and charity shops. Small/medium sized towns have been forced to consider 'repurposing' to a mixed-use future, whereby many of the empty shops are replaced with housing and leisure.

The wider explanation for the expansion of out-of-town retailing in its various forms relates to the long-term increase in real incomes. It resulted in a rise in the demand for goods to buy and hence more shops. The demand could not be met by the physical constraints of traditional high street shopping centres. But around the millennium, the tide began to turn, supply had outpaced demand, and there were beginning to be too many shops. The recession following the global financial crisis and the falling real incomes that followed led to a shake out of retailers. There were now clearly too many shops, and vacancy rates in small-town centres rose to unprecedented levels, in some cases over 20% by 2009.

Online sales

At this time, online sales were just beginning. The growth of British online retailing started in the second half of the 2000s when broadband connections became available in homes. During 2008–2009, broadband connections became the norm and the effective platform for the online sales revolution. It was the start of fundamental change in shopping habits.

Physical shopping for comparison goods typically involves browsing through the range and prices of goods available in stores. The larger the shopping centre, the greater the benefits of agglomeration economies and the choice of goods. However, shopping on the Internet enables people to search through potentially a wider range of goods with their prices without leaving their home. It is this choice and convenience that has stimulated its rise.

Online sales in the UK at the beginning of 2009 were just over 5% of sales, but by the end of 2019, they were just over 20%, the highest share in Europe. Food sales were least affected by online sales, while the clothing and footwear sector has been significantly impacted. However, the biggest impact of the use of the Internet during this period on shopping centres was seen in closures of many bank branches, and most travel agents, and book and music shops. These closures affected the reasons for visiting high/main streets and hence reduced the revenue from passing trade of the remaining shops.

Online purchases have now become an integral part of shopping choices in the UK. Physical stores are still central to shopping. Many British retailers have promoted a facility to order online and collect in a store known as 'click and collect'. Consumers benefit from the choice provided by the Internet and the convenience that they can pick the purchased item up at a shop with the certainty it is in stock. As a result, almost 90% of UK retail sales still involved physical stores in 2019. This extensive use of click and collect has favoured out-of-town shopping centres for collection by car. Online sales have therefore supported the decentralisation of shopping.

The rise of online sales, taking revenue from physical stores, has also exacerbated the surplus of shops. This issue has been exacerbated by little growth in wages for over a decade after the global financial crisis, with average incomes in 2019 still below that in 2007. By the beginning of 2020, many national fashion shopping chains had begun closing branches. Many of these stores acted as essential shopping centre attractions for customers. The national average vacancy rate had now reached 13% and much higher in small towns in peripheral regions. The vacancy rates were higher than in 2009 after the global financial crisis.

The rise of online sales has been boosted by the pandemic when many stores were closed for months in 2020. The only way to purchase any goods at that time was online. The full ramifications may take some time to be seen. Nevertheless, it is clear that some national retailers have decided to extend and accelerate previous plans to reduce store networks. In the recession after the global financial crisis, it was the small towns that were mainly affected. The expanding redundant excess space from the continuing rise of online sales is likely to be focused on shopping malls and retail parks (power centers).

The pattern of retail change expounded above is the British story of decentralisation. The experience of other countries will be different. There will be common ingredients including the long-term rises in real incomes in western economies after World

War II. Recent stagnant or falling real incomes in many countries are also a mutual thread. Similarly, the suburbanisation of the population/urban decentralisation has inevitably led to pressure for the development/expansion of retail centres away from traditional central locations.

Furthermore, new technologies and products linked to food preparation/storage have meant that lower-order shopping trip patterns have adapted and can be less frequent. The expansion of car ownership has altered distances households are prepared to travel. Together, these developments have brought increased flexibility in shopping patterns.

The result has been a rearrangement of the retail hierarchy, including functions of existing centres at different levels. This is reflected in a changing mix of shops over time and vacancies in shopping centres. The dominant influence until the last decade has been changes in transport costs in influencing suburbanisation and the pattern of shopping trips. Online sales are now in the vanguard of retail change.

These retail changes, encompassing decentralisation and online sales, have been occurring around the world. The actual physical form of the developments and the timing of change have varied. In particular, retail change is also influenced by planning policies of an individual country. The UK has a comparatively restrictive planning system in international terms that has sought to restrict out-of-town shopping centres. In contrast, the United States has had a more liberal approach to suburban shopping malls that has resulted in a clear surplus.

A key issue is how urban planning policies have sought to accommodate the rise of the car. In some countries, strategies have been designed to restrict car usage, while in other cases, the opposite has occurred, providing more city/town centre parking spaces. Planning policies are generally not classified into this simple dichotomy but do influence the spatial development and form of retail centres. This issue is discussed further in Chapter 13.

Evolution of warehousing

Warehouses have existed for thousands of years to provide storage for agricultural produce. As commerce and trade grew, warehouses were often located in central locations at ports to store goods to be loaded on to or received off ships. With the arrival of railways and canals, warehouses were also located in central city locations near the rail head or canal terminus.

Warehouses increased dramatically in size in the 1800s. The arrival of the forklift truck together with standardised pallets led to a configuration step change in the 1930s. As a result, the warehouse abandoned the historic multi-storey model to become a single-storey 'shed' from the 1960s onwards.

Containerisation created a further revolution in the distribution of goods from the 1970s. The standardisation of container sizes significantly lowered the cost of freight transport and shipping times. Part of the changes to the distribution process involved the removal of manual sorting at warehouses and ports. It thereby contributed to the expansion of global trade and flexibility for the locations of warehouses.

As roads became more efficient, particularly with the arrival of motorway or freeway networks, warehouses began to decentralise to be adjacent to road junctions. But the location of warehouses was also conceived on a regional and national scale. A newly improved road link could lead to a reappraisal of where to locate warehouse depots. For example, an improved road between two cities could lead to the closure of a depot at one city when previously there were two. Alternatively, it could result in the closure of both depots a new one opened on a junction halfway between the two cities.

Location decisions are then essentially about cost minimisation and the optimum location changes as the transport network is modified and accessibility changes. The expansion of the motorway network has meant that large distribution centres can serve national/regional markets. Indeed, in Europe, one central location can distribute to a number of countries.

In recent decades, technological change has continued to shape the formulation of the warehouse. From the 1980s, warehousing began to develop from a passive storage system as the use of barcodes permitted a sophisticated inventory management. But the changes to warehousing are not simply about their internal management. The nature of the distribution of goods has changed dramatically. Warehousing is now less about storage and much more about distribution. It has created a new form of warehouse, a distribution centre.

This change is part of a wider use of 'Just in Time' in manufacturing and retailing through supply chain management. In the 1970s, manufacturers used to supply shops directly from their factories. In this distribution framework, goods were stored either in the factories or in the backrooms of shops. Shops had to wait for deliveries by their suppliers. During the 1980s, retailers took more control of the process by setting up central distribution centres to which suppliers would deliver to. The retailers then delivered goods directly to their shops.

In the 1990s, as global trade brought the importation of many goods, the process was marginally adjusted. To encompass global sourcing, an importation centre as an additional stage was added prior to goods reaching the distribution centre. As a result of these distribution changes, retailers could hold less stock, but paradoxically, the use of national or regional distribution centres led to larger and larger warehouses. It also led to a substantial increase in the number of warehouses as well as the amount of space.

This continuing expansion of warehousing has seen the explosion of mega-sheds around the world over the last quarter of a century. These mega-sheds are over 10,000 sq metres and have become possible with automation and improved management systems, and sizes are increasing rapidly. These changes are supported by internal racking systems getting taller that require 'high bay' warehouses between 10 and 40 or more metres high. The warehousing is designed for sorting and picking rather than storage and so there is rapid turnover of 'stock' and heavy traffic of lorries in and out.

The latest stage in the warehousing saga is the reformulation of the distribution of goods that has centred on online sales and home delivery. While the distribution process outlined above has not fundamentally changed and remains, online sales required an alternative strategy. In addition, online sales have expanded exponentially since the late 2000s as the previous section noted. Ecommerce has been the stimulus for a dramatic rise in the amount of warehousing to meet the demand.

The growth of online sales has made distribution processes more complex including the need to incorporate returns. Instead of the traditional delivery to stores, deliveries can be to homes or collection points (including stores and offices). Hub and spoke networks have been established with regional/national warehouses in conjunction with local urban depots. These depots are parcel hubs, where goods can be transhipped between vehicles for local delivery. This local delivery is known as 'the last mile', direct to consumers. The result is that smaller warehouses have been built, and many factories and light industrial units converted to parcel storage and distribution.

Determinants of the location of warehousing

Given the changes to distribution processes that have occurred, it is useful to recap on the locations of warehousing. Despite the improvements in information technology, logistics and supply chain management accessibility to road transport infrastructure is still the key to the location of warehousing. Freight travel, certainly in western countries, is principally by road so that the locations of individual warehousing are primarily chosen by reference to motorway/expressway networks. There are two dimensions to this decision, location within and between cities. These are now considered in turn.

Analysis of the changing intra-urban location of distribution warehousing has coalesced around the term, 'Logistics Sprawl'. It relates to a trend for new warehousing development to locate away from inner urban areas towards more suburban localities on the edge of cities. Studies have reported this occurrence in cities in Belgium, France, Germany, Spain, Sweden, the United States and Germany. However, a study in Paris noted that local parcel distribution is usually subcontracted and that long-standing firms continue to use central historic depots. It means that parcel depots are less dispersed than general logistics warehousing.

The locations outside city centres have many advantages, not just good transport infrastructure connections to the motorway network or airports. There are large areas of readily available land at a relatively low price, offering expansion potential and freedom from conflicts with neighbours. A study in Flanders finds that land price is the most important locational factor. At the same time, central urban locations suffer from a lack of sites with large areas of available land and congestion.

At the inter-urban level, the location of warehousing the importance of the wider accessibility is essential to distribution networks. This applies at the regional and national level, and even cross-national, for example, in terms of European logistics networks. In Europe, there is consequently a concentration of distribution centres in Belgium and the Netherlands.

A study in the United States of 143 metropolitan counties shows that it was strongly correlated with accessibility to air and motorway networks, but to a much lesser extent to rail networks. In the UK, the area centring on the East Midlands and Birmingham is sometimes referred to as the 'golden triangle of logistics'. The area is at the heart of the nation's motorway network providing access to over 90% of the UK population within four hours' drive.

Although a strategic point on the national motorway network is a major factor in distribution location, it is not the only consideration. Distribution warehouses are also located to deliver to large hinterlands. Proximity to large population centres to deliver goods to consumers as quickly as possible is also a significant locational influence.

The result is that a map of distribution centres in the UK shows concentrations at all the major urban cores, not just in the Midlands triangle. Further, the growth of online sales has seen a recent expansion of large satellite warehousing in outlying medium sized urban centres across the country.

The evolution of warehousing has therefore seen dramatic recent changes, facilitated by technological change. Originally, the role of warehouses was for the storage of goods, but they are now collection/transfer points in the distribution process. As part of this transformation, the optimum locations of warehousing have moved to be accessible to motorway/expressway intersections. Accessibility is considered both in terms of the immediate location to the road network but also with regard to national/regional and even continental deliveries.

The changing nature of western economies with the extensive importation of manufactured goods has resulted in a substantial increase in warehousing. This trend has been amplified by the growth of online retail sales. Distribution warehousing has become an established real estate form in its own right.

Cities in the information age

The chapter has so far mapped out how individual real estate forms and the location of land uses have altered the urban landscape since the 1980s. In particular, the changes can be seen as resulting from a combination of the motor age and the emergence of the information age. It is possible to think in terms of these influences in temporal terms as cars first and then computers, although they overlap. In one sense, there are two long-term urban development cycles: first, that driven by adjusting to car usage, and second, stimulated by the development of ICT. This section focuses on the latter as it is still not complete.

The continuing ICT revolution has seen a growing pace in change and began to influence first the nature of office building design in the 1980s, then communications and travel, and the location of land uses. The nature of office work has changed through the adoption of personal computers for virtually all staff, although this is still evolving. As a result, office buildings have had to adapt to the new work environment. The 1980s saw significant obsolescence as many office buildings built only in the 1960s could not be adapted to the heavy cabling required for a new computer age.

Historic offices too suffered during this period. However, the subsequent emergence of improved fibre optic and wireless technologies has reduced the cabling space requirements. On the other hand, the use of computers has led to a shift to large open-plan offices. The types of office buildings required are changing, and arguably, the emphasis is not on building design per se but in the accommodation of equipment. The latest buildings are often light and functional – structures determined by services with short life/cheaper buildings providing flexibility and cheap to replace.

The implications are not confined to the office buildings themselves. It is not just that the office requirements have been transformed but ICT has also contributed to the obsolescence of city centres. The need for redevelopment/renewal initiated decentralisation of offices in the 1980s and promoted the expansion of office parks. More generally, ICT has meant the substitution of information transfer for transport of goods and people. It also enables teleworking from home, hot-desking and teleconferencing.

At the extreme, it creates the concept of electronic villages/cottages with the opportunity for greater spatial freedom. This is reflected in the population of some deep rural areas rising, such as parts of the highlands of Scotland after a century of long-term decline in this peripheral region.

In retailing, ICT has stimulated home delivery of goods after shopping on the Internet, television programmes down telephone lines, etc. It has increased choice and flexibility for consumers. Initially, online shopping was for books and DVDs, a small percentage of sales on the high/main street. However, it is now beginning to have ramifications across all types of shopping centres as the retail industry adapts to 'omni-channel' shopping. The percentage of consumer expenditure accounted for by shops has fallen substantially, affecting traditional small centres in particular.

All sectors of the urban economy are experiencing fundamental change. In particular, there is a reduction in the demand for commercial buildings. The spatial pattern of land uses within cities is also evolving and historic locational relationships changing. Cities are arguably in the process of reformulation. It may take decades for changes to occur as land and buildings are recycled but there are many signs of changes to traditional land use locations. Housing, for example, is replacing offices and shops in city/town centres. The decentralisation forces may ultimately lead to a continuing dilution of the city centre. It leads to questions about the future shape of cities and their very survival.

Cities have so far survived the Internet revolution but do the original reasons for cities still exist? There are some positive signs – the population of cities that were in decline have begun to rise again. Most Internet sales still involve a visit to shop to collect the order. It is useful to revisit the reasons for cities. Agglomeration economies created cities, but do they still apply? The answer must be yes in the sense that cities are still the location for specialist business services. They are still the primary locus/ concentration of labour skills, and there are still household and social agglomeration economies from living in cities.

Summary

The resurgence of cities as services centres has also been associated with continued urban decentralisation. This decentralisation included movement of offices to both smaller centres and suburban locations. Historically, offices were focused on city centres which in turn were demarcated into prestige addresses for different professions.

Now, there is a spatial fragmentation of office locations with a polycentric pattern within a city region. As well as in the traditional central business district, there are a range of other office centres. These can include a converted residential quarter, central fringe localities developed on former industrial land, edge cities and office or business parks. Office centres now represent a diffuse spectrum located from the city centre to the suburbs.

This extensive splintering began with the flexibility provided by the car use, and it accelerated with improvements in ICT. The associated reforms to business practices freed up the spatial separation of business functions and introduced new activities such as call centres. It has led to the redesign of offices and the removal of spatial constraints. The reduction in the necessity of F2F contact has permitted the take-up of decentralised locations with less congestion, cheap land and space to expand.

At decentralised locations, office users can still benefit from urbanisation and localisation economies but only the CBD enables complexity activity economies. The centre serves the needs of industries or business functions that require the immediate spatial proximity through F2F contact. It is therefore the focus of a range of financial services activities.

Labour is also attracted to city centres by the extensive range of social amenities, which people can frequent in breaks or after the working day. Dispersed office locations can counter this city centre attractiveness to workers by the establishment of edge cities with shops and local amenities.

Shopping was for centuries a central hub of cities, latterly as a high or main street. The advent of the family car led to fundamental changes in shopping patterns.

This occurred because the population has been suburbanising and the greater flexibility provided by the car. The resultant decentralisation differs from country to country. In the UK, it has been portrayed as three waves from the 1970s. It was first seen in supermarkets opening up outside town centres and was subsequently followed by new out-of-town retail forms from the 1980s.

The most common out-of-town format is the retail park concept where units have evolved to increasingly look like shops. Many of the units are also now occupied by national brands that were once only to be found in town centres. Brands have been attracted to the larger floorspace units available in retail parks. City/town centres have also changed with the introduction of in-town centre shopping malls but have been disadvantaged by car parking constraints in contrast to out-of-town centres.

These changes have undermined the historic intra-urban retail hierarchy. However, the city centre remains at the top of the retail hierarchy with a focus on leisure shopping, together with cafes and restaurants. The inter-urban hierarchy has also been reformulated. Households have switched from shopping in local towns to travelling to regional centres to buy clothes, etc.

Small towns have been squeezed by competition from sales being diverted to regional centres and to out-of-town retail parks and supermarkets. The result is high vacancy rates in these towns following on the global financial crisis as demand fell.

Online sales have subsequently grown substantially to just over one-fifth of all sales in the UK by 2020. The Internet offers people the opportunity to assess a large range of goods with their prices without leaving their home. The initial impact saw closures of bank branches, travel agents, and book and music shops. In some cases, people can order online and collect in a store through a facility known as 'click and collect'.

The continued rise of online sales, together with stationary real incomes, has exacerbated the surplus of shops. Many national fashion shopping chains had begun closing branches by the beginning of 2020. Vacancy rates in towns and cities were then higher than in 2009 after the global financial crisis. The rise of online sales was boosted by the pandemic in 2020, accelerating plans to reduce store networks, and this is likely to be focused on shopping malls and retail parks.

This account of retail change in Britain may be unique, but other countries have experienced parallel decentralisation with common underlying threads. The result has been a rearrangement of the retail hierarchy, with a changing distribution of store types over time in centres. The car has been the driving underlying force bringing suburbanisation and revolutionising shopping patterns. However, over the last decade, the expansion of online sales has been propelling retail change.

Planning policies also impinge on the nature of retail decentralisation. A key issue is how urban planning policies in different countries seek to contain car usage. The success of these policies affects the spatial development and form of retail centres.

Warehouses have a long history in city centre locations or at ports. Over the years, technological advances have been transformed them from multi-storey buildings to single-storey sheds. Containerisation from the 1970s not only contributed to the expansion of global trade but also brought locational freedom to the siting of warehouses.

Locations of warehouses became linked to road networks to minimise costs. More recently, technological change has brought a new role for warehousing as part of reformulated distribution processes. A just in time approach to distribution means retailers hold less stock in stores and use national or regional distribution centres. It has led to a substantial increase in the number of warehouses, and they have also become larger and larger. The new mega-sheds are designed for sorting and rapid turnover of 'stock' with heavy traffic of lorries in and out.

The emergence of extensive online sales has seen a further dramatic rise in the amount of warehousing to meet the demand. There is now a more complex distribution process direct to homes. Hub and spoke networks have been established with regional/national warehouses in conjunction with local urban depots. These depots enable the transfer of parcels for local delivery known as 'the last mile'.

There is an international prevalence of 'logistics sprawl' as new warehousing development moves away from inner urban areas towards more suburban localities on the edge of cities. However, some parcel depots continue to use central historic premises. The reasons lie in accessibility to transport connections and also readily available cheap large land sites. At the inter-urban level, there are regional concentrations of warehousing chosen to facilitate distribution of goods across whole countries, even continents. In addition, closeness to large urban centres is also important to ensure efficient delivery to customers.

The growth of online sales has led to a further step in the increased role for warehousing and the expansion of demand for these buildings. The consequences are seen in the substantial growth of large satellite warehousing in outlying medium sized urban centres.

These changes in offices, shops and warehousing are part of a wider configuration of the urban system. Since the end of World War II, cities have experienced and have been impacted by two long-term urban development cycles. The first centred on car usage, while the second has come from the thrust of changes brought about by ICT.

The arrival of personal computers has led to a revolution in office work with the implications for the design and use of buildings. ICT has led to some offices becoming obsolescent and created much more locational freedom, whether it be in office parks or working from home. In retailing, online sales are taking hold as a normal element of shopping and consumer expenditure accounted for by shops is being reduced. The result is that there is falling demand for offices and shops, while at the same time, warehouse space is expanding.

The locations of land uses within cities have also seen long-term changes with pressures for decentralisation. Town and city centres have been subject to the greatest threats to their original functions. There are questions about the continuing importance of agglomeration economies, the original reasons for cities. While some aspects are now diluted, certainly in relation to a city centre location, there are still economic benefits of spatial concentration for some industries. The populations of cities also provide a supply of labour and ensure there are household and social agglomeration economies.

Learning outcomes

The renaissance of cities as services centres has also been associated with continued urban decentralisation.

There is now a polycentric pattern to office centres in cities, encompassing the city centre and office parks.

The motor age and improvements in ICT have led to a rethink on the design and location of offices.

At dispersed office locations, firms can still benefit from urbanisation and localisation economies but only the CBD offers the opportunity for complexity activity economies.

Decentralised office locations can counter the city centre attractiveness to workers by the establishment of edge cities with shops and local amenities.

The historic role of central urban locations for shops has been undermined, as a result of suburbanisation and the greater flexibility provided by the car.

The form of retail decentralisation varies from country to country and began in the UK in the 1970s.

The most common out-of-town format in the UK is the retail park concept where units are clustered around car parking.

Many units in retail parks are now occupied by national brands that were once only to be found in town centres.

The city centre continues as the peak of the local retail hierarchy with an emphasis on leisure shopping supported by cafes and restaurants.

The inter-urban retail hierarchy has been restructured. Many households have been drawn away from small towns to regional centres to buy clothes, etc.

Small towns have also lost out to competition from out-of-town retail parks and supermarkets. The result is high vacancy rates in these towns.

Online sales have grown substantially contributing to a surplus of shops, accelerating plans to reduce the store networks of national retailers.

Planning policies influence the form of retail decentralisation, depending on their approach to accommodating car usage.

The traditional role of a warehouse has been subject to fundamental change, impacting on their design and location.

Regional and national warehouses can now be described as mega-sheds to enable rapid sorting and delivery as part of a just in time approach to distribution.

Online sales have resulted in an explosion in the demand for warehousing.

With the distribution process for online sales requiring transportation direct to customers' homes, hub and spoke networks have been established. It involves depots where there is the transfer of parcels for local delivery known as 'the last mile'.

The effective revolution in warehousing has led to 'logistics sprawl' as they are increasingly located on the edge of cities. These locations permit benefits from the accessibility to road networks and the availability of cheap land.

There are also regional concentrations of warehousing at locations that expedite the national/continental distribution of goods.

There have been two long-term urban development cycles caused by the motor age and then the information age.

The extensive use and power of computers have had fundamental implications for every aspect of life, including work, shopping and leisure. The consequences are seen in falling demand for offices and shops, while at the same time, warehouse space is expanding.

Decentralisation has threatened the traditional functions of the centres of urban areas.

The importance of agglomeration economies, the original reasons for cities, can be questioned.

There are still economic benefits of spatial concentration in city centres for some industries and attractions for people to live in cities.

Bibliography

Dablanc L and Browne M (2019) Introduction to special section on logistics sprawl, *Journal of Transport Geography*, 88. https://doi.org/10.1016/j.jtrangeo.2019.01.010

Hall P (2001) Global city-regions in the twenty-first century, in A.J. Scott (ed) *Global City-Regions: Trends, Theory, Policy*, Oxford University Press, Oxford, 59–77.

Hall D and Frodsham M (2019) *Occupational Drivers of Investment Performance in the Logistics Sector*, IPF, London.

Jones C (2013) *Office Markets and Public Policy*, Wiley-Blackwell, Chichester.

Jones C and Livingstone N (2018) The 'online high street' or the high street online? The implications for the urban retail hierarchy, *The International Review of Retail, Distribution and Consumer Research*, 28, 1, 47–63.

Jones C (2021) Reframing the intra-urban retail hierarchy, *Cities*, 109.

Schiller R (1986) Retail decentralisation: The coming of the Third Wave, *The Planner*, July, 13–15.

13 Real estate investment, planning and urban economic change

Objectives

The spotlight in this chapter is on the relationships between urban change, planning and real estate investment. The premise of the chapter is that urban change in its broadest sense can be moderated by land and property owners as well as planning policies. It draws on knowledge from previous chapters. Chapter 6 explained the logic of planning. In Chapter 8, it was noted that much of the commercial real estate stock in cities is owned by financial institutions and large property companies known as RE-ITs. In particular, this chapter examines to what extent and how land and real estate ownership, together with planning policies, can mould the configuration and the pace of urban economic change. It is primarily concerned with the relationship between urban development, new property forms and property investment.

The structure of the chapter is as follows:

- Land market and urban development
- Planning policy responses to urban change
- Investment, urban change and new real estate forms
- Establishment of new investment classes
- Longer-term implications for real estate investment in retailing
- Financial institutions and other real estate innovations – distribution warehousing, green offices and mixed use
- Wider implications for the real estate market

Land market and urban development

In Alonso's model as set out in Chapter 4, the market is assumed to be perfect in the sense that there is a free market with perfect information. Land rent is determined in the here and now, and there are no leases that constrain the market. Firms are presumed to maximise their profits in deciding the rent to bid for different locations. Owners accept the highest bid for their land. As a result, land is allocated efficiently by the market to the most profitable use based on the rental value.

There are great insights from the Alonso model although it is based on restrictive assumptions. In the analysis here, two assumptions are relaxed. Given that the concern here is with urban development or urban change, the land market is taken not to be static. Second, the behaviour of landowners/developers is seen from a more

DOI: 10.1201/9781003027515-15

behavioural perspective. In other words, land is not sold or let to the highest bidder necessarily as it depends on a landowner's objective for or reason for holding land.

A good example is a farmer who owns agricultural land on the edge of built-up area of a city that is expanding. Developers are looking to buy agricultural land to build housing to meet the growing demand. The farmer's land could be worth far more for housing than for agricultural use. But the farmer may not want to sell. A farmer may hold out for a land value higher than the going rate for greenfield housing because of the potential loss of way of life. Another farmer may be more willing to sell, say because of approaching retirement.

Similarly, some wealthy landowners may view the pleasantness and enjoyment of living of their estates as far more important than very lucrative offers made by developers. In these examples, individual preferences about lifestyles can trump monetary values. These landowners may also see it as part of their long-term family heritage. In other words, the motives and circumstance of landowners vary and they are important in shaping the dynamics of urban change. The result is that the actual urban growth may not follow an orderly pattern.

These examples illustrate that there are different types of landowners. The two above are users of land as part of a production process and consumers of land, respectively. To these two groups can be added those who have explicitly bought land as a financial investment. In addition, land can be purchased for public or community ownership. It is also possible to distinguish between active and passive owners. This dichotomy does not necessarily map to the groups above and relates to the behaviour of owners rather than why they hold the land.

Active owners try to overcome site constraints to make land marketable/suitable for development. Such activities could include interfacing with the planning system to improve land value perhaps through a change of use classification. These landowners may also seek to improve the development potential of their land by resolving physical/infrastructure constraints, possibly by buying adjacent land. On the other hand, passive owners make no effort even if they plan to sell.

Land market constraints are often seen as a particular issue in the inner parts of cities, subject say to the closure of industrial plants brought about by deindustrialisation or decentralisation. These obsolescent industrial buildings ultimately can become large tracts of derelict land awaiting redevelopment.

The problem is often not that these individual surplus land plots are not advertised for sale but the asking price is too high, given the broader state of the surrounding environment and potential contamination. In fact, the true value of this land may be actually negative given the capital expenditure required to redevelop to an alternative use. This in turn justifies some form of state intervention as discussed in Chapter 23.

To summarise, the operation of the land market therefore impinges on urban change, both at the periphery and in terms of redevelopment. The nature of the relationship between owner and land is likely to influence decisions about making land available for development. Active landowners accelerate or promote urban developments, while passive ones may constrain change. Irrespective of landowners' inclinations, there can be also barriers to the redevelopment of obsolescent buildings.

Once the land has been purchased for development, property companies have an important role in the (re)configuration of a city. Their decisions are crucial to moulding a city's streetscape and the spatial pattern of land use. In theory, property developers seek to predict and meet the needs of the community as a way of maximising profits. Nonetheless, it is possible within a behavioural perspective to see property development as not necessarily tuned to meet the demands of the local economy and to produce the most efficient use of urban land.

There are a number of reasons, including poor decision-making perhaps based on inaccurate information. Some authors argue that development is driven by investment needs of property companies/institutions rather than the user demands of a city. The result can be 'white elephants', such as shopping centres in the wrong place, overbearing tall office blocks or half-empty housing developments. In any case by choosing the design and locations for new development, developers shape future urban change for better or worse.

Planning policy responses to urban change

Planning is visualised in Chapter 6 as a means to regulate real estate markets and the urban economy. But planning policy is also a value-driven activity with social goals and a view about the nature of urban form and change. Historically, planners have viewed themselves as physical or environmental planners and social engineers in terms of shaping neighbourhoods and cities, or through creating new communities.

Chapter 6 also refers to planning as intervening in the land use market and in the process of shaping, regulating and stimulating real estate markets. From this perspective, planners can embrace their role as market actors in order to manipulate the market to achieve social goals. The chapter now considers the interface between specific planning policies on the one hand and market forces on the other.

The view of planners as market actors implies some engagement with market forces. However, this view is not necessarily recognised by practitioners. For example, the promotion of 'urban sustainable development' usually has only limited reference to market forces. This issue is discussed in Chapter 16. In fact, many practitioners take the view that market forces have to fit in or be expressed within the plan set out for the city. In some cases, this means that planning policies are designed not to shape, regulate or stimulate the market but to stop market forces.

In effect, this happens with regard to retail planning in many countries that are designed to maintain traditional urban retail centre hierarchies. While trying to support and revitalise high/main streets, these policies seek at the same time to deter 'out-of-town' shopping centres. These policies are often justified on the basis that existing centres should be safeguarded, as they offer distinctive qualities including access to public transport.

The UK is a good example of where these policies have been applied. After a decade in the 1980s when planning acquiesced to the development of out-of-town shopping planning policy was reversed. It shifted in the mid-1990s to one of defending the status quo of the existing urban retail hierarchy, especially town centres.

To achieve this goal, there is a sequential test for site selection that new development must pass. To be successful, out-of-town retail planning applications have to demonstrate that there will be no adverse impact on the vitality and viability of existing town centres. This impact is measured by potential falls in rental and capital values (as well as other criteria).

The evidence is that the implementation of these policies in the UK has misfired. The successful establishment of out-of-town shopping centres described in Chapter 12 has progressed in hand with the decline of many traditional shopping centres. Despite continuous policy over more than two decades aimed at restricting out-of-town retail centres, they have flourished. These centres often offer an equivalent range of stores as a city or town centre.

This raises important generic questions about the goals of planning. First, there are serious question marks over the efficacy of policies that aim to maintain a historic retail hierarchy when cities are dynamic entities. This is particularly true as there is continuing and long-term suburbanisation and decentralisation across the world.

Second, in western economies, retail planning has to embrace the motor age (and online sales) and changing shopping patterns and hence shape market forces to ensure a sustainable urban future. Overall, the evidence points to the limitations of planning in seeking to arrest or stop market trends, in particular towards decentralisation.

Investment, urban change and new real estate forms

With much of the commercial real estate stock in many countries owned by financial institutions and property companies (REITs), much economic power is vested in these investors. For brevity, these different types of investors are all referred to as financial institutions in this chapter. They have a similar investment perspective in terms of their choice of individual real estate assets. This chapter now considers how their investment strategies respond to the challenges brought about by urban change. In particular, it considers how they embrace the arrival of new forms of real estate.

It is useful at the outset to illustrate the importance of investment by financial institutions. The influence of financial institutions is illustrated by the London centric investment strategies of financial institutions in the UK. The majority of their investments are in London and the South East of England where most institutions are based.

This preference is reflected in the concentration of new office development in the London area. The area receives a greater share of private office construction than its national proportion of banking, finance insurance and business services employment. The result is that office development in provincial cities may not be keeping pace with the growth of business services.

As a consequence, insufficient new office supply is acting as constraint on their local economic development. It means that London has an advantage as it is replacing its older offices quicker and improving the quality of its stock compared with provincial cities.

These observations are important for an understanding of real estate investment by financial institutions. Investment decisions are logically determined by reference to a combination of risk and return. The risk is perceived to be much lower in London than in provincial cities, hence the focus on investment in the capital.

The spatial pattern of investment reflects that financial institutions are risk averse. More generally, financial institutions choose to purchase, as a general rule, only 'prime property' that are located so that there is consistently excess demand for it (see examples later). This safety first behaviour is highly relevant to looking at their response to urban/real estate change.

The relationship between urban growth and returns to real estate investment was more broadly highlighted by Henry George who writing in 1879 (previously quoted in Chapter 6) during a period of dramatic urban expansion noted,

> *"Our (land owner).... is now a millionaire. Like another Rip van Winkle he may have lain down and slept, still he is rich .... from the increase in population."*

The forebears of the Duke of Westminster in London did just that. The family owned some farm pastures in Mayfair on the edge of the London built-up area during the 1700s. The subsequent growth of London engulfed this agricultural land, and its value rose dramatically. Its impact is demonstrated by the traditional British Monopoly board game in which Mayfair is the most expensive property (in London). Moreover, the Duke at one time was the richest man in Britain.

Real estate investment is no longer so simple or as lucrative. The market conditions of the 1800s were growing cities around a central core or node. These circumstances no longer necessarily prevail, and as seen in previous chapters, urban growth patterns have become more complex in recent decades.

The initial decentralisation that occurred left the spatial structure of land use in cities unchanged until relatively recently. However, from the 1970s, as noted in Chapter 12, there was a new urban development cycle brought about by the motor age. This cycle was followed by a second, centring on the development of information and communications technology (ICT). The result is that cities have experienced a revolution in the shape and organisation of land uses.

Technological change has also always led to the evolution of real estate forms. It is useful to see this in historical perspective. The invention of the electric lift led to multi-storey office blocks from the 1880s. The requirements for offices were transformed from the 1980s with the ICT revolution, especially with the universal adoption of the personal computer. Decentralised business space was also enabled by ICT and car usage.

The factory originated with steam power and the industrial revolution. Industrial estates were set up with the arrival of electricity in the 1890s. The industrial shed became part of the urban landscape with the emergence of production lines and motorways. Distribution warehouses were supported by developments in ICT.

In the retailing sector, the high/main street was the locus of activity from at least the eighteen hundreds in western cities. Suburbanisation has always created demand for new shops, but until recently, they have taken the same high/main street form. Ultimately, out-of-town developments followed suburbanisation and the flexibility provided by cars for shopping. In the UK, the primary form of out-of-town centre was a retail park (power center) as explained in Chapter 12.

Substantive direct investment in real estate assets by financial institutions began only in the 1950s. At that time, the choice was simply between high/main street shops, city centre offices and factories or industrial units. These can be viewed as real estate investment classes. In the UK, shops were initially the most common form of real estate investment, but they were overtaken by offices in the 1970s. Shops regained their popularity in the 1990s as offices suffered from obsolescence with advances in ICT. There is therefore a degree of fluidity in institutional portfolios.

The picture is more complex than this simple story as Table 13.1 records. The table chronicles both the changing nature of real estate assets and their contribution to the stock owned by financial institutions for the period 1981–2010. These three decades cover the period of substantial upheaval noted above associated with new urban development cycles.

The continuing fall in the proportion of the aggregate 'institutional portfolio' accounted for by standard office investments from the 1980s is not matched by a growth in standard shops. Instead, new investment classes emerge and populate the institutional portfolio.

Table 13.1 The Changing Aggregate Institutional Portfolio in the UK by Asset Class,
1981–2010

Property Segment	1981 %	1985 %	1990 %	1995 %	2000 %	2005 %	2010 %
Standard Shop	15.2	18.7	16.8	16.8	13.8	10.9	10.3
Shopping Centre	9.4	12.1	12.7	17.3	18.4	19.6	17.9
Retail Warehouse	0.6	0.9	2.9	7.8	12.4	18.1	19.4
Department/Variety Store	1.9	1.8	1.9	2.2	1.8	1.5	1.3
Supermarket	0.7	0.9	0.5	2.1	1.5	1.4	3.8
Other Retail	0.3	0.4	0.5	0.7	0.7	0.6	0.8
Standard Office	56.2	52.2	50.5	36.8	33.6	28.5	27.8
Office Park	0.0	0.5	2.2	2.9	4.6	4.1	3.4
Standard Industrial	15.0	11.9	11.1	11.1	10.4	12.3	11.9
Distribution Warehouse	0.5	0.5	1.0	2.4	2.8	3.0	3.3

Source: MSCI/IPD (2010).
Note: After 2010, this database begins to lose its comprehensiveness.

Comparison of 1981 and 2010 reveals a dramatic transformation in the pattern of investments by financial institutions. There is a decline in the role of traditional real estate asset classes. Standard offices fall as a proportion of the total by more than half, standard shops by a third and standard industrial by just over a fifth.

There has been an expansion in the new real estate forms, notably shopping centres and retail warehouses/parks, and to a lesser extent office parks and distribution warehouses. Shopping centres doubled their share of the 'institutional portfolio' from 9% to 18%. Institutional investment in office parks grew from zero to almost 5% in 2000 but had fallen away a little in the subsequent decade. The largest increase in the aggregate portfolio share is that of retail warehouses: from almost zero in 1981, and still under 3% in 1990, to 18% in 2005. In 2010, it is almost one-fifth of the value of real estate assets owned by financial institutions.

These statistics in Table 13.1 demonstrate the momentous upheaval in land use patterns and real estate forms in cities over 30 years. It appears that financial institutions have adapted their real estate portfolios to the new urban reality. However, closer examination gives a suggestion of a lag between new development of new real estate forms and actual investment. This is most evident with retail warehouses that emerged in the 1980s (see Chapter 12) but only gain a noticeable foothold in institutional portfolios during the latter half of the 1990s. This phenomenon is not surprising because institutions are cautious investors while new investment forms and new locations suggest risk.

The emergence of new investment classes

The process by which new real estate forms are accepted by financial institutions as suitable investments can be slow. It is useful to examine this process as a pathway in the form of stages by which new real estate forms can become established in the institutional portfolio. Three potential stages can be identified as follows:

• Initially, new real estate forms are built by specialist development companies. This is because conservative institutions only invest once the market for new investment forms has 'matured'. At this point, profits from these new innovative real estate forms are weak because rent levels are low to attract demand.

How long this stage lasts depends on the degree of newness, the extent to which developments meet consumer needs and the acceptance by the planning system. During this stage, the product may evolve substantially, and capital values are very low to reflect the risk of obsolescence.

- The next stage is a growth period that sees rapid market acceptance and increasing rents. At this point, potential competitors who are watching developments in the initial stage join a potential bandwagon and enter the market. The property company's task is now to sell not just the new product to tenants, but also its particular scheme.

Growth may still be constrained by planning regulations that seek to maintain the *status quo*. There is increased interest in purchasing properties as an investment by financial institutions. Nevertheless, there are continuing perceived risks of obsolescence as the product develops and uncertainty about long-term acceptability to occupiers.

These concerns ensure that investors/purchasers require a return that factors in a high-risk premium. In turn, capital values remain relatively low. The relative prices of a new real estate form in comparison with the equivalent standard or traditional asset class are therefore lower because it is seen as more risky.

- If there is a period of sustained letting activity and the product achieves credibility as an investment, then a mature or sustainable market is established. The latter marketability is demonstrated by an active resale investment market. Such market acceptability implies the achievement of a critical mass. The asset class needs also to possibly demonstrate its ability to sustain lettings through downturns in a real estate cycle.

The acceptance of the real estate form as a mainstream investment medium leads to a fall in the associated risk premium and a rise in capital values. This process illustrates that the development of a mature market for an asset class requires product acceptance in the occupier and investment markets.

At the beginning of this process, the risk premium is higher for the new asset class than for comparable standard property forms. As the market matures, this differential should disappear and may even be negative.

This process can be seen with the emergence of retail warehouses/parks in the UK. The first wave of these developments in the 1980s was undertaken by a range of specialist and local property companies. In the 1980s, there is an investment 'risk premium' differential of approximately 2% between retail warehouses and standard shops. This risk premium rises during the recession of the early 1990s reflecting the surety of traditional assets in a downturn. The risk premium subsequently reduces to 1.3 by 1993, but a low differential lingers through the 1990s.

Only in 2000, does the risk premium on retail warehouses fall below that of standard shops. By then, all retail parks are predominantly owned by financial institutions. This suggests that the process of market establishment for retail warehouses took almost 20 years. The process was distorted and slowed by the macroeconomic/property cycle.

In comparison, office parks that were introduced as a new asset class at broadly the same time have a different investment maturity experience. The risk premium differential between office parks and standard offices is much more short-lived. Within six years of office parks appearing, the differential disappears.

The difference partly reflects that a retail warehouse is a distinct new real estate form rather than an office park that can be described as a traditional form in a new location. Additionally, the retail warehouse evolved over time to be located in a retail park including much obsolescence. There were also uncertainties created by retail planning policy discouraging retail warehouses.

The conclusions of this analysis are that the major reshape of intra-urban land use patterns that brought the introduction of new real estate forms has had significant implications for investors. Financial institutions ultimately embrace the new real estate forms within their property portfolios. However, this adoption of new real estate forms required the establishment of investment markets for these innovations. For retail warehouses, this process took two decades.

The longer-term implications for real estate investment in retailing

The changes in the retail sector have broader fundamental implications for investment by financial institutions that go beyond simply the purchase of new asset forms. Before out-of-town retailing, investment was confined to city/town centres where the traditional prime shopping pitch was at the central location within the high/main street.

This central location was at the tip of the retail hierarchy in a town or city. It was the most accessible location with always excess demand for its location. When a shop tenant left, there would be another waiting. If a particular type of shop was no longer viable, other shops selling different goods would be waiting to take their place. Investing in shops was a very safe option.

Today, the UK retail investment environment has changed. To recap, the emergence of shopping centres has reduced supply constraints. There has been a vast expansion of space which has not been in traditional town centres. As Chapter 12 notes, the arrival of out of town centres has contributed to excess shops. In fact, following the global financial crisis, it was the first time that there had been excess shops/high vacancies. A trend that continues with the growth of online sales.

The excess shops inevitably increase the risks for investors. Initially, it meant that individual investments in small-town centre shops were under threat from the lack of demand. In 2006, the number of standard shops owned by financial institutions was less than half that of the mid-1990s. But there are also changes to the underlying dynamic of retail investment.

For a standard shop in a town centre, the retailer is responsible for fitting it out, not the landlord. In contrast, shopping centres require periodic upgrading to ensure they are attractive to customers and maintain footfall. It therefore introduces obsolescence to retail investment. On top of that, the excess shops mean increased competition between shopping centres for custom and a greater pressure on periodic upgrade spending by investors.

Financial institutions and other real estate innovations

Distribution warehousing

Chapter 12 records the long history of warehousing from passive storage to its present formulation as distribution centres. The overall impact has been a revolution in the occupational requirements of and the demand for industrial real estate. There has also been a dramatic rise in the amount of logistics warehousing in recent decades.

Given the rapid expansion of the sector, it has a high proportion of new stock. With online sales continuing to surge as part of a long-term upswing, there is a consequential high occupancy rate in warehousing. Many of these distribution centres have tenants on particularly long leases compared to other real estate sectors. However, only a limited percentage of this real estate class is owned by financial institutions in the UK.

They represent only a small proportion of the institutional real estate portfolio. As Table 13.1 indicates, distribution warehouses represented only 1% in 1981. Despite their soaring importance, these warehouses only represented 4.4% of real estate owned by financial institutions in 2014. This lack of interest contrasts with what occurred to retail warehouses as they became established in the urban retail landscape.

A comparative analysis of the evolution of distribution warehousing and retail warehouses reveals some initial similarities to this process. Initially, just like retail warehouses, speculative and specialist developers primarily build these warehouses. However, there was subsequently a divergence with a trend towards bespoke building (often with a pre-let to a retailer) and owner occupation. The lack of interest by financial institutions almost certainly lies in the role of product development

The essential long-term shape of retail warehouses coalesced at the beginning of the 1990s with the arrival of retail parks as a standard format. In contrast, distribution warehousing as an investment class is clearly subject to perpetual obsolescence with the impact of technological change, while there is now a convergence to a broad norm of the characteristics required. The rethinking of the optimum distribution processes, including locations, has meant that often businesses choose new facilities over vacant warehousing.

At best, these innovations lead to long letting periods and lower rent for second-hand properties. With a relatively short development timescale, logistical innovations accelerating there is investment uncertainty that suggests a high-risk premium for this investment class. All these factors contribute to a lack of investment attractiveness to financial institutions.

Green offices

There is increasing concern about sustainability of buildings given that they are a major source of greenhouse gases. The wider issue of urban sustainability and the housing market is taken up in Chapter 16, but here, the chapter is concerned only with green offices and the attitudes of occupiers and investors. The existence of green offices is relatively new, and there is no universal definition with classifications varying across the globe.

Reference to green buildings in the United States relates generally to the Leadership in Environmental and Energy Design (LEED) classification, whereas in the UK, it is to British Research Establishment Environmental Assessment Method (BREEAM). BREEAM was launched in 1990, and the LEED rating system began life in 1994.

The higher environmental standards of green offices implicitly imply higher costs. Historically, there have been barriers to the greening of the office stock involving who should pay these extra costs. The issue has been encapsulated traditionally by what is known as the 'vicious circle of blame' in which developers, investors and occupiers all blame the others for the lack of progress. The challenge for change is exemplified by the fact that the existing established (non-green) office norms were attractive to occupiers and investors.

The established standard form was the ubiquitous architecturally fashionable air-conditioned offices found around the world. These offices generate significant energy demands. There are a generation of air-conditioned offices built from the 1980s with other physical specifications to meet the development of ICT. They quickly became accepted as the model for 'prime' offices by financial institutions.

This 'institutional standard' became seen by investors as essential for long-term attractiveness. At the same time, occupiers seeking prestige office space saw these buildings as meeting their needs. Nevertheless, they are at odds with the general world policy commitment to promote green buildings.

One potential avenue to the greening of the office stock is via the existence of a green premium. If there is a premium rental value for green-labelled offices in a city logically, it should lead to the new development of these types of buildings if land is available. In other words, increased rents (and hence capital values) could make green developments (and refurbishments) profitable.

The reality is more complex because choosing a new office involves normally considering a range of substitute properties in which the choice balances a range of attributes such as rent, location and size with green features. The tenant's choice is likely to involve the trade-off between the 'positive' and the 'negative' characteristics of the different offices.

The greening of the office stock can be seen as a process where at the beginning the greenness of offices is a secondary demand characteristic subservient to location and the suitability/functionality of space. In most countries, the green credential of an office is still not a key demand characteristic. Over time, if the benefits of green offices to tenants become widely accepted, then they would become more attractive than the older non-green stock. These views would be reflected in the relative rents.

Ultimately, the culmination of this process is that non-green offices become obsolete and require refurbishment or complete redevelopment. Such 'green obsolescence' is an end or a near-end state to the greening process, but at present in most localities, non-labelled offices constitute the larger proportion of the stock.

It could take some years and be dependent on the supply of new green offices. The dynamics of change within real estate markets is slow, influenced by prevailing lease structures and the state of the national economy. It is also dependent on any restrictions on new development, including planning and conservation regulations.

An 'external' driver is the threat of the green policy agenda that much of the world has signed up to address climate change. Increasingly, since the millennium, governments, including the European Union collectively, are enforcing stricter green standards, not only for new offices but also for the existing stock. These policies inevitably accelerate green obsolescence. New greener building technologies have also supported these initiatives.

In the UK, it appears that it is the obsolescence threat that persuaded many investors in new and refurbished offices to go green. This has happened even though there was no evidence of a green rental premium. Instead, there is a belief that green buildings are easier to let, thereby enhancing future rental income. The result was an explosion of green developments during the 2000s. Even so, the proportion of green office stock is still in the minority and varies significantly by city. There are clearly important local real estate market factors at work.

Mixed-use development

Planning policies around the world have embraced a desire for mixed-use neighbour-hoods. The basic tenets can be traced back to the arguments of Jane Jacobs in her 1961 book, 'Life and Death of Great American Cities'. Instead of the then planning ortho-doxy designed for the separation of land uses to avoid externalities, mixed use is seen as the basis for vibrancy, innovation and enterprise. Jacobs emphasises the intensity and character of street life and the public realm, created by the intermingling of land uses and high urban density.

Today, mixed land use is also seen by planners around the world as contributing to ur-ban sustainability (see Chapter 16). It is normally linked to high residential density which is seen as increasing the viability of local shops and services, and the take-up of public transport. There is an intensification of land uses in a city. Mixed land use is also often closely associated with diversified (tenure/social class) residential communities in plan-ning policy. Much of these policies are advocated under the banner, 'New Urbanism'.

The precise definition of mixed-use development is in practice blurred or at least subject to interpretation. In the traditional model, a mixed-use high-density neigh-bourhood was a traditional high/main street. In these neighbourhoods, there was a mix of retail and business premises intermingled with residential dwellings. Housing was either above the commercial uses or in close proximity.

Mixed-use development today can take a number of forms. It can be an individual mixed-use building, a large development scheme or a mixed-use neighbourhood/town centre. A comprehensive mixed-use development scheme would comprise buildings each with separate if interrelated uses and perhaps linked public space.

Despite the planning consensus towards the desirability of mixed-use development, there are barriers to its adoption. There can be a clash between the policy and accepta-bility to investors and homeowners, and hence market viability. There may be a lack of depth to the demand for mixed-use neighbourhoods. New house buyers may not want to live above shops and be unhappy about industrial use nearby with environmental health disadvantages.

Many speculative housing developers, with an inevitable conservative perspective, would tend to avoid mixed-use developments. These obstacles are more likely to be significant in the suburbs, whereas in new city centre housing developments, it is more acceptable to residents to incorporate offices or retail units.

Traditionally, city centres have always contained a combination of retail and offices who have flourished in buildings side by side or with offices on top. Indeed, there is a mutual synergy between the two uses. However, mixed-use developments, combining in particular offices and housing, can be seen as a new real estate form. It faces many questions about potential demand, management and long-term value.

The incorporation of owner-occupied housing into mixed-use schemes brings frag-mented ownership and use. This creates issues not just about the day-to-day manage-ment of the estate but also to potential long-term returns. Attention to the design of a mixed-use development is arguably the key to managing any potential use conflicts. One common approach to address this issue is the separation of entrances for each activity in the scheme. It thereby ensures greater security and identity say for both households living in the residential units and occupiers of the offices.

Where there are communal internal and external spaces, then a single manage-ment company model is necessary funded by service charges on the occupiers. This inevitably may be the cause of some friction. The resources, interests and priorities of different types of occupiers and owners can vary.

The general need for intensive management underpinning mixed-use development compared with standard offices also seeks to undermine the investment case. For this reason, residential units in a mixed-use scheme may be organised in distinct (adjacent) buildings that can be sold independently.

Given these investment barriers, mixed-use buildings comprising offices and residential apartments are very much the exception to the rule. Nevertheless, many of the latest generation of multi-storey towers around the world combine offices, residential, hotels and retail. More generally, investors normally seek to meet the planners' mixed-use vision through schemes rather than individual buildings.

While in development terms they can be conceptualised as an integrated mixed-use scheme, ultimately on completion they are likely to be reconstituted as separate investments. As development projects, they may require a complex funding vehicle. The subsequent decomposition of the finished buildings means that a distinct investment class is unlikely to emerge.

Wider implications for the real estate market

The pace of change in urban economies over the last 30 or so years has had implications for the real estate market, beyond new property forms and investment consequences. The continuing urban transformation has inevitably put pressure on tenants to be able to respond to ongoing changes.

More flexibility is demanded to respond to market forces, and it has been translated into short leases. At its extreme, it has spawned the advent of serviced office suites or pop-up shops, short-term retail sales spaces. These can be rented for a minimal period, from days to weeks.

In addition, the greater choice of properties in the retail sector has led to the power in negotiation switching to tenants from landlords. The leases lengths, reported in Chapter 8, reflect these changes. Historically, financial institutions in the UK have required a 25-year lease with five-year upward-only rent reviews. These began to be less prevalent at the turn of the millennium, and terms have consistently fallen since then to an average of less than ten years.

The substantial upheaval in the urban environment has also meant that economic functional obsolescence of individual buildings has become commonplace. The phenomenon brings with it associated redevelopment and refurbishment. It has contributed to shorter-term real estate investment horizons by financial institutions and the greater turnover of stock.

The traditional institutional investment perspective of real estate as long-term assets (inflation hedging) is no more. The average holding period for an asset has fallen significantly mirroring the march of urban land use changes.

Summary

The pattern of land uses is explained using the Alonso model in Chapter 4 based on a perfect land market. However, this chapter has focused on change and the behaviour of key players in this process. In doing so, it has examined how the motives of individual owners of real estate, tenants and planning authorities have influenced the process of change. These motives are shown to relate to the specific activities of these key players.

In the land market, there are different types of landownership. Land can be bought for private consumption to be used to produce goods or simply to be consumed as a place to live. It is also bought as an investment, for example, by financial institutions. Land can also be acquired for public or community ownership. These motivations map into different approaches to ownership.

It is possible to distinguish between active and passive owners. Active owners try to overcome site constraints to improve the development potential of their land. On the other hand, passive owners make no effort even if they plan to sell. Even where there are active attempts to sell vacant land, especially in inner parts of cities, there may be substantial barriers to the take-up by alternative uses. The workings of the land market therefore shape the contours of urban (re)development.

Real estate companies too can influence the configuration of urban change and the spatial pattern of land use by their choice of developments. Their decision-making and demand assessments may not identify the actual needs of a local area. The result can be new developments that fail financially but still overshadow the urban form for years to come.

Planning policies can also nurture and regulate urban form and seek to control new development. Specific policies are designed to establish desirable physical built forms that meet environmental standards and create communities. In the process, planning intervenes in real estate markets to achieve these social goals. Planning as market actors therefore shapes, regulates or stimulates the market and in some cases seeks to stop market forces.

Retail planning in many countries designed to maintain traditional urban retail centre hierarchies is cast also to stop decentralising market forces. At the same time as supporting and revitalising high/main streets, these policies attempt to deter 'out-of-town' shopping centres. The evidence from the UK is that there are serious limitations to such a strategy.

Despite long-term restrictions on out-of-town retail centres, they have blossomed. Britain has seen an extensive spread of out-of-town shopping centres with a parallel decline of many traditional shopping centres.

This retailing example provokes questions about to what extent planning can obstruct market forces. In particular, the motor age is driving long-term suburbanisation and decentralisation in cities across the world. Planning needs to find ways to meet its social objectives within this market imperative.

Financial institutions, including REITs, own much of the largest commercial real estate stock in cities and have a strong influence on the market. Traditionally, they have invested in only 'prime property', for example offices in city centres. They are generally risk averse in their investment strategies, so that they take a conservative approach to real estate innovations.

Real estate investment returns have historically benefitted from the expansion of cities. It generally led to rising values both of existing properties and in the surrounding areas that were engulfed. However, recent urban development cycles centring on the car and ICT have brought a transformation in real estate forms and the decentralisation of land uses. For financial institutions, this emergence of new real estate forms together with locational changes is an investment challenge.

Nevertheless, in the three decades from 1981, there is a transformation in the type of real estate assets held by financial institutions in the UK. Ownership of standard offices falls as a proportion of the total by more than half, standard shops by a third and standard industrial by just over a fifth. These declines are balanced by purchases of shopping centres and retail warehouses/parks, and to a lesser extent office parks and distribution warehouses. The rise of retail warehouses is most dramatic: their percentage rose from almost zero to a fifth of the value of all real estate assets owned by financial institutions in 2010.

This upheaval does not necessarily occur overnight, and there can be a lag between the arrival of new real estate forms and institutional investment. The process of acceptance can be seen through a series of stages. In the first stage, a new real estate form is built by specialist development companies. Initially, the demand is unproven and rent levels are low to attract tenants. Capital values are low too because the product may be evolving with the risk of obsolescence.

The following stage in the development of a new real estate form is a period of market acceptance by tenants leading to increasing rents. While there may be rapid take-up of the scheme, there are likely some ongoing concerns about their long-term prospects. As a result, the purchase of these assets is subject to a risk premium above the equivalent traditional assets.

Once there has been a period of sustained letting and resale activity, then the product achieves credibility as an investment. This acceptance leads to a fall in the associated risk premium and a rise in capital values. At this point, financial institutions widely incorporate the new real estate investment class into their portfolios.

This process can be seen with the emergence of retail warehouses/parks in the UK from the 1980s. Initially, there is an investment 'risk premium' differential of approximately 2% between retail warehouses and standard shops. It takes until 2000 for this risk premium to fall below that of standard shops. By then, retail parks are predominantly owned by financial institutions. The process of establishing a mature investment market for retail warehouses took almost 20 years.

Office parks can also be shown to have an equivalent investment maturity process if much more concertinaed. Overall, the arrival of new real estate forms has had substantial implications for financial institutions. Ultimately, these new asset classes are incorporated within their real estate portfolios, but the process to achieve acceptability was not overnight.

The out-of-town retailing revolution has also meant more than simply a switch in the structure of real estate assets in the portfolios of financial institutions. There are wider consequences for investment in retail as a whole. It has increased the risks of investing in retail by creating a surplus of shops. The problem has been exacerbated by the rapid acceptance of online shopping. The focus on shopping centre investment necessitates their periodic upgrading to address the pressures of competition between centres.

The real estate sector continues to evolve generating further questions for the investment decisions of financial institutions. This ever-ending change is exemplified by distribution warehouses. The rapid expansion of this real estate asset class has also been associated with design changes brought about by technological change.

It has also led to rethinking the distribution process and the location of warehousing. One consequence of the apparent perpetual innovation is that financial institutions in the UK have shown a lack of investment interest.

In the office sector, climate change has led to a growing pressure for green buildings. However, the higher costs associated with improved environmental standards have been an obstruction to the greening of the stock. At the same time, the problem is exacerbated by established prime office forms, architecturally fashionable air-conditioned office skyscrapers, that are energy guzzlers.

One path to the greening of the office stock is if tenants are prepared to pay a green rental premium. Higher rents (and hence capital values) could therefore make green developments (and refurbishments) profitable. At present, this green premium does not occur in most countries. The choice of a new office involves normally considering a range of substitute properties in which the choice balances a range of attributes such as rent, location and size with green features.

The greening of the office stock can be seen as a process where at the beginning the greenness of offices is a secondary demand characteristic subservient to location and the suitability/functionality of space. Over time, the benefits of green offices to tenants are recognised compared with the older non-green stock. Eventually, non-green offices would be unwanted and need refurbishment or complete redevelopment.

It will take a long time for such 'green obsolescence' to materialise naturally through market forces. However, since the millennium, governments are enforcing a green agenda requiring stricter environmental standards, not only for new offices but also for the existing stock. In the UK, it is the continuing expectation of the march of higher required standards that has convinced developers and investors to go green.

A further policy impinging on the real estate market is the rethinking of planning orthodoxy to promote mixed-use developments. The policy sees not only mixed commercial and housing uses in close proximity, but often high residential density. There is no universal definition, and it can be interpreted in a number of different ways. A large mixed-use development scheme could comprise offices, shops and residential blocks perhaps linked public space.

There is a potential clash between the desires of planning for mixed-use neighbourhoods and market forces. While the market impediments are likely to be stronger in the suburbs, there is more scope for city centre housing developments encompassing offices or retail units. However, such developments still face questions in relation to potential demand, management and long-term value.

The management of potential conflicts can be moderated by design including separate entrances for different uses. To further avoid ongoing management issues, a mixed-use scheme may be decomposed into distinct (adjacent) buildings, including residential blocks, that can be sold independently. Even so, there are many examples of multi-storey towers around the world combining offices, residential, hotels and retail uses.

The chapter has demonstrated how the transformation of the spatial economy has brought new real estate forms and investment consequences. The pace of change has also contributed to short leases. The power in lease negotiations has also switched more to tenants from landlords.

Economic functional obsolescence is becoming commonplace changing the nature of investment decisions. Traditionally, 'Location, Location, Location' has been the mantra of the real estate market. However, 'prime' locations stagnate to be replaced by the new 'prime' and new property forms. Real estate investment has had to adapt and contributed to shorter-term real estate investment horizons by financial institutions and a greater turnover of stock.

Learning outcomes

Different types of landowners may have diverse approaches to ownership and development on it.

There may be substantial financial barriers to the development of vacant land, especially in inner parts of cities.

Developers do not necessarily identify the true real estate needs of a local area.

Planning policies are designed to meet social goals by intervening in real estate markets. In doing so, planning shapes, regulates, stimulates or arrests market forces.

Retail planning in many countries seeks to maintain traditional urban hierarchies and stop decentralising market forces. The evidence from the UK is that there are shortcomings to such a strategy.

There is an important question as to whether planning can stem long-term decentralising market forces.

Financial institutions that own many of the retail and office buildings in cities are risk averse and take a conservative approach to real estate innovations.

The advent of new real estate forms, together with locational changes in land uses, is an investment challenge for financial institutions.

The three decades from 1981 heralded a transformation in the type of real estate assets held by financial institutions in the UK. The contributions of standard offices, shops and industrial units to their investment portfolios fell away to be replaced mainly by shopping centres and retail warehouses/parks.

The rise of retail warehouses saw their percentage increase from almost zero to a fifth of the value of all real estate assets owned by financial institutions in 2010.

The adoption of new real estate classes as investments can be seen as a slow process, through a series of stages incorporating product development, acceptance by tenants and evidence of marketability.

Through this process, progress can be measured by the changing risk premium financial institutions apply to the purchase of a new type of asset.

It took almost 20years for retail warehouses to be established in the UK as an investment class by financial institutions.

The evidence from the UK is that the out-of-town retailing revolution, followed by online sales, has increased the risks of investing in retail by creating a surplus of shops.

There has been a rapid expansion of distribution warehousing with technological change generating design changes. This apparent ceaseless upgrading has been unattractive to investment by financial institutions in the UK.

Climate change is pushing an imperative to make offices more environmentally friendly, namely greener. One channel to achieve this is if tenants are prepared to pay a green rental premium, but this is not evident in most countries. It could take a long time for such a green premium to emerge naturally through market forces.

However, since the millennium, governments are enforcing a green agenda requiring stricter environmental standards for offices. In the UK, the clear direction of policy travel towards higher required standards contributed to a surge in new green office developments.

The reframing of planning conventions to promote mixed-use developments is also impacting on real estate development. There are market obstacles to large mixed-use development schemes. A solution is to disaggregate a scheme into distinct single-use elements that are sold independently.

Economic functional obsolescence is becoming commonplace bringing shorter time horizons for real estate investment and tenancy decisions. Prime locations are no longer set in stone as spatial land use patterns change and new real estate forms are established.

Bibliography

Adams C D and May (1991) Active and passive behaviour in land ownership, *Urban Studies*, 28, 5, 687–705.

Jones C (2009) Remaking the monopoly board: Urban economic change and property investment, *Urban Studies*, 46, 11, 2363–2380.

Jones C (2013) *Office Markets and Public Policy*, Wiley-Blackwell, Chichester.

Jones C (2014) Land use planning policies and market forces: Utopian aspirations thwarted?, *Land Use Policy*, 38, 5, 573–579.

Oyedokun T, Dunse N and Jones C (2018) The impact of green premium on the development of green labeled offices in the U.K., *Journal of Sustainable Real Estate*, 10, 81–107.

14 Urban public finance

Objectives

A key attraction of living in a city is the availability and range of public services for households. Similarly, firms benefit from many of these services. Most of these services are usually provided by local authorities. This chapter looks at the financing of these services through in particular a local tax base. In many parts of the world, the level of local property taxes, dependent on capital values, is a crucial part of the financial equation of funding services.

A further influence is the spatial variation in the demand for public services. The chapter examines the problems that arise and solutions through a prism that distinguishes the service needs and tax resources of the urban core and its suburban rings. A final ingredient to the problem of financing public services is the impact of continuing suburbanisation.

The first part of the chapter explains the context in terms of the services offered by public authorities and why. The nature of local government is also examined. The next sections consider the use of local property taxes to fund these services and their impact on the property market. The urban fiscal problem is then explained and why central government grants are required to address the issue. The unit then addresses some of the difficulties that arise in the equalisation of resource needs across different areas.

The structure of the chapter is as follows:

- Urban public services
- Structure of local government
- Financing of public services
- Property and land taxation
- Influence of tax on local property values
- Urban fiscal problem
- Central government equalisation
- Impact of urban change
- Multiplier effects

Urban public services

Some of the basic reasons for living in cities can be seen as social agglomeration economies in the form of range of services. Many of these are provided by or regulated by the public sector. The reasons for public services provision can be seen in a wider

DOI: 10.1201/9781003027515-16

perspective. Some services can be seen as public goods including the police, the fire service, public health, electricity supply infrastructure, roads, street lights, public transport and parks.

The fire service and public health can also be regarded as addressing negative externalities. Other public services effectively resolve negative externalities. Examples encompass water, refuse collection/waste management and planning in its broadest sense. There are also market failures to be addressed such as education, public transport, library services, recreation and cultural activities. In addition, the local public state can provide welfare support in the form of social housing, social services, etc.

Not all these services are provided by local government. The precise mix of suppliers between local, regional and central governments and the private and public sectors vary by country. The range of services and the priorities of local government have also changed over time as the following illustrates. Public health in the high-density living of cities has been historically very important. Looking back in time, cities have experienced outbreaks of cholera, typhoid and other contagious diseases. It was a policy imperative to avoid such episodes.

From the mid-1800s, many cities set about ensuring clean water, to fight cholera in particular, that encompassed civil engineering projects with a system of aqueducts and treatment plants. A system of sewers was also built to remove and then treat human effluent. However, these tasks are not now undertaken by local authorities in many countries. Instead, they are the responsibility of national/regional public agencies or private companies to standards regulated by the state. The same applies to the provision of electricity generation.

Besides, clean water public health concerns saw local municipal by-laws on building standards in the 1800s and efforts to reduce overcrowding by slum clearance. Hospitals serving cities were also built in order to support the ill. They also reduced the negative externalities associated with disease, isolated the ill and reduced the spread of contagious diseases. Hospitals were first operated by churches and charities. They can now be operated by city, state/regional and central government agencies, as well as private enterprises. The precise system of health care varies by country.

Urban public transport is often the responsibility of a publicly elected body. Traditionally, local government ran public transport. Today, city authorities do not normally operate public transport. Instead, they may oversee bus, tram and (underground) train operators by, for example, granting franchises. This can include its coordination, and subsidies to fares and services at times and in areas where no commercial services are provided.

Public school education is usually devolved to the local level. However, in some cases, it can be the responsibility of a higher tier of government such as provinces in Canada and Länder (states) in Germany. Churches operate schools in some countries, but funding is from the state. There are local school boards in the United States. While school education may be organised at the local level, it is usually subject to national standards in relation to the curriculum and examinations.

This discussion illustrates that there are a range of public services that urban dwellers benefit from but that different countries deliver and fund them in diverse ways. It is difficult to say even that there are core services provided by a local authority. Over time, the core services have often been reduced as responsibilities have been shifted to national public agencies or private companies. In the UK, the core services currently incorporate roads, parks and recreation, local social services and community development.

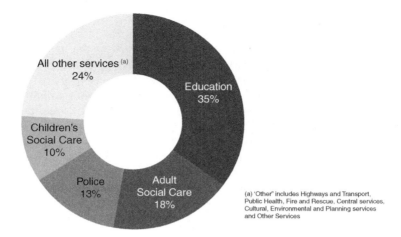

Figure 14.1 Proportions of Budgeted Expenditure on Services in England 2020–2021.

A breakdown of expenditure for local authorities in England is given in Figure 14.1. The largest share is education, followed by adult and children's social care. The figure shows also the additional disparate range of activities of English local authorities that encompass fire, planning, cultural, environmental health services. However, the contribution of the cost of these activities is low relative to total expenditure by local authorities. While these figures apply to England only the relative dominance of education and social services is likely to be broadly applicable to other countries.

The precise activities are partly determined by central government and local policies. Local authorities effectively act as agents of central government. A wide range of national legislation requires local authorities to have a legal responsibility to deliver certain services. Local services provision is also subject to central government financial constraints on local authorities, for example capping total expenditure.

Even when there are legal requirements for a local authority to meet, it may choose to provide a service above the minimum set out in law. This is because there is often local scope to determine the level and range of services. These local preferences are determined by the democratic process in which priorities determined at elections by voters selecting councillors/representatives.

Structure of local government

The range of services provided by an individual local authority within a given country normally depends on its population size and position in a tier of local government. In most countries, there are tiers of local authorities with different responsibilities and autonomy. In Germany, for example, there are 16 states and beneath that Landkreise and over 10,000 municipalities. In some countries, large state or regional governments control most urban services. The highest tiers also tend to have strategic planning responsibilities.

Local government structures in the UK are complex. There are a combination of unitary authorities, and a two-tier system – county and district councils in England. In fact, there are five types of local authority – county councils, district councils, unitary authorities, metropolitan districts and London boroughs. In all, there are 343 councils in England of which district councils represent more than half.

Local government in Scotland and Wales comprises only unitary authorities. Local authorities in Northern Ireland are also unitary and they have less responsibilities. Although the UK has hundreds of local authorities, it has one of the most centralised structures in terms of limiting the power of local authorities.

The United States has a tradition of decentralised political authority with a huge number of local governments, more than 80,000 of them. Of this total, approximately 19,000 are municipal governments. There are also some 3,000 county administrations. There are differences between states in the responsibilities allotted to these different types of authorities. There are also single purpose authorities such as school districts.

The picture of local government areas and responsibilities is therefore very diverse. The pattern varies from great fragmentation to regional administration. Where local government predominates, it tends to be constituted in areas that are less than the size of a functional urban area. It is common for central cores and suburban areas to be located in different local government areas.

Financing of public services

In principle, there are a number of different ways of funding local public services. They could be funded out of general national taxation. However, as a general rule, there is an element of local taxation to reflect localised democratic accountability and choices in services provision. The choice of local tax is then between ones on incomes, expenditure/sales and property/land values. Revenue can also be derived from user charges and financial support from central government.

A locally determined income tax at the urban level is not usually collected, although there are notable exceptions such as New York and Philadelphia. The problem with local income taxes is that higher paid workers can simply move to a lower taxed area to avoid payment. Sales/expenditure taxes also have a limited take-up at the local level. Some cities in the United States apply a local tax rate on top of their state's sales tax. However, there are problems about collecting the tax on mail-order goods or online purchases originating beyond the administrative boundaries.

The most common form of a local tax is one on the value of individual real estate properties located within a local government area. There are different forms of property tax. Many centuries ago, there was a windows tax in some European countries. Today, taxes are proportionate to real estate values. The UK has a distinctive residential property tax called the 'council tax' that essentially applies the tax based on value bands.

Some countries have in the past applied land value taxation. However, it is difficult to identify the split between the value of buildings and land. In the United States, unusually, some cities tax the value of land at a higher rate and the value of the buildings and improvements at a lower one.

The dominance of the collection of property taxes reflects, in part, the fact that real estate is fixed in comparison with income for households or firms. It is impossible for property occupiers to avoid paying it. Nevertheless, there are issues about the relationship between a tax on the value of properties and the individual use of public services.

High value properties generate more taxable income. However, the occupants do not necessarily use the equivalent proportion of local services. This could occur when large properties are owned by empty nesters, but of course, they may have done at an earlier stage in their family life cycle. A property tax is also not designed to be related to usage but also rather to the ability to pay.

This issue raises the question of user charges for local publicly provided services. These charges could be applied, for example, to sports facilities, leisure amenities, through to waste collection. The significance of these fees varies between countries. There are limits to user charges as the state may be seeking to encourage the use of services for the common good. These include education and in particular services to support low-income households.

Local redistributive services, including welfare support, in some cases can be financially supported by specific ring-fenced (or 'conditional') grants from a higher tier of government. Direct grants from central government, for example, are often given for public transport. More widely local authorities can receive a block grant from 'central' government towards general services provision. The proportion accounted for by this support grant varies substantially between countries.

At one end of the spectrum is the UK where funding from central government amounts to the order of 80% of local authority expenditure. This high figure for the UK reflects the fact that property tax on commercial real estate is set and collected by central government. This is unusual. In contrast, the figures for other large western countries are much lower, approximately 45% in Germany, 33% for the United States and France 25%. However, there is no clear norm.

Influence of local taxes on property market

Local taxes are by definition set locally. This means that there are variations between local authorities' property tax rates. In turn, it is important to note that local taxes influence property values. The reason for this is based on the 'surplus rent' theory that is the basis for the Alonso model set out in Chapter 4. Essentially, it is because rent is determined by the sum available to pay after all other costs, including property tax, are accounted for.

We can look at this in detail by considering the introduction of a local commercial property tax (it did not exist before). This tax will reduce the amount a tenant will be prepared to pay as it increases their costs. It affects all tenants so rents will ultimately be reduced across the whole local authority area (subject to lease review dates for commercial property). If the market is very efficient, then the reduction in individual rents will equate precisely to the property tax. These lower rents will then be capitalised into lower capital values.

The analysis above applies to commercial real estate rents, but it could also be applied to residential property. The introduction of a tax on residential capital values will lead to households taking this into account when they decide how much to bid for a house. The tax, say collected, on a monthly basis, will reduce the income available to pay mortgage repayments. It thereby constrains a household's capacity to service a mortgage loan. The aggregate effect is to lower house prices.

Differential rates of tax between local authority areas should also be capitalised in property values. Consider two identical houses on either side of a road whose centre represents a local authority boundary. If the property tax rate is higher in one local authority, then market forces should result in a lower price for the house in that area. The same logic broadly applies to the local impact of differential sales taxes between areas on property market values.

Urban fiscal problem

There are wider issues about the impact of differential local property tax rates. Again, we can see this by considering an increase in the tax in one area, while other areas' rates are unchanged. Consider an increase in residential taxes in a central local authority compared with authorities in suburban locations. This could lead to movement out to lower taxed suburbs. The result would logically be an increase in house prices in the suburbs until the differential in taxes is reflected in values.

The process as described above is a simplification because household residential choice is not usually just between the level of taxes in different areas. There is, for example, also a choice between services offered in different local authorities. More services mean more expenditure and that implies higher taxes. Each local authority offers a mix of tax and services to a household.

This choice led Tiebout in 1956 to postulate the idea that these differences may create communities of like-minded households. If the taxes/expenditures mixes of local authorities differ, then individuals may move to the local authority community that best matches their own preferences. Some people may move to a high tax area because they want better services such as good schooling. This process would be reinforcing, grouping together people of like minds.

The scale and veracity of such a process are limited by the cost of moving, and there would need to be a strong desire. It also ignores other locational preferences and the role of accessibility to work set out in Chapter 5. This dilutes the Tiebout effect. People may also have difficulties finding an area with their desired mix of local service provision unless it is simply based on one factor such as education. It is also dependent on sufficient local authorities with a variety of policies to enable a real choice. Only in a country such as the United States with a very dispersed local governance is this realistic.

More importantly, some members of a community are more free to move than others – owner occupiers rather than social housing tenants, rich rather than the poor. Indeed, there is one scenario that highlights the barriers to this process. The poor could be 'trapped' in inner-city cores where incomes and housing/property values are low. At the same time, local public expenditure per capita is high because of the local social problems associated with low incomes and a concentration of the elderly.

If there is no tax revenue directly from commercial real estate, as in the UK, there is a potential fiscal problem. The combination of low residential property values and high social care expenditure (more than a quarter of total local authority expenditure on average in England) means that local inner-city tax rates are relatively high.

In contrast, the rich live in suburbs where incomes and house/property prices are high. Suburban public expenditure is relatively low because there is less demand for social services. Local suburban tax rates are therefore low reflecting low demand for social care and the high tax base.

This fiscal problem could be exacerbated through the spatial spillover of services. The tax base is related to residential location, but the use of services is not confined to the local area in which a person lives. At the same time, as suburbanites pay lower taxes, they can benefit from services at the urban core. They can commute into the city centre or travel in during the evening or weekends to make use of cultural facilities and public services in the city centre. In some cases, cities address this freeloader problem by trying to limit services or apply user charges to those from outside the boundary.

If cities are able to collect tax on commercial real estate, then the problem could be to some extent ameliorated depending on the spatial distribution of the revenue. However, if it is distributed on a per capita basis across the city, this does not resolve the problem. The relative differences between core and suburbs remain.

The nature of the problem also varies dependent on the nature of the spatial scale of local authorities. It is of course possible to have citywide local authorities (with upper and lower tier authorities). This allows some internal equalisation within a city region. However, the evidence presented earlier is that it is the norm for separate (and multiple) local governments for inner core and suburban areas. There is therefore a clear urban fiscal problem in providing services to core areas based on local authority tax revenues alone.

Central government equalisation

The analysis of the urban fiscal problem up to now has looked at variations in taxes and expenditure and has implicitly assumed no central government intervention. Now, we introduce a central government grant(s) to local authorities to address the inequalities identified in the previous section. In fact, the inequalities noted between urban core and suburbs are only a specific example of a wider set of differences between local authorities.

Accepting that government should intervene, what are the principles by which financial support should be allocated? To assess this task, we need to reconsider the fundamentals that have created these inequalities. First, there are the differences in spending on social care needs by local authority highlighted in the core suburbs dichotomy. More generally, needs depend on the characteristics of a local population and area.

The result is that the expenditure per capita to provide the same level of services varies by local authority. Central government grants to equalise for these variations in expenditure needs can be allocated by a formula. It would necessarily take into account the demographic structure of local populations and also physical differences in areas, such as the lengths of main roads. There are different ways that this can be achieved. The process below is a simplified example of what has been applied in the UK. It is a useful way of explaining the issue.

In this equalisation system, grants to local authorities are based on a standard cost of services for an average area. Examples include the standard national cost of educating a school pupil or the maintenance of a kilometre of road. These standard costs would be applied to the local characteristics of an area, for example the number of pupils in primary and secondary schools.

The grant would also be adjusted for variation in costs between authorities, such as to take account of say sparsity of population or high labour costs. In this way, a cost assessment is undertaken for each service provided by a local authority.

For each authority, the grant will apply these standard costs to the services needs to estimate a 'Standard Spending Assessment' for that area. This is equivalent to the cost of providing a standard level of services (expected by central government) by the local authority to account for local needs.

These spending assessments treat individual services as distinct. The next stage is a block grant that reflects the aggregate needs of a local authority. Alternatively, individual services could receive a targeted grant based on the characteristics of the community. Local authorities can still choose to a degree spend more or less on a particular service if funding is not ring fenced.

The process so far equalises for needs but not the tax resources of each area. The final stage adjusts for these latter differences. One way to do so is for the overall grant paid to a particular local authority to be calculated by subtracting the amount central government deems should be collected from local property tax. This sum is calculated by applying a standard multiple to its local property tax base.

This last stage in the process is a recognition that any government equalisation scheme needs to go beyond just differential local needs. Clearly, different local area property tax bases mean that it could be cheaper or more expensive to a household/ taxpayer to receive the same service. The cost per taxpayer depends on where you live or operate as a firm. The desire to create territorial equity between local authorities in this regard can be defined formally as equalising the fiscal capacity of each local authority area.

In some countries, a government provides a specific grant directly to take account of these spatial variations in (property) tax income. To fully equalise, the specific aim of such a grant would be to enable the same rate of tax in each local authority to buy the same level of service.

To summarise, the variations in the characteristics of local authority areas offer a strong case for central government equalisation grants. These equalisation grants can be a combination of needs and resources. There are different ways of achieving these goals, and the basics of one approach have been presented as an illustration. In theory, then the conflict between suburbs and core city outlined above should disappear, but in practice, it is only lessened.

There are a number of reasons why there may not be full equalisation. First, in relation to the example set out above, the estimates underpinning the standard spending assessments may not be accurate. The calculations would be based on complex statistical analysis, and there would be inevitably a continuing debate about their correctness. Second, the funding made available by central government may not be sufficient to provide for full equalisation.

A third reason is that the formula to allocate grants to local authorities may be inadequate. In the UK, for example, the formula appears to have been influenced by the political complexion of central government. The formula has been politicised and is not a purely technocratic exercise. Labour Governments are accused of favouring cities, while Conservative Administrations are similarly arguably biased towards 'counties' or rural areas.

Impact of urban change

A huge issue for public authorities is how the funding and provision of local services are affected by urban or spatial change. This is particularly important for areas suffering urban decline and population loss. The reasons for this decline are explained in Chapter 9. Detroit was also posited as an extreme example where between 1970 and 2010 the city's population halved. There were also punishing consequences for the city's finances.

The implications for public finance in the city were first on the property tax base. As people left the city, many properties were abandoned in some neighbourhoods and real estate prices collapsed. Manufacturing industry in the city experienced severe deindustrialisation and job losses through decentralisation, especially of car industry plants. The city's government found it difficult to afford the provision of services as it accumulated a large budget deficit. In 2013, the city government eventually filed for bankruptcy. New York had almost suffered a similar fate in 1975.

The problems are not simply a falling tax base. The residents and firms who remain expect and require the services for which their taxes contribute to. It is useful to look more closely at the problems of services provision by taking education as an example. Education is one of the highest components of local government expenditure.

Population decline means falling school rolls and hence lower than optimum class sizes. In financial terms, it means that the cost per pupil increases because of the fixed costs incorporated in schools and the need to meet academic standards.

The local authority in the urban core employs the same number of teachers and schools for less and less pupils. The solution is to close and amalgamate schools, but this inevitably is a slow and costly process. Just like education, the provision of other local services costs more per capita with falling residential density, and neighbourhoods of almost empty houses. These services also require rationalisation of fixed supporting infrastructure costs such as depots.

Urban decline also incorporates the selective flows of suburbanisation that also lead to a greater concentration of the poor and the elderly as noted earlier in the chapter. In the case of Detroit, it has been characterised as the 'white flight' to the suburbs leaving behind marginalised African-Americans and the old trapped. The result is further demands on local services and greater fiscal pressures on the city.

In theory, these additional costs could be accommodated by adapting the formula for central government equalisation funds (if they exist) as discussed above. However, the suburban local authorities, the recipients of households moving out from the urban core, also have extra costs. New schools, for example, may be required to meet the rising number of pupils. These authorities can argue that they require additional central support for such needs.

The problem is that if a central government accedes to requests to adjust the equalisation formula for these extra costs what is achieved? If the total funds available for equalisation remain unchanged, then the modifications to the formula could simply cancel each other out. If this happens, both the core and suburban authorities have to absorb the additional costs of migration. The precise financial impact of migration depends on the extent of equalisation, and its formula in terms of both needs and fiscal capacity.

A related issue that follows from financial equalisation is the potential lack of incentive for a local community accepting housing development. This is important when there is a shortage of housing. Fiscal capacity equalisation between local authorities, in particular, can reduce incentives for new housing development. The reason is that the associated rise in property tax base could be diluted through a reduction in central government grant/transfer.

Instead of new housing leading to new tax resources for a community to draw on, there may be simply just more expenditures. These perceived negatives exacerbate the views of residents in areas that are reluctant to accept new housing, often referred to as development not in my back yard (NIMBYISM). Assuming equalisation continues, the solution could be a specific grant or incentive to accept new housing. In this vein in England, the central government has provided a direct grant, 'New Homes Bonus', to local authorities linked to the number of new houses that are built.

Table 14.1 Notional Breakdown of a Local
Authority Budget

	£m
Local Authority Budget	100
Total Central Grant	80
Local Property Tax	20

Multiplier effects

The use of central government grants to support local authorities has further potential ramifications. The accuracy of needs element assessments is difficult given the dynamic nature of urban systems. Equalisation funds may also not be sufficient. If the assessments are insufficient, there is a multiplier effect on the amount paid by a property taxpayer. This can be illustrated with a simple numerical example in Table 14.1.

In the table, the local authority budget is funded by a grant from central government and a local property tax in the ratio 80:20. These proportions are just indicative, but are similar to the budget funding package in the UK as a whole.

Now, assume the government underestimates the local authority's needs by 5% or the actual budget rises to £105m and the government refuses to increase its aggregate grant(s). This means that the property tax collected has to rise to £25m to fund the budget. In these circumstances, a 5% increase in expenditure has led to a 25% tax rise for the local community.

The multiplier in this instance is 5. The actual multiplier depends on the proportional funding by central to local government. The less proportional support from central government then the lower the multiplier. Nevertheless, when cities are suffering population decline, an underestimate of their expenditure needs is quite likely with a multiplier effect further exaggerating their fiscal problems.

In these examples, an increase in local expenditure is seen as underestimated by central government. It could be an error by the government, and if needs expenditure is underestimated, then the effect is magnified on local property taxpayers. This is of general importance if there is systematic underestimation for particular types of local authority areas because the formula is flawed. The multiplier effects also apply if central government consciously reduces its financial contributions to local government.

Summary

The provision of public services provided by local authorities contributes to the positive social agglomerations of locating in a city. Many of these services are urban public goods that people benefit from and cannot opt out of. In some cases, these services can be seen as resolving the negative externalities of high-density urban living. Other public services are designed to address market failure or welfare support.

The priorities of city government have changed over time as have the mechanisms. Services such as clean water were historically pioneered by city governments, but can now be provided by the private sector. Similarly, other services such as transport once developed by local authorities are no longer directly operated by them. Overall, there is no agreed core range of urban public services provided by a city. The services are also delivered and funded in disparate ways in different countries.

In England, services provided by local authorities encompass education, social care, fire, planning, cultural activities and environmental health. The largest share of this local expenditure is education, followed by adult and children's social care. For many of these services, local authorities effectively act as agents of central government to at least national standards. However, there is local autonomy to adopt local priorities.

Local government is often organised in tiers. The systems can be complicated; for example, England has unitary authorities together with others set within tier structures. There can also be single purpose local authorities. The number and size of local authorities vary substantially with the United States having a very decentralised system. Even in countries with highly consolidated systems, urban cores and suburban rings tend to be in different local government areas.

Local government is funded at least in part by local taxation as an integral part of democratic accountability in the provision of services. Revenue support can also be obtained from user charges and via grant from central government. While local taxation can take the form of income or sale taxes, the most prevalent tax is on property. Property tax can take different forms but is the least avoidable local tax.

Property tax can be queried as the payment does not necessarily relate to the use of local services, but it does link to the ability to pay. User charges can also be drawn on to pay for local services, but these are limited where low-income households are required to benefit. The role of central government in financing local authorities can be through grants to support individual services or a block grant to cover a range of activities.

The percentage accounted for by central grants varies enormously between countries. At one extreme, the percentage in the UK is around 80%, whereas it is normally much lower, only the order of 25% in France.

Local variations in property tax rates have consequences for real estate values. The payment of the tax logically leads to a fall in property values as it reduces the capacity of occupiers to pay rent or mortgages. Areas that have high tax rates will see a larger impact on real estate values than low tax areas. Differential local property tax rates could possibly also see households and firms moving between areas.

Potential migration between local authority areas is not just about variation in taxes but also encompasses the services provided. Tiebout in 1956 hypothesised that households could move to a local authority area that offers the best match in terms of the balance of tax and service levels. This potential Tiebout effect is likely to be swamped by the costs of moving and many other local preferences on migration. It is also dependent on a range of local authorities offering a real choice.

Most important low-income households are least able to move. They can be 'trapped' in inner-city cores where the property tax base is low. At the same time, there are high costs associated with social care, resulting in inner-city tax rates being relatively high. The opposite is true in the suburbs where the rich live and there is less demand for social services so that tax rates are low. Citywide local authorities could equalise the imbalance between resources and needs, but usually, there are separate local governments for core and suburban areas.

The fiscal differences between local authority areas in terms of expenditure needs and tax bases, and associated problems, can be addressed by central government grants. Needs equalisation can be achieved by a grant(s) that takes into account the demographic structure of local populations and also physical differences in areas. The starting point for such a grant could be the standard cost of services for an average area. These standard costs would be applied to the local characteristics of an area and also adjusted for variations in costs between authorities.

The process could lead to an aggregate block grant or a series of grants for individual services. Resource equalisation could then take account of the local tax base, and how much a local authority can afford to pay towards the recognised costs. This can be achieved by subtracting an amount based on a standard multiple of its local property tax base. In this way, there is territorial equity between local authorities, whereby the same rate of tax in each authority buys the same level of service.

There are a number of reasons why in practice full equalisation may not occur. It could be that estimates of local needs may not be statistically accurate, and in any case, the amount of funding allocated by the government may not be sufficient. A third reason is that provision of grants is in the realms of political economy and is not a simple technocratic process.

Urban decline and population loss is a particular challenge to the funding and provision of local services. It affects the local property tax base and the cost of providing services. The problem has affected the cores of many cities that have experienced de-industrialisation and decentralisation. Detroit as a severe case ultimately succumbed to bankruptcy in 2013.

The problems are twofold. Taxes collected fall. Population loss also leads to the increased cost of services derived partly from a legacy of fixed costs and surplus facilities. There are also adjustment costs, while also service provision to the remaining residents becomes more expensive, reflecting lower residential densities.

The selective flows of suburbanisation also lead to a continuing concentration of the poor and the elderly in the urban core. There are the concomitant extra demands on local services and greater fiscal pressures on the city. At the same time, the suburbs have extra costs providing facilities with the rise of their population. Given that there may be a ceiling on central government funding, the additional costs in the core and suburbs may have to be absorbed internally.

Fiscal capacity equalisation between local authorities can create negative incentives towards a local community accepting new housing development. This is a particular problem if there is a housing shortage that is exacerbated by NIMBYISM. The solution can be specific grants to promote new urban development.

Failure by central government to fully support local authorities can lead to a multiplier effect on the property tax paid by local residents. The precise multiplier depends on the relative contribution to a local authority budget from central government. If central government contributes 80%, the multiplier effect is 5. A 5% increase in unsupported expenditure has led to a 25% tax rise for the local community.

Learning outcomes

Services provided by local authorities contribute to the positive social agglomerations of locating in a city.

There is no agreed core range of urban public services provided by a city.

Urban public services are delivered and funded in disparate ways in different countries.

For many of their services, local authorities effectively act as agents of central government to at least national standards.

Local government is often organised in tiers, but structures in individual countries are diverse and can be complex.

In most countries, urban cores and suburban rings tend to be in different local government areas.

Local government is funded at least in part by local taxation, grants from central government and user charges.

Property tax can take different forms but is the most common local tax.

The percentage of local government finance accounted for by central grants varies enormously between countries, with high figure of around 80% in the UK.

Local variations in property tax rates have impacts on real estate values.

Tiebout argued that households could move to a local authority area that offers the best match to their preferences in terms of tax and service levels. This effect is likely to be swamped by the costs of moving and many other local preferences on migration.

Low-income households may be 'trapped' in inner-city cores where property taxes can be high. In contrast, high-income households live in the suburbs where tax rates are low.

Needs spending of local authorities can be equalised by central government grants. They can take into account the demographic structure of local populations and also physical differences in areas.

Resource equalisation can also address the variations in the local tax base of local authorities. Territorial equity would enable the same rate of tax in each local authority to purchase the same level of service.

Full equalisation is unlikely to occur in practice.

Urban decline affects the local property tax base and the cost of providing services. In the extreme case of Detroit, it brought bankruptcy in 2013.

The selective flows of suburbanisation lead to a continuing concentration of the poor and the elderly in the urban core and a rise in the suburban population. Both bring extra demands on local services and greater fiscal pressures on each component of the city region.

Solving the local fiscal equalisation issue reduces financial incentives for urban residential development.

Differences between central and local government perspectives on required expenditure can lead to a multiplier effect on the property tax paid by local residents.

Bibliography

Friedrich P, Gwiazda J and Chang W N (2003) Development of Local Public Finance in Europe, CESIFO Working Paper 1107, Ifo Institute for Economic Research, Munich.

Tiebout C M (1956) A pure theory of local expenditures, *Journal of Political Economy*, 64, 5, 416–424.

Vickerman R (1989) *Urban Economies: Analysis and Policy*, Philip Allan, London.

15 Transport policies

Objectives

The rising use of cars in cities was facilitated by the expansion and improvement of urban road networks. New bypasses were built, roads upgraded to dual carriageways or simply widened as well as urban motorways created in some cases right into the heart of cities. Nevertheless, it has led to continuing and increasing congestion over recent decades.

As noted in Chapter 11, large cities, in particular, are now suffering from congestion that also contributes to pollution. Commuting travel times by car are getting longer as average vehicle speeds fall. It is a global problem as cities increase in size around the world. Cities have sought different ways to accommodate and tame car use. The range of approaches is now evaluated in this chapter.

The structure of the chapter is as follows:

- Scale of the problem
- Congestion road pricing
- Investment in roads
- Traffic management
- Enhancement of public transport
- Low emission zones and car bans
- Investment in cycling
- Implications for land use patterns and real estate market impacts
- A future postscript

Scale of the problem

Chapter 10 charts the rise of car usage over the last century, and it is useful to reread the statistics presented to be reminded of the scale of change. From the 1950s on, the growth in the use of cars in western economies accelerated. The number of cars registered in the UK is now more than six times that in 1960 with a parallel decline in the use of buses by people of working age.

The impact on road usage can be discerned from the following statistics for the UK for the 25 years to 2019:

- The distances travelled by cars and taxis increased by almost 30% to 278 billion vehicle miles over this period.
- This mileage is at a record high.

DOI: 10.1201/9781003027515-17

- Lorry traffic increased by 12.8% to 17.4 billion vehicle miles over the period, although it is below the peak seen in the mid-2000s
- Bus and coach traffic fell by 16.1% to 2.4 billion vehicle miles.
- Van traffic has exhibited the fastest percentage growth of any vehicle, more than doubling over the period
- Van traffic is at an all-time high of 55.5 billion vehicle miles.

These aggregate figures hide cyclical and structural influences, with traffic for example falling with the recession following the global financial crisis. The growth in car traffic has subsequently outstripped population growth, indicating an increase in the average car driver distance travelled. The growth of van traffic reflects the growth of online sales and home deliveries.

Overall, there is a long-term, if slowing, upward trend in the mileage of cars and vans in the UK while that for buses continues to fall. The reduction in the use of buses reflected falling subsidies and rising fares, although it is important to note that falling bus use is not universal. Other countries have applied different approaches to subsidies (see below).

Cars consistently account for around four-fifths of all distances travelled by motor vehicles in the UK. Mileage by cars equates to five times that of vans. The predominant use is of cars on urban roads rather than motorways. Only one-fifth of all car travel is on a motorway.

Cycling has experienced what can be described as a rebirth in popularity in the UK. It fell dramatically during the 1950s and 1960s, and there then followed a period of relative stability. From 2000, there was a steady growth, and cycling use in 2019 was just over a third above that in 2000. The pandemic then saw a surge in cycling in 2020 as more people took it up.

These statistics and travel patterns will differ around the world. As an illustration, just over a quarter of trips made by Dutch residents are by bicycle, compared to under 1% in the United States. The Netherlands and the United States are extremes, but the rise of the car is almost ubiquitous while cycling is also rising in popularity worldwide.

Similarly, the use of the car for commuting to city centres varies considerably not just between countries, but also between cities in a given country. City size is also an important determinant. At one extreme, commuting by car into central areas is as high as 90%, but in most cities, public transport is the norm. In many European cities, the use of public transport is heavily subsidised; for example, in Oslo, the public purse covers approximately 60% of its costs.

The pattern of vehicle use and city size are critical to the level of congestion. Nevertheless, congestion is perceived as a worldwide problem facing cities. An important influencing factor is the role of travel costs and subsidies. They are also relevant to the potential solutions that are now considered.

Congestion road pricing

The long-term rise in the use of cars in many urban areas has led to arguably excessive congestion in most large core cities. It can be seen as acting as a constraint on their urban development. It has led to the introduction of charging cars to enter the central areas of some cities around the world. This is referred to here as road pricing.

Tolls or taxes for the use of a road have a long history stretching back many centuries. Tolls are often applied to recoup the costs of a road, bridge or tunnel construction and maintenance. Many countries currently charge tolls on selected long-distance routes, hence primarily on inter-urban road networks or roads skirting large city regions.

The logic of road pricing is very different. It is about attempting to reduce congestion on existing intra-urban roads. The idea for road pricing was developed in the 1960s, but at that time, the technology was not available to put it into practice. Toll booths would have been the only potential mechanism, and this would probably have increased congestion by causing queuing.

The technology is now no longer a constraint on the adoption of road pricing, and schemes have been introduced in cities around the globe. However, they are not very popular and are often rejected by the local electorate. Some of the schemes in operation are now briefly considered to illustrate the different approaches.

Singapore

The city introduced the world's first congestion pricing scheme in its central area in 1975, and the zone was extended in 1995. It amounts to 7 sq kilometres and includes the CBD and main shopping area, Orchard Road. In 1998, the system became electronic with a network of gantries. Sensors on each gantry automatically deduct the toll charges when a vehicle passes through via a tracking device in each car. The scheme was also extended to travel on the city's expressways.

There are now two types of charges. The first relates to a charge for travel into this central cordon. The second type comprises 'radial' toll charges on the use of major expressways in the city. These charges fluctuate based on expressway traffic flows/congestion levels, so vary by time of day.

London

The scheme was introduced in 2003. It takes the form of a flat daily fee for entering a central zone of the city between 7am and 10pm. The zone covers an area of 22 km sq within the inner ring road. It encompasses the City of London financial centre as well as the 'West End' that includes retail, offices and entertainment centre. Drivers only pay once, no matter how many times you drive in and out of the zone on the same day. Just over 130,000 live within the central congestion charging cordon, while total employment exceeds 1 million.

The charge is based on automatic recognition of vehicle registration by cameras. Drivers pay online, and it can be automatically deducted from an account. Residents living within or very close to the zone are eligible for a 90% discount. A plan to extend the area covered was abandoned when there was a change in political control.

Santiago

The city introduced a road charge in 2004. Motorists using a network of highways pay according to the distance travelled and the time of day. The charging is calculated by an electronic device, known as tag that needs to be purchased for each car. Drivers can buy a monthly tag/pass usable on all the toll roads in the city region.

Stockholm

The scheme was introduced in 2006, and the charge applies between 6.30am and 6.30pm within an inner zone of the city. This zone covers around 35 km sq km and has the order of 600,000 residents. The cordon is crossed on weekdays by more than a quarter of million commuters living outside the zone, and approaching 100,000 commuters living within it.

The tolls are collected electronically via automatic number plate recognition cameras on the entrances and exits to the zone. Vehicles pay a fee dependent on the time of day, with the highest rates during rush hours. Drivers receive a monthly statement of fees. In 2016, the congestion tax began to be charged on the Essingeleden motorway that passes through the western part of the city.

Overview

These examples demonstrate the different ways that road pricing can be implemented. London focuses only on the city centre, while Singapore, Santiago and Stockholm take a more holistic approach encompassing the city. In the Singapore, London and Stockholm schemes, vehicles are charged for entering a central cordoned area. Charges vary by time of day/congestion levels except in London.

Logic of road pricing

The analysis now sets out the reasoning for justifying road pricing. The urban transport problem can be specified within an economic framework based on the demand for and supply of road (space). From this perspective, the core of the problem is that motorists only perceive and count the partial costs of travel for themselves. This personal perspective could focus only on simple fuel costs, ignoring a whole gamut of other costs, including maintenance and depreciation.

The social cost also includes the pollution and climate change gases emissions, noise, accidents to self and others. In addition, congestion particularly increases pollution and there are delays induced on others. Drivers only value their own trip costs, ignoring the impact of their decision to add an extra vehicle to the route on other motorists. The relationship between marginal social costs (MSC) and marginal private costs (MPC) of travelling an extra unit distance is shown in Figure 15.1. Both are dependent on the number of vehicles on the road. MSC is consistently above MPC, and the difference between the two increases as congestion rises.

These differences in marginal private and social costs can be set out as follows:

- Marginal private costs (MPC) = cost of travelling an extra unit distance
- Marginal social costs (MSC) = MPC + pollution/environment costs, etc.

The importance of this distinction is shown in Figure 15.1. MPC and MSC per unit distance both begin to rise faster when traffic flows reach a level that there is congestion. The cost curves represent supply curves, and the demand for car use is a function of price in the normal way. Demand falls with the rise in congestion, but without any policy intervention, the equilibrium level of traffic flows is at **A**, the intersection of demand and the MPC.

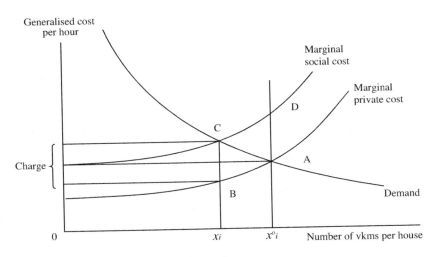

Figure 15.1 Economics of the Urban Transport Problem.

The problem is that the traffic flows at **A** are much higher than a socially optimal one that takes into account the social costs. In economics, the traditional argument is that congestion/pollution is an externality and the solution is to 'internalise the externality' by charging the true social cost. To achieve this, the introduction of a tax, CB, for the use of the road (road pricing) will mean that demand should be choked off. At first, the total costs to the motorists in terms of private costs and the tax will be at **D,** but some motorists will opt not to travel and the flow will fall to **C**. The tax collected will then be **CB** x traffic flow number.

Issues raised by road pricing

The logic of road pricing requires that there needs to be a charging system that is determined by the amount of congestion/pollution a vehicle creates (i.e. marginal social cost of a vehicle). This means that the charge should be linked to vehicle size, speed and area. The examples around the world demonstrate this is not always the case.

There are issues about the efficiency of road pricing linked to how motorists respond to these schemes. In theory, it is based on responding to price signals linked to variable social costs which in turn are related to traffic flows. As noted, not all schemes have a pricing variability built in. Those that do use a standard rate based on time of day rather than actual congestion. On the other hand, such an approach offers clear signals to motorists to make decisions.

Motorists gain by having less congestion but lose by paying more, and this gives rise to distributional questions. Some motorists will be rationed out, that is the whole point of the policy but is it fair? Who will it be that is rationed out? The poor use the buses and the rich can afford the tax so it is middle incomes who logically suffer most.

This effect is pronounced because the charge has to be sufficiently high to discourage drivers. The fairest way is for everyone to suffer congestion. It may be the most equitable, but arguments for road pricing are based on improving the efficiency of a road system not equity.

Given that the charge is negative for most drivers, the net positive effects of road pricing on well-being depend on what the toll revenues are spent on. In most cases, the revenue is promised to improve public transport. It is therefore interesting to note that the first cities to introduce road pricing also have well-developed public transit systems. In these cities, there is a relatively easy choice of switching to public transport.

Impact of road pricing

A key question is how responsive is demand for car travel to price, in other words the shape of the demand curve with respect to price. If demand is totally inelastic, an increase will have no effect on travel, and all motorists will be worse off. In London, a study found that there was a 30% immediate fall in cars in the central area in the first year after its introduction. In Singapore, numbers fell by much more, while in Stockholm, it was a little less. In general, congestion fell by more than the drop off in car numbers.

In London, 40% of drivers initially rationed out by the congestion charge transferred to the buses and up to a half to the underground or rail. However, the responsiveness to switch from a car is partly based on the availability of public transport as a substitute. Not all cities have sufficient public transport infrastructure in place.

The potential introduction of road pricing schemes in cities often falters because of the perceived insufficiency of public transport to bear the transfer of traffic. The revenue from a scheme is normally promised to be used to support public transport and improve its services. Nevertheless, there can be a potential time lag between the introduction of the charge and the upgrading of the public transport system.

Electors are often unconvinced that the promised improvements will materialise and vote against the disruption caused by the scheme. Road pricing is therefore not a universal panacea for congestion, and the chapter now considers alternatives.

Investment in roads

Arterial roads

Road pricing seeks to reduce the traffic on existing urban roads to reduce congestion. An alternative is to increase the amount or supply of road space. This perspective has seen the promotion of new networks of urban roads and motorways into the heart of cities. However, there is a high cost of restructuring urban road systems as cities are not easily adapted. New roads require extensive demolition of housing and the displacement of communities.

The approach has also failed to reduce congestion. Decisions to build roads were often based on cost-benefit analysis that assessed the social benefits and costs over time. These appraisals consistently failed, by underestimating traffic flows generated by new roads. Despite extensive new roads and improvements in cities over decades, the level of congestion has risen.

The inability of the capacity of the road network to cope with the rising demand from road vehicles has had its own direct effects on travel. The increased congestion and travel times choke off demand without recourse to road pricing. In some cities, the solution to the shortage of road space has simply been to formally ration vehicle use. One way is to differentiate vehicles say into two groups, on the basis of their registration number. Only one set of vehicles are permitted on the roads one day, and the other set the next day.

Bypasses and ring roads

Most cities have ring roads or bypasses that circumnavigate the urban area. Invariably, there are inner and outer ring roads. Beijing is the ultimate example where there are seven concentric rings or orbital roads. This solution logically diverts demand from urban centres and is primarily aimed at long-distance traffic.

These ring roads also support suburb-to-suburb commuting, and as Chapter 11 notes, there is now often increasing suburban gridlock with heavy congestion on these roads. A further potential problem is that the increase in circumferential traffic may mean that there is space created on arterial roads. Bypasses could therefore encourage more car users to travel to city centres.

Traffic management

An alternative to reducing demand via direct pricing or increasing the supply of road space is using roads more efficiently by managing the traffic. Urban traffic management can divert and reduce demand by a range of measures:

- Parking restrictions
- Park and ride
- Route restrictions
- Restricting loading and unloading times

These attempts to control traffic in different ways and their effectiveness are now considered in turn.

Parking restrictions

Parking restrictions can encompass local prohibitions on particular streets, and metered short time constraints on others. Residential parking permits in central districts by limiting curb spaces for residents and their guests can also prevent commuter parking. Charges for parking in car parks are usually on an hourly basis.

The advantages of these restrictions are that they are cheap to administer and politically acceptable. The disadvantages are that they tend not to restrict traffic during peak times. In some cases, parking restrictions on arterial roads are designed to improve traffic flows. There is also the possibility that reducing the number of local trips encourages through traffic (unlike road pricing).

A further problem is that parking charges are often determined administratively or commercially, for example if the car park is owned by or run for their benefit of a shopping centre. Some car parks owned by retailers set charges designed to maximise shoppers rather than constrain road traffic. If these charges are below the market rate needed to equate demand and supply for parking space in the locality, there is excess demand. This can result in cruising as cars search for non-existent places.

Central area car parking charges are almost equivalent for travellers into the area as road pricing, where there is a flat fee for entering the control zone. In addition, parking charges on top of a congestion charge amplify the price effect. On the other hand, the impact of car parking charges is diluted if there is widespread free car parking for commuters at their place of work.

Studies in the 1990s of the United States found that at least 90% of car commuters had free car parking at work. In one city, just over half of car commuters into the CBD had a free car parking space. This CBD proportion is much lower in other parts of the world. Nevertheless, most commuters around the world have free car parking at work even if it is on the street. Although these workplaces may be in decentralised locations, free parking can be seen to encourage commuting by car.

Constraints on car parking at work can be via planning restrictions on new developments and a tax on existing car parking spaces. Planning restrictions on car parking are a relatively new phenomenon. Historically, planning policies required a minimum number of car parking spaces based on the floor space of developments. Today, the position has been reversed in many cities and there is a low cap on the number of spaces.

A tax on car parking spaces at the workplace is a relatively new idea. In the UK, only one provincial city, Nottingham, has brought in such a tax, known as the 'The Workplace Parking Levy'. It was introduced in 2012, and the tax is paid by businesses with 11 or more car parking spaces within the city's administrative boundaries. In most cases, the tax is passed on to the car user.

A workplace parking levy has the potential to significantly increase the cost of commuting like a central congestion charge. The pricing effect in terms of rationing out commuters is different. Whereas the congestion charge is designed to shift commuters to public transport, this substitution is not so straightforward. First, in decentralised locations, commuters could simply shift to parking on a street if it is available. Second, workplaces in decentralised locations tend not to be well served by public transport.

In many cases, commuters to decentralised locations may travel from well beyond the urban fringe. They have chosen their residential locations assuming a car journey. Even suburb-to-suburb commuters are unlikely to turn to public transport. If it is available, it is likely to involve a more complex time-consuming journey. Even if the revenue from workplace parking levies is used for extending public transport, there may be only a limited effect. It is possible that the workplace levy could encourage car sharing/pooling.

Park and ride

These facilities are parking spaces with public transport connections that are adjacent. They permit city centre commuters, visitors and shoppers to leave their vehicles. They can then transfer to public transport for the remainder of the journey. A park and ride facility can link to a bus or rail station.

Park and rides are generally located in the suburbs or on the outer edges of large cities. The success of park and ride arguably depends on other policies, for example road pricing or limitations on central parking places. Some people argue that they shift travelling from making a whole journey by public transport. That case is not proven.

Route restrictions

These restrictions are aimed at manipulating the use of urban roads. They include designing a one-way system on an existing network to enable a more efficient road system. These routes could be more appropriate in terms of width. Alternatively, the restrictions could be designed to make it more difficult for cars to travel through or to reach a city centre. Such an approach attempts to increase barriers for car users and contributes to a shift to alternative public transport modes. It is often combined with bus-only lanes or routes.

Restricting loading and unloading times

The aim is to restrict deliveries by commercial vehicles on public roads, especially at peak rush hours. In some cases, the restrictions apply during the whole of the normal working day, so that deliveries take place in the early evening. There is a problem of enforcement. Cruising may occur as lorries wait until the appropriate delivery time periods, creating further congestion. Reduced demand for commercial use may in the longer term also generate more private demand.

Enhancement of public transport

Many of the initiatives discussed above are designed to 'persuade' car travellers on to public transport. The aim is to divert demand from cars to reduce congestion. The presumption is that the public transport is available and of a standard to meet the greater demands on it. For this reason, normally, the revenue from congestion charges (and the workplace parking levy) supports the improvement of public transport.

The longer-term perspective is that public transport is a classic example of an inferior good. As incomes have risen, people have shifted from public transport to using cars. Nevertheless, without public transport, cities cannot function. The need to subsidise public transport to address this problem therefore emerged in the second half of the last century as users declined while services and costs did not.

The financial difficulties are heightened by the twin peaks in daily demand. The social benefits of public transport (rather than the social costs of cars) can justify subsidies to transport operators. These social benefits are in terms of the reduction of pollution, the lower demand for car parking and the provision of travel for low-income households and the elderly.

Different countries have various approaches to subsidies. Some countries have experimented with free public transport. In the UK, there are only limited subsidies to urban public transport. In 1986, local buses were 'deregulated' and provided in the main by private companies. There are only subsidies on a minority of routes, on the basis of social benefit criteria.

Public transport policies cannot be viewed in isolation as car congestion reduces the efficiency of buses. Bus routes/lanes are one way of potentially improving efficiency, and reducing travel times (if not costs). In parallel, cycle lanes are a similar approach to making alternatives to the car more attractive, and persuading people to switch travel mode (see later). These lanes also contribute to the relative attractiveness of commuting modes.

Low emission zones and car bans

Cities around the world are looking at ways to control air pollution. These policies are not directly about controlling congestion but addressing climate change and improving urban health such as the incidence of asthma and lung diseases. The reduction of diesel emissions is particularly important for health.

Low emission zones have recently been introduced across Europe based on common pollution standards. They have similarities with road pricing as the mechanisms applied overlap. The aims of the policy are to directly control pollution by reference to vehicle usage in cities. The essential idea is that the worst polluting vehicles are banned or charged a fee for entering a zone. These zones can be a single road, part of a city or a whole city region.

They are called different names such as 'clean air zones' in the UK, 'Umweltzonen' in Germany and 'Zone à Circulation Restreinte' in France. Most low emission zones restrict buses and coaches and heavy goods vehicles. In some cases, there are constraints on cars and vans. Many of these zones are still at the proposal stage.

The actual restrictions vary. A low emission zone was introduced in Stuttgart in 2008. There is a citywide zone that bans vehicles that do not meet environmental standards. These restrictions, in particular, apply to older vehicles with diesel engines. Vehicles require a green sticker for their windscreen, and without it, drivers can be fined by the police. Delivery vehicles are exempt.

In Greater Lyon's low emission zone, the focus of restrictions is on diesel heavy goods vehicles and vans. Again, the scheme operates on the basis of a vehicle having the appropriate sticker based on its emissions. Older commercial vehicles are excluded from the zone that covers the whole city. These exclusions are becoming more restrictive over time by encompassing newer dates of vehicle registration. Cars are generally exempt from the restrictions.

The London Low Emission Zone began in 2008 with a charge on commercial vehicles in the city that do not conform to high emission standards. The zone covers most of Greater London. In addition, from April 2019, there have been tighter emission standards in an 'Ultra Low Emission Zone' in the Central London congestion charging zone. This affects all petrol and diesel vehicles including cars. Within the existing congestion charge zone, there will be an extra charge for vehicles that fail the emission test.

Low emission zones vary across cities and countries but are part of a wide movement to limit motor vehicle use in cities. So far, it has seen increasing moves towards banning (selected) road vehicles. It can be seen as a mechanism to reduce pollution but also as a part of a grander scheme towards banning petrol and diesel road vehicles altogether. Some national governments plan to ban diesels and motor cars in due course.

But there are moves afoot to bar cars altogether from city centres. From November 2018 on, Madrid has barred most non-resident vehicles from driving anywhere in the city centre. The only vehicles that are allowed are those cars that belong to locals, as well as zero-emissions delivery vehicles, taxis, and public transit like buses.

Pontevedra, a small city (town) in Spain, with a historic centre, has similarly stopped cars crossing the city and got rid of street parking. It has closed all surface car parks in the city centre and opened underground ones and others on the periphery. In a similar vein, Oslo has not banned cars but virtually all street parking.

City governments across Europe it seems are falling out of love with the car. In some ways, it is the culmination of a long process over many decades. Copenhagen and Brussels have large city centre car-free zones. This is equivalent to extending the use of pedestrianisation that began for shopping centres. This policy started in 1953: the Lijnbaan, in Rotterdam, was the first purpose-built pedestrian street in Europe.

Car and motor vehicle-free zones in cities raise many questions about how a city functions. There may be knock-on consequences for the urban economy and possibly congestion in areas beyond the car-free zones. There are also potential implications for the long-term spatial structure of cities and the real estate sector. These issues are discussed in detail in a later section.

Investment in cycling

Many cities that have had a strong cycling culture have a network of segregated lanes. Copenhagen, for instance, is perceived as one of the most cycle-friendly cities in the world. It has invested in infrastructure with an aim to reach 50% of all commuting to be undertaken by bicycle.

However, in the majority of cities, there is limited cycle use for commuting. The rise in cycling in recent decades in the UK, for example, has been mainly for leisure, shopping, education and social trips. Many of these are for short journeys that replace the car. In some cases, cycle trips are to catch public transport. But leisure cycling increases the demand for road space and can create more congestion.

The rise in cycling has led to plans for more designated cycling infrastructure to further promote it as a cheap, healthy and pollution-free mode of travel. There was a stimulus to cycling with the pandemic that also instigated temporary cycle lanes in cities. The introduction of segregated lanes to promote safe cycling has also created resistance through conflicts with motor vehicles.

Besides investment in cycle lanes, many cities have schemes for public cycle hire on a short-term basis. The schemes can be supported by the latest tracking devices. They enable people to have access to a bicycle on an impromptu basis but are unlikely to be used regularly by commuters. The facility could replace short taxi rides or travel by public transport. Many of the schemes, in the UK at least, have not proved to be ultimately financially viable.

Despite the obvious health benefits, commuting by bicycles in most cities remains very much a minority activity. The emergence of electric cycles (and scooters) is likely to boost take-up. (Electric) cycles also offer the potential to replace delivery vans in high-density areas. These could contribute to a reduction in congestion.

The policy problem is that in a country like the UK urban cycling infrastructure is in its relative infancy. Substantial and costly adaptations are required to the urban streetscape to promote safe cycling. However, the likely resultant increase in cycling in absolute trip terms is likely to be very modest. It also serves a cycling community that is predominantly male, young and in professional groups.

With a fixed amount of road space, cycle lanes invariably restrict the amount of space for other vehicles or pedestrians. It is therefore debatable given the costs whether cycle lanes always represent value for money, compared to investing in public transport. But such a financial analysis ignores the health benefits.

Implications for land use patterns and real estate market impacts

Transport restrictions have potential consequences for the working of the urban economy. Changing transport infrastructure and costs must have specific implications for the real estate market. The structure of house prices and commercial rent structures are dependent on transport costs as set out in Chapters 4 and 5. Reformulating the cost of travel/transport means that there will be issues about which locations benefit and lose out. These impacts apply to policies, whether they are road pricing or parking restrictions or other constraints. This section examines the implications for different real estate sectors.

Retailing

In the city centre, for example, pedestrianisation or new route restrictions could change the flows of pedestrians at the micro-level. These modifications could have distributional implications for sales by retailers with some gaining and others losing. It ultimately leads to a revised spatial pattern of rents and prime sites with the resultant locations of different types of retailers evolving. While some retail locations gain in popularity, others are in decline.

Parking restrictions too can contribute to the loss of attractiveness of retail locations. In small towns, parking constraints have led to car-borne shoppers looking elsewhere as noted in Chapter 12. In the suburbs, retailers have found that parking restrictions in front of their stores reduce passing trade.

There is an instructive case study of the impact of the introduction of road pricing in central London on the John Lewis departmental store within the cordon. It found that there was a small negative impact on the number of shopping trips in the first year of the charge. However, there was a diversion to other stores in the group in surrounding centres.

There was therefore a short-term negative impact, but customers may learn to adjust, as happened when pedestrianisation was introduced in many town centres. John Lewis adapted by opening on Sunday when road pricing did not apply at that time. The issues raised by this store apply more broadly to shopping centres.

In general, travel constraints in certain retail centres are almost certainly going to benefit those without constraints. A congestion charge or a car parking fee can add considerably to the cost of a shopping trip. The picture is complicated by the other relative attractions of centres such as size. Nevertheless, the main long-term beneficiaries, in particular, are decentralised shopping centres such as retail parks that have readily available (and free) adjacent car parking.

Offices

The urban transport policies discussed above to reduce congestion are aimed at improving travel times to city centres. They are thereby seeking to make the CBD more attractive to office workers and businesses. However, the long-term impact, particularly of road pricing and pedestrianisation, on office location is not necessarily so straightforward.

It is probable that the demand for high-order services in the CBD is not affected because of the benefits from complex-activity economies for many businesses at a central location (see Chapter 3). The picture is less clear for offices providing low-order services, for example professional services. Central travel constraints could expedite long-term decentralisation trends to suburban locations for these businesses. Restrictions reduce the rents/profitability and new office building sizes in directly affected areas while encouraging development in unconstrained (or less constrained) localities.

Schemes to mitigate congestion in cities therefore work at two levels for the office market: first, directly on congestion by inducing modal shift from the car to public transport, and second, indirectly by potentially reducing the demand for offices (and businesses generally) in the city centre. In theory, addressing congestion should arrest some of the causes of decentralisation but the indirect impacts work in the opposite direction. The induced decentralisation takes the form of the displaced firms locating in peripheral office parks.

Logistics warehousing

Low emission zones and parking restrictions logically lead to a fine-tuning of the location of distribution warehousing. In particular, the location of local depots just outside the perimeter of a zone is encouraged. At these depots, the contents of large heavy goods vehicles are discharged into small pollution compliant vehicles for local delivery.

Housing

The precise impact of road pricing on the housing market depends on how it is implemented. If there is little housing within the pricing cordon, as in London, the effect is minimal. On the other hand, if say the cordon is drawn along say the ring road, it does impact on houses prices. The cordon then differentiates between types of commuters. The tax on commuting varies by residential location. This is the position in Stockholm.

The access-space model developed in Chapter 5 can be used to assess the long-term impact on the spatial pattern of house prices in these circumstances. Crossing a congestion charge cordon by car rise has consequences for households' housing decisions. The price households are prepared to pay for housing is a function of commuting costs.

In a city where most people work at its centre, the access-space model derives a negative exponential house price gradient. This gradient is 'distorted' by a congestion charge, with a downward step at the point of the cordon. Prices on the inward side of the cordon are relatively higher, and there may be some movement back into the city. Overall, the dominant commuting costs at each location are logically embedded in the spatial pattern of house prices.

Using the access-space model, it is also possible to project what would happen if a free bus service is introduced. The only travel cost would then be travel time. The reduction of travel costs in the model as Chapter 5 explains implies households moving further out. It also brings a flatter house price gradient and more urban sprawl in the long term. However, the caveat to this conclusion is that the dispersal is dependent on the availability of bus routes.

A future postscript

The long-term rise in car ownership fuelled the congestion and pollution problems of cities. It has also stimulated various transport policy approaches to resolve them considered in this chapter. However, there are signs on the horizon that the role and significance of commuting are about to change. In the process, the issues associated with car use will be moderated.

Actions to address climate change have now got a greater sense of urgency. Sooner rather than later, petrol and diesel engine motor vehicles will be replaced by electric cars and freight vehicles. Indeed, this is implicit in the periodic raising of pollution standards within low emission zones. Implicitly, the strategy is working themselves out of business once electric vehicles are universal.

Commuting problems could be diluted before the arrival of electric cars by improvements in ICT. There has been a trend for at least a decade of part working from home in combination with hot-desking. The pandemic has accelerated this trend and brought 'the future' closer. The result has been that many office workers forced to work at home have found themselves in the world of virtual meetings. Students too have studied using virtual classes. Work patterns are unlikely to turn back the clock as efficiency has not suffered.

Reduced commuting has consequences for the housing market. Looking through the prism of the access-space model, commuting costs fall resulting in people living further out (see free bus travel in the last section). Working from home brings greater residential locational flexibility. Such freedom enables not just a move outwards to a house in the suburbs with a garden but also beyond. It encourages decentralised urban forms.

Summary

There has been a long-term growth in the use of cars in western economies. More recently, van traffic has experienced an even faster rise. At the same time, the use of buses in many countries is falling. Cycling on the other hand is gaining in popularity. While there are differences in commuting patterns by travel modes between cities, the problem of congestion is a ubiquitous urban problem.

Congestion represents an excess demand for road space. The solutions are seen as a combination of road pricing/improved traffic management and subsidies to public transport, even banning cars. Road pricing is the charging of cars to enter (the central) areas of cities.

The idea for road pricing can be traced back to the 1960s, but the technology was not available to implement it.

The chapter has reviewed four examples of congestion road pricing – Singapore, London, Santiago and Stockholm. These case studies illustrate different approaches to its implementation. London focuses only on a city centre cordon and applies a flat rate charge. The other three also apply taxes on urban expressways. Charges vary by time of day/congestion levels except in London.

The logic of road pricing is that motorists only consider some of the marginal private costs when deciding to travel. While they underestimate the private costs, they completely ignore the social costs. Social costs encompass the pollution and climate change gases' emissions, noise, accidents to self and others, together with congestion delays. Both MPC and MSC increase with the number of vehicles on a road.

Road pricing at its simplest involves charging motorists the MSC rather than the MPC. As the MSC varies with congestion levels, the theory sees motorists responding to price signals linked to traffic flows. In practice, this is difficult.

There are distributional questions about road pricing as it is designed to reduce demand. It affects middle earners most. However, the case for road pricing is based on efficiency not equity. It is also dependent on there being ready alternatives in the form of public transport. Provided there are alternatives, the evidence suggests that demand can fall by the order of 30%.

Instead of rationing road space, one alternative is to build more urban roads. However, the social costs are high, given the necessary destruction of residential communities. In addition, new roads have simply generated more road traffic and do not ultimately address the congestion problem.

Building a ring road is a way of diverting demand from urban cores and is primarily aimed at long-distance traffic. However, many ring roads are themselves suffering congestion as a result of suburb-to-suburb commuting. If successful in diverting inter-urban motor traffic, bypasses could encourage space for more car users to travel into city centres.

Traffic management is an alternative potential solution to congestion by diverting or reducing demand. Parking controls, in particular, are cheap to administer and reduce the number of local trips. However, there are a number of disadvantages including the possibility of encouraging through traffic (unlike road pricing) and cars clogging roads searching for spaces.

There are increasing constraints on car parking at work which can be via planning restrictions on new developments. A workplace parking levy is also a policy option under active consideration with Nottingham having already introduced a scheme. There are a number of practical and theoretical questions over this policy. Commuters could just shift to parking on the street or just pay the cost as there is unlikely to be suitable public transport to shift to.

Other ways of managing travel into city centres include park and ride facilities. They provide parking spaces with public transport connections into city centres. Route restrictions too can be aimed at reducing central car use. Commercial vehicles can also face restrictions on when they can unload to improve peak hour traffic flows.

Many urban transport policies aim to shift car travellers on to public transport, and without it, cities cannot function. But there is also a need to subsidise public transport. That can be justified by the social benefits, but subsidy regimes vary by country. Besides subsidies bus travel can be made more attractive by designated routes/lanes that potentially improve efficiency, reducing travel times.

Low emission zones in European cities have recently been introduced to directly control pollution rather congestion per se. The worst polluting vehicles are banned or charged a fee for entering a zone. Those charged are mainly heavy goods vehicles or old diesel cars, but the restrictions vary by city.

These zones can be seen as part of a longer-term strategy to phase out petrol and diesel motor vehicles in favour of electric engines. But in some cases, all motor vehicles are banned from city centres and large swathes of central areas pedestrianised. This latter trend can be seen as a culmination of a process started in the 1950s.

Cycling is growing in popularity as a means of urban travel. In some cities, it has a strong historical tradition and there is widespread use for commuting. However, in other cities it is a new phenomenon and the necessary infrastructure is lacking. As a result, segregated lanes to promote safe cycling have been built and more planned. There are also schemes for public cycle hire on a short-term basis in many cities.

In countries where urban cycling infrastructure is limited, there are barriers to building segregated cycle lanes. Adapting the existing streetscape can be difficult and costly. The increase in cycling is likely to be only modest in comparison with those travelling on other modes. It raises questions about whether such lanes are valued for money based simply on the extra trips generated.

Restrictions on travel patterns, modifications to transport infrastructure and increasing or subsidising different transport modes must have implications for real estate. In the retail sector, parking availability and pedestrianisation, in particular, shape the level of demand and spatial flows of customers. Revisions to any restrictions have consequences with the spatial pattern of shops and rents evolving. Travel constraints in general benefit decentralised shopping centres that have readily available (and free) adjacent car parking.

Constraints on city centre travel for offices are unlikely to impact high-order services in the CBD. However, for low-order services, the restrictions could expedite long-term decentralisation trends to suburban locations such as office parks. Schemes to mitigate congestion potentially reduce the demand for offices in the city centre.

The precise impact of road pricing on the housing market depends on the location of the pricing cordon. If the cordon differentiates between types of commuters, then the impact will be embedded in the spatial structure of house prices of the city. Prices on the inward side of the cordon are relatively higher. There may be some movement back into the city to avoid the congestion charge.

As a postscript, it is interesting to postulate that the congestion problem could be resolved by itself in the long term. Improvements in ICT have already led to more working from home and less commuting into cities. Working from home was essential for many workers in the pandemic. It acted as a demonstration of the benefits, and an accelerator for the trend.

Working from home enables increased flexibility about where to reside. Living in a village and the countryside becomes more possible and with it a move to more decentralised urban forms.

Learning outcomes

Although commuting patterns by travel modes vary between cities, the problem of congestion is a universal urban challenge.

Congestion can be characterised as an excess demand for road space.

Road pricing is the charging of cars to enter (the central) areas of cities.

The basic logic of road pricing is to charge motorists the MSC rather than the MPC.

Road pricing rations out motorists who are unwilling or unable to pay the congestion charge.

The case for road pricing is based on improving the efficiency of the road system, but it is also dependent on public transport alternatives for motorists.

An alternative to rationing space is to build more urban roads.

New urban arterial roads simply generate more traffic at high social costs by bulldozing communities.

Building a ring road is a way of diverting demand from urban cores and is primarily aimed at long-distance traffic.

Traffic management, including parking restrictions, can divert or reduce demand in city centres.

Increasing constraints on car parking at work can be through planning restrictions on new developments or a workplace parking levy.

A workplace levy has a number of practical and theoretical difficulties, notably the inability of many commuters to switch to public transport.

There is also a general need to subsidise public transport that can be justified by the social benefits.

Low emission zones in European cities ban high polluting vehicles or charge them a fee for entering.

Cycling is growing in popularity but requires more segregated lanes. However, the cost of providing this infrastructure is subject to a debate about value for money.

Restrictions on travel patterns, modifications to transport infrastructure and increasing or subsidising different transport modes must have implications for real estate.

Central travel constraints in general favour the expansion of decentralised shopping centres with adjacent car parking.

Schemes to mitigate congestion potentially reduce the demand for offices in the city centre.

The impacts of road pricing on commuters are embedded in the spatial structure of house prices of the city.

The congestion problem in the long term could be addressed by ICT enabling working from home, and households moving to more decentralised urban forms.

Bibliography

Santos G (2005) Urban congestion charging: A comparison between London and Singapore, *Transport Reviews*, 25, 5, 511–534.

Whitehead T (2002) Road user charging: And business performance: Identifying the processes of economic change, *Transport Policy*, 9, 3, 221–240.

16 Urban sustainability

Objectives

There is an increasing awareness of the need to address the global warming of the planet. This involves the reduction of the emission of greenhouse gases. Chapter 15 noted that one way is to develop new technology to replace petrol and diesel motor vehicles. The use of these motor vehicles is predominantly in movement around cities. As a result, cities are widely acknowledged as the main sources of pollution and environmental degradation. It is, therefore, appropriate to address the sustainability issue at an urban level. Many commentators extend these arguments to consider desirable sustainable urban forms to limit pollution.

This chapter is about the nature of what are sustainable urban forms. To do so, it addresses the following aspects of urban sustainability:

- Nature of urban sustainability
- Global policy context
- The compact city, smart growth and urbanism
- Decentralised eco-settlements
- Polycentric sustainable model
- Walkable neighbourhoods
- Reformulation of sustainable city problem
- Interaction with real estate market

Nature of urban sustainability

Taking a purely ecological approach to sustainability suggests that the goal at the city level is not possible. Cities depend on imports and exports of resources and waste, so the nature of the urban environment is such that it has very little assimilative capacity. Cities, whatever their size is, impact not only on their own environment but also on surrounding environments too.

An ecological view of sustainability is too simplistic in terms of implementing urban change. Urban sustainability can be seen more broadly as comprising three dimensions – environmental (including transport), social and economic aspects. The physical characteristics of an urban form can be seen to influence these three dimensions.

DOI: 10.1201/9781003027515-18

Environmental dimension

The environmental dimension encompasses emissions from vehicles and the provision of open, especially green, space. The environmental benefits of open green spaces include the amelioration of an urban heat island effect and 'free' cooling to buildings.

An urban form can logically influence travel demand as more compact and higher density cities encourage more sustainable modes of travel, such as walking, cycling and public transport. However, the socio-economic characteristics of households also have an important influence on travel distance per person. The cost and availability of different transport modes are also relevant to the level of urban pollution as considered in the previous chapter.

Social acceptability

Social sustainability embraces issues of both quality of life and social equity. The role of an urban form in social sustainability is quite complex. Socio-economic characteristics again can have a strong influence on the quality of life. Higher densities and mixed-use urban forms have been argued to lead to a better quality of life owing to more social interaction, community spirit and cultural vitality.

These impacts follow from assuming less social segregation and, hence, better access to amenities such as shops for disadvantaged groups. However, there are competing arguments that high density has negative impacts such as poorer access to the benefits of green spaces, poorer health and mental well-being, potential bad neighbour effects and reduced living space. These negativities give rise to questions of the social acceptability of high density.

Economic viability

The economic benefits of different urban forms are central to debates about the nature of urban sustainability. Sustainable urban form solutions need to be economically viable. Commentators have suggested that higher density forms support more diverse local service provision, by making local businesses and services more viable. Higher density mixed-use central areas are claimed to encourage more interaction, thereby promoting innovation and economic growth.

There are counter arguments that the viability of services depends on income as well as density, and deprived urban neighbourhoods may lack services even when densities are high. At the same time, as previous chapters demonstrate, much economic activity has a preference for dispersed urban locations.

Overview

The key to improving or achieving urban sustainability is the relationships between these environmental, social and economic dimensions. In choosing a sustainable future, there are potential trade-offs between these aspects of cities although environmental improvements are paramount. Nevertheless, cities are fundamentally economic entities, and urban sustainability solutions need to meet a test of viability.

There is a debate about the specification of sustainable urban forms that has two principal 'extreme' alternatives. One side of the argument advocates a high-density, mixed-use centralised urban form. The other side proposes low-density dispersed urban forms. This debate provides the structure for this chapter.

Global policy context

The Brundtland Report, the World Commission on Economic Development, in 1987 was the first global attempt to address the sustainability problem. In essence, it stems from the finite number of resources in the world but infinite number of human wants. To achieve sustainable development, this report argues it is necessary to meet present needs, while ensuring the needs of future generations are also catered for. The report warned that significant changes need to be made to ensure a sustainable global future.

Despite the debate noted above between high density and decentralised urban forms, policy makers have in the main focused on urban containment and more compact forms. The policy was taken up vigorously by the planning profession around the world. The European Union was an early sponsor. In the UK government strategy documents have consistently promoted compaction and sustainable communities.

Similar concepts can also be found in North America in the form of 'New Urbanism' and 'Smart Growth' initiatives. Both believe in urban forms that are of high density and mixed use, and which are contained to reduce travel distances and the dependence on private transport. New Urbanism also focuses, in particular, on pedestrianised neighbourhoods. These neighborhoods encompass public spaces that provide a supportive environment for social connectivity.

Smart growth proponents emphasise the infrastructure costs of developments on the suburban fringe. The strategy of smart growth is to channel new development into existing urban areas and away from undeveloped areas. In doing so, it seeks to see shifts towards travel on to public transport and the promotion of mixed use.

The planning policy in the European Union has evolved towards a more complex variant of compactness, namely polycentricity. The polycentric sustainable urban model sees a city region or a megacity comprising a number of centres linked together by fast efficient public transport. At one level, polycentricity can be seen as a combination or compromise of the compact city and dispersed urban form. However, a planned polycentric region is designed to ensure that a system of centres is compact and dense.

Urban sustainability from these perspectives can be seen as part of a wider policy agenda including land use patterns, energy use/savings of buildings, minimization of waste and water use and efficient transport. In terms of the urban form, the key characteristics are as follows:

- Intensive use of urban land
- Networks of green corridors
- Mixture of land uses
- Provision of affordable homes
- Local identity

These characteristics are encapsulated in the umbrella term of the 'compact city'. Urban sprawl is viewed as a completely negative outcome.

From a global perspective, there are many questions about the meaning of high density. What is high density in a European or a North American city is low density in Asian cities. There are, therefore, issues about the potential for policy transfer of ap plying the compaction model of relatively small western cities to other parts of the world. Many rapidly growing Asian cities are seeking to reduce densities not increase them.

Within these fast-growing cities and megacities, there have been smaller scale at tempts at creating the sustainable city. These sustainable or ecocities are small newbuild settlements built to master plans that focus on green energy use and the promotion of public transport. Stand-alone demonstration projects like these have only a limited role to play as exemplars to adapt cities to more sustainable urban forms.

Despite the universal planning consensus that supports a high-density, mixed-use contained urban form as discussed earlier, the dimensions of the sustainable city are a complex issue. Planning arguments tend to emphasise the environmental and social benefits of high-density cities. There is a lack of critical appraisal of the disadvantages and constraints and the importance of economics.

The urban economy is a fundamental influence on sustainability. Sustainability pol icies have to balance the fundamentals of the local economy with meeting social and environmental objectives. The next sections of the chapter consider the underlying economics of potential sustainable urban forms.

The compact city, smart growth and new urbanism

Although compact urban forms have much in common in terms of their characteris tics, it is the compact city that has the most developed economic underpinning. Many of the economic arguments in favour of the compact city develop from agglomera tion economies or benefits and the fundamental reasons why cities exist as set out in Chapter 3.

The economic case for the compact city builds on the role of agglomeration econo mies by stressing the role of density. Higher employment densities in a city are seen as enhancing potential agglomeration economies. A part of the reason is that knowledge spillovers between firms in an industry in close physical proximity can lead to innova tion. Technological change in this way promotes urban economic growth by reducing production costs or stimulating product development.

The higher the residential density of a city, it can mean the closer people are to their place of work. High residential density further supports good public transport provi sion. The less time people spend commuting it is argued they are then more productive. This view sees labour productivity as negatively related to urban sprawl.

High residential density incorporating mixed land use is also deemed to have ben efits for the urban economy. It is seen as increasing the viability of services such as shops as there is a high concentration of demand, enabling the viability of a range of shops. It means that there is greater choice of services for households, making it an attractive place to live and enhancing the quality of life.

Counter arguments

These interlocking arguments in favour of the compact city have a series of coun ter reasonings. The overriding logic of the compact city argument derives from the importance of agglomeration economies as the basis for cities. This is the driving force for the growth of cities and that this impact is enhanced by high urban densities. However, there is a contradiction between these arguments that the compact city encourages economic growth and the implicit assumption that the compact city should be tightly defined.

A high-density constrained compact city has difficulties coping with significant population growth. There is less potential for expansion, particularly, if urban development is already at a high density. It can also be argued that beyond a certain size, negative agglomeration economies, such as congestion and pollution, occur. It follows that cities can become too big to be sustainable.

Higher residential densities could lead to overcrowding, a lack of green space and traffic noise that will not be socially acceptable. In the compact city, environmental quality may suffer from the loss of open spaces as they are used for development. In terms of residential preferences, a compact form may be less desirable for some individuals. Households with children may prefer to locate further away from the city centre where they have a garden (see later in this section).

Furthermore, the role of household incomes is important in determining the viability of services not just density. The compact city model simplifies this relationship by assuming mixed income neighbourhoods and implicitly assumes a constant urban density. However, as Chapter 5 demonstrates, market forces generate an urban spatial structure with density falling away from the centre. In general, too, there is a spectrum of household incomes from low to high with distance outwards from the city centre.

The ultimate logic of the proponents of compact city economics is that market forces would create compact cities, yet the opposite is occurring. Chapter 12 sets out why these phenomena are occurring across all commercial land uses. Urban systems are decentralising driven by market forces and both the choices of firms and households.

Urban containment and compact city policies are, therefore, working against spatial economic forces. Chapter 13 discusses the wider issues that this creates. In a sense the compact city model is saying to the real estate market here is the urban 'plan', and development should be guided by it. Assuming the compact city works, it creates a number of 'unfortunate' consequences that follow from the operation of real estate markets.

The compact city involves limiting land supply through development constraints on greenfield sites on the edge of an urban area. This could be achieved, for example, by green belts of countryside surrounding cities. If there is continuing economic growth within the city boundaries, the increased demand for land will lead to higher densities as the proponents of the compact city seek. However, this will happen because land values will be bid up, and as a result, land is used more intensively.

In the long term, these higher land values have potential negative implications for the local economy. It could be followed by demands for higher wages. The higher commercial real estate and labour costs reduce the competitive advantage for firms in a compact city. Economic growth could be siphoned away to other areas by these increased costs. In other words, it encourages moves to alternative cheaper locations.

High housing costs lead to lower to medium households potentially priced out of the market in the city. Households then have to commute from cheaper locations in neighbouring towns, beyond say a green belt. At these locations, they can afford to live in a house with a garden. The result is high travel costs and carbon emissions. In these circumstances, the compact city does not minimise travel.

The compact city, therefore, intuitively or superficially promotes sustainability, but in practice, there are many underlying queries or barriers to its application. Arguably, to date, its promotion by the planning profession has had limited success. The same conclusion applies to New Urbanism and Smart Growth. A part of the problem is that these new urban forms are not intuitively attractive to many households.

There is, therefore, market resistance to change, particularly in suburban areas. Although many households enjoy living in decentralised suburbia or aspire to it, many are unconvinced by the benefits of high-density living. Suburban house buyers are attracted to homogeneous neighbourhoods. The introduction of new high-density development in an area can fuel fears of lower house prices in an area and stimulate NIMBYISM.

Households are also reluctant to give up the use of their car. Car ownership is primarily a function of real incomes not residential density. Introducing higher density developments into affluent suburban areas could simply mean greater congestion and noise. This could occur even with improved public transport.

The promotion of the planning ideas linked to mixed-use and high-density living has primarily been successful in inner city areas as discussed in Chapters 9 and 10. Mixed-use neighbourhoods incorporating residential, retail, leisure and offices appeal to young adults partly as part of a lifestyle choice. In city centres or locations just off the centre, travelling by car is less attractive than cycling or using public transport.

In the UK, for example, the government has successfully encouraged the building of small flats in central urban locations on brownfield sites. These flats are mainly occupied by childless households, including students. As purchases, they are attractive as starter homes. Many of these households move out to low-density suburbs as they move to the child-bearing stage of the family life cycle.

Decentralised eco-settlements

Given that there are obstacles to compact cities, the answer to a sustainable urban form may lie in a different direction. In this regard, it is important to emphasise there are market forces that are driving decentralised urban forms. The optimum location for many commercial and industrial firms is no longer a central location. At the same time, as incomes increase in the long term, households seek out low-density housing. Arguably, there are unstoppable market forces that planned compact urban forms cannot stem.

These pressures have led arguments towards the promotion of self-contained decentralised settlements. These would offer the benefits of a low-density 'rural' or 'semi-rural' lifestyle with low development costs with low energy consumption and congestion. This can be viewed as very much an ideal (rather like compact cities). These settlements have a parallel with the freestanding towns and small cities that the urban system is naturally generating as considered in Chapter 11. These islands of growth or urban archipelagos are located in former rural areas or at the urban and the rural interface.

A key potential issue is the population size of these dispersed communities. The freestanding towns/cities that are flourishing are much larger than the eco-settlements that are intuitively more like the size of a village. The proposed 'eco-towns' initiative, for example, promoted by the UK Government in 2007 envisaged populations of 5–10,000. Despite these differences, the planning system could come to an accommodation with urban development market forces.

The main argument against such a decentralised approach is that the residents of small urban settlements can only, in reality, be part of a polycentric urban network with commuting linkages. There will still need to be large centres/central places of economic activity (and population) for manufacturing, retail and administrative services. Rather than these eco-settlements being self-contained travel to work areas, commuters may travel long distances to work. Inter-urban commuting is the norm with the long-term growth of car ownership.

Nevertheless, as noted in Chapter 15, looking to the future improvements in information and communication technologies (ICT) have seen part-working from home become common place. It is a long-term trend likely accelerated by the pandemic. The pollution from commuting by car will also be addressed by electrification. These developments point to the ultimate viability of decentralised sustainable urban forms if not at the time of writing.

Polycentric sustainable model

With the limitations of the compact city and dispersed settlements as sustainable urban forms, a polycentric solution has been proposed as a related alternative. The polycentric sustainable urban structure has also been described as 'decentralised concentration'. As noted earlier in the chapter, the planning in the European Union has promoted polycentricity as a way forward for sustainability.

The term polycentricity at one level is simply a description of a subregional urban system as set out in Chapter 11. Nevertheless, polycentric urban regions take a number of different forms. They can have a dominant centre such as the Stockholm or Frankfurt Rhine-Main regions. Alternatively, they can have two or more cores such as Rotterdam/ The Hague. There is also no definitive number of centres to a polycentric urban region.

As a normative policy tool, it is rather fuzzy. The key to the polycentric sustainable urban model is the linking of centres within a city region by fast efficient public transport. It is argued that the polycentric urban system enables agglomeration benefits to be gained in each of the centres. At the same time, it avoids the risk of negative agglomeration effects associated with large urban structures.

The European Union has made great claims for the model. In this way, it is claimed to promote economic growth and equality across Europe. The polycentric structure is asserted to promote links between industrial clusters and encourage innovation and economic growth. There are some doubts expressed as to whether or not this occurs in reality. A key argument is that competitiveness and cohesion are encouraged through developing connectivity between the various centres within a city by good transport links.

In one sense as a planning strategy, it is about adapting and shaping existing urban development to make it more polycentric. In that sense, such a strategy depends on the historic urban forms to be manipulated. A common theme is that public transport systems can be used to adapt urban change within city regions to make them more compact or at least constrain urban sprawl.

It includes the promotion/configuration of (new) attractive centres with amenities that are well connected by public transport, for example, through new rail stations. In this way, the strategy seeks a potential reduction in the dominance of a central core.

The improvement of public transport is key to addressing the extensive car commuting flows between centres that already exist within polycentric urban regions. Given the diversity of commuting flows between centres, it requires an extensive and efficient public transport system. It will need to encompass the entire city region and be very expensive to develop from scratch.

The polycentric sustainable solution is to a degree working with decentralising market forces. However, to the extent it is trying to increase the densification of smaller centres, it is steering against the market, just like the compact city. This is a barrier. Yet, the primary success of the polycentric strategy depends on the ability of commuters to abandon their cars in favour of the upgraded public transport.

Walkable neighbourhoods

One potential micro-planning solution to the urban sustainability problem is to reduce the use of cars by creating walkable neighbourhoods. The idea is to reconfigure low-density neighbourhoods to make them more pedestrian friendly. The objective is encapsulated in the concept of providing a range of services within a 15- or 20-minute walk from home. At its extreme work, home, shops, entertainment, education and healthcare are all reachable within this time frame.

A less extreme view sees urban and transport planning ensure that residential, business, and other land uses are within walking distance of public transport. The strategy not only is about the reformulation of public transport routes and the location of bus or tram stops but also often includes the densification of neighbourhoods.

There are inevitable issues about such a strategy. It would be theoretically possible, at least partially, to physically create such neighbourhood relationships. Such neighbourhoods existed in pre-car days. The problem is that human behaviour is not determined by its physical environment. As noted earlier, households are stubbornly reticent to give up the use of their car.

Households do not simply shop at the nearest (neighbourhood) shop as a car allows the choice of a range of retail centres. Neighbourhood shops in any case are likely to stock only a limited range of goods. Accessibility to public transport does not necessarily substantially increase its use for commuting. It is difficult for public transport systems to meet the diverse commuting flow demands within a city region. The car is just more efficient if more polluting.

Reformulation of urban sustainability problem

The potential sustainable urban forms described above can be characterised as normative and theoretical and supported by minimal and disputed evidence. To different degrees, they run counter to market forces. Nevertheless, compact cities/high-density planning policies have become very pervasive. The problem is that these policies face a range of challenges that mean that they suffer impracticalities.

This section develops an alternative perspective on urban sustainability. It begins by noting that the five main elements that make up an urban form are as follows:

- Land use patterns
- Density – number of sub-elements – gross population, net residential, commercial and industrial employment densities
- Transport infrastructure
- Characteristics of the built environment
- Layout

The operation of the real estate market together with planning determines the spatial pattern of land use, the density of development, the characteristics of the built environment and layout. To be more precise, the **real estate market** and **transport infrastructure** are the key determinants of the spatial structure and, hence, sustainable urban form. The real estate market is, therefore, at the heart of achieving urban sustainability.

An alternative, less presumptive, approach is to express the problem differently as **maximising potential urban output or productivity subject to a series of sustainability constraints**. These would encompass social, environmental and economic factors. From an economic perspective, these constraints would include the **viability** of sectors of the local economy. This approach can be viewed as guidelines for adapting existing city regions.

A sustainable urban system has to ensure that the components of the urban economy are viable. These encompass manufacturing and services, the labour market, the public transport system, and public administration. Perhaps most important of all land use or real estate markets generate sufficient offices, shops, industrial units and housing to meet the needs of the urban economy.

A linked social constraint would be an **adequate supply of housing** for the workforce and their families and full employment. Otherwise, workers will need to commute long distances with the associated pollution. All constraints would need to be met to satisfy sustainability. The precise 'sustainable' urban form will also vary depending on factors such as the industrial mix or structure of the city and the incomes of households.

The formulation assumes that a sustainable urban form depends on the nature of the local economy. A sustainable urban form with a high-income population could look very different from a city of poor households. This can be explained with the following comparative example.

Consider two cities with the same population, household numbers and demographic structure but with different income levels, one poor and one rich. The poor city has no planning constraints, so the market determines the extent of the built-up area within which everyone commutes. As the demand for housing is a function of income, the rich demand more space, so without planning constraints, the rich city would cover a larger area.

If the rich city was constrained to the same area as the poor by planning constraints, then many households would have to commute long distances from surrounding settlements. Based on simply travel patterns, the poor city is more sustainable than the rich city – reinforced by the likely higher use of cars in the rich city. However, this would ignore the important income variable and its influence on the urban form. It also highlights the difficulty of generalisation and the limits of physical constraints on the urban form as a sustainability solution.

A plan for making a city more sustainable must take into account local circumstances/ characteristics, the existing urban form and the operation of the local real estate markets. The precise form of the solution that this formulation of the problem generates is not definitive. There are too many variables, and in any case, a city is an individual and a dynamic entity. The approach is best seen as a set of guiding principles.

Given the central role of the real estate market, planning to achieve a sustainable urban form must centre on transport infrastructure and shaping the property market. But as the example of the rich city above shows it requires an adequate supply of housing for the workforce and their families (affordable housing constraint) otherwise, there is commuting from surrounding areas.

The sustainability of cities is not simply driven by economic factors, but desired urban forms will need to be subservient to economic viability. As an example, many households prefer low-density living, and this preference drives market forces. If there is no accommodation with market forces, it will mean that a 'sustainable' urban form will be not be stable in the long run. The next section considers the potential for adapting residential land use patterns in this regard.

Economic viability of adapting cities

Housing is the primary urban land use and, hence, is the principal land use focus for improving the sustainability of a city. The task is then to examine the potential for adapting a local housing market to a more sustainable form by building more houses/flats for sale in the right locations. These houses need to be built on 'brownfield' land within the existing city boundaries to ensure against urban sprawl. The analysis here draws on a study that estimates the spatial patterns of local housing development viability within five provincial cities in the UK.

For each city, viability maps were derived from data on individual house prices and on construction costs for individual development projects. By examining viability, it is possible to assess how easy and where it is possible to build new housing. These intra-urban patterns of viability are found to be determined by the spatial structure of existing house prices. In other words, development viability is best in neighbourhoods where house prices are highest.

In terms of urban sustainability, this is a problem. These high-priced neighbourhoods are also those with the lowest density. These neighbourhoods benefit from high demand and where there is most resistance to change. At the same time, in some cities, there are large swathes of negative viability even without accounting for the additional costs of brownfield development.

It is these latter neighbourhoods that offer the best opportunity to improve the sustainability of a city. New housing will be seen as upgrading the physical environment of these neighbourhoods. However, the negative viability of these areas means that it will require public funding to support residential development. In addition, in all probability investment in new public transport, infrastructure will also be necessary, at least in some areas, to encourage inward housing demand.

There are, therefore, major constraints to the reconfiguration of urban housing markets. The challenge of adapting whole neighbourhoods means that it cannot be undertaken in a piecemeal way. Adapting cities will arguably require grand designs and public expenditure to fundamentally reconstruct intra-urban house price structures and neighbourhoods.

The reshaping of cities will require positive planning policies to achieve a specific more sustainable urban form. The actual form will vary by city. Urban strategies must centre on transport infrastructure and redrawing the residential market. The process of adjustment must include the establishment of attractive new housing, involving perhaps new property forms in non-traditional parts of a city.

The restructuring of a city in this way is not necessarily a smooth process and could take decades. They will also have to ensure that the housing provided is inclusive, in the sense that it will be affordable. Strategies to adapt the city form will, therefore, require a consensual long-term public policy framework to ensure confidence for private real estate investment.

Summary

From a simple ecological perspective, given the level of urban pollution, the sustainability of cities is a misnomer. However, urban sustainability can be seen more widely as comprising three dimensions – environmental (including transport), social and economic aspects. The physical characteristics of cities can be seen to influence all three dimensions.

Environmental sustainability is affected by the scale of green space in a city and the degree of compactness impacts on travel demand. The relationship between the urban form and social sustainability is more complicated.

On the one hand, it is argued that high densities and mixed-use urban forms bring a good quality of life with social interaction, community spirit and cultural vitality. On the other hand, high density is seen as having weak access to green spaces, poor health and mental well-being, potential bad neighbour effects and reduced living space.

The economic sustainability of different urban forms is also disputed. However, crucially sustainable urban form solutions need to be economically viable to ensure a long-term legacy. The goal of addressing urban sustainability is particularly linked to the relationships between the environmental and economic dimensions.

There is a dispute about the nature of sustainable urban forms that has two radically different options. One solution proposes a high-density, mixed-use centralised urban form. The alternative way forward supports low-density dispersed urban forms.

The global context to this policy debate was set by the Brundtland Report of the World Commission on Economic Development, in 1987. The message of the report was that to ensure a sustainable global future, significant changes were required. Policy makers have accepted this sustainability challenge by extolling the merits of high-density and compact urban forms.

The policy has been promoted in different guises by planners across the globe. In Europe, it took the umbrella terms of the compact city and polycentricity. In North America, there are variants known as 'New Urbanism' and 'Smart Growth' initiatives. They have different internal logics, but all emphasise a shift towards the use of public transport.

These policies encompass land use patterns, energy use/savings for buildings, minimization of waste and water use and efficient transport. A sustainable urban form incorporates densification, green space, mixed land use neighbourhoods with local identity and access to affordable homes. Urban sprawl is seen as the antithesis of sustainability.

These western planning policies are not easily transferred to high-density cities in much of the rest of the world. The priority of many rapidly growing Asian cities is seeking to reduce their very high densities not to increase them. There are, therefore, issues about the universality of these policies, as well as about their economic viability.

The economic case for the compact city centres on the role of agglomeration economies and their relationship to density. The closer firms are together the greater the innovation through knowledge spillovers and, hence, the higher urban economic growth. Similarly, short journeys to work promote labour productivity. High residential density encompassing mixed land use enables the provision of a greater range of services and shops, thereby enriching the quality of life.

There are a number of counter arguments. By definition, a compact city has to be tightly defined so it will ultimately struggle to manage economic and population growth. It will bring associated congestion and pollution. Higher residential densities could lead to overcrowding, and environmental quality may suffer from the loss of open spaces. The compact city also runs counter to market forces as urban systems are decentralising.

The compact city involves limiting land supply and if as a result market forces increase demand for land. It will lead to the higher densities and the intensive use of land as desired by the policy. However, land and property prices also rise with negative consequences. Firms and households could be priced out of the city, with the latter having to commute from surrounding areas.

The policy of compact urban forms has had limited success and is intuitively unattractive to many suburban households with families. There is stiff resistance to abandoning the use of a car. Mixed-use and high-density living is attractive to young adults and has been primarily successful in inner city areas.

An alternative sustainable urban form has been proposed in the shape of self-contained decentralised eco-settlements. Such settlements have the advantage of similarity to larger urban forms, namely, small towns, being spawned by market forces. The primary difficulty of these settlements is the problem of self-containment. Realistically, most residents of such small urban settlements can only be part of a polycentric urban network within which workers commute to larger centres.

A polycentric sustainable urban structure has been promoted by the European Union planning policy. The vital element to the polycentric sustainable urban model is the connection between centres within a city region by fast efficient public transport. An implicit goal is the reduction in the dominance of a central core so that negative agglomeration effects of size can be moderated.

The underlying premise is that public transport systems can be used to adapt the urban structure within a city region to make it, as a whole, more compact and limit urban sprawl. The policy depends on providing a viable solution to car commuting. It, therefore, requires an extensive and efficient public transport system.

At the micro-level designing walkable neighbourhoods have been identified as a way of improving urban sustainability by diminishing the use of cars. The target is to ensure a range of services within a distance from a home, say 15 or 20 minutes. There are different versions of this idea including requiring residential, business and other land uses are within walking distance of public transport.

Although it is possible to achieve this aim of ensuring public transport is located to meet this goal, there are questions over whether it can achieve its ultimate goals. Households are stubbornly reticent to give up the use of their car. Accessibility to public transport does not necessarily substantially increase its use. The flexibility of the car enables households to choose from a range of retail centres and optimise travel times to work.

An alternative perspective on the sustainable urban form is to see the challenge as maximising the potential urban output or productivity subject to a series of sustainability constraints. In addition, an urban form is defined as the spatial pattern of land use, the density of development, the characteristics of the built environment and layout. The real estate market and transport infrastructure are the key determinants of spatial urban form and, hence, sustainable urban form.

A sustainable urban system has to be viable and encompass an adequate supply of housing for the workforce. This approach does not have a definitive solution as it depends on the characteristics of the individual city and its economy. A plan for making a city more sustainable needs to take into account local circumstances including the existing urban form and the operation of the local real estate markets.

Reshaping cities will require positive planning policies to achieve a more sustainable urban form and centre on transport infrastructure and redrawing real estate markets. Housing is the primary urban land use and, hence, is the principal land use focus for improving the sustainability of a city. New housing needs to be built on 'brownfield' land within the existing city boundaries to ensure against urban sprawl.

Unfortunately, development viability is highest in low-density wealthy neighbourhoods where there is most resistance to change. As a result, the best opportunities to improve the sustainability of a city is through the upgrading of low-income neighbourhoods with the most potential for physical improvement.

The desired sustainable urban form will vary by city. The process of adjustment could take decades and is likely to focus on affordable housing in less established residential neighbourhoods. In doing so, strategies will entail sustained public investment within a long-term plan to support private real estate investment and new transport infrastructure.

Learning outcomes

Urban sustainability depends on three dimensions – environmental (including transport), social and economic aspects.

Environmental sustainability is affected by the scale of the green space in a city and the degree of compactness impacts on travel demand.

Social sustainability involves a good quality of life with social interaction, community spirit and cultural vitality.

Sustainable urban form solutions need to be economically viable to ensure a long-term legacy.

Policy makers have addressed the urban sustainability challenge by endorsing a move towards high-density and compact urban forms.

These planning policies have taken different guises across the world, from the 'compact city' through to 'New Urbanism' and 'Smart Growth'.

Urban sprawl is seen as the antithesis of sustainability.

These high-density western planning policies are not easily transferred to rapidly growing cities that are seeking to reduce their very high densities not to increase them.

The economic case for the compact city centres on the notion that agglomeration economies are enhanced by high density.

This argument is disputed, for example, a compact city runs counter to market forces that has seen urban systems decentralising.

One particular issue about the compact city is that if successful in stimulating demand, the result would be rising land and property prices. Firms and households could be priced out of the city, with the latter having to commute from surrounding areas.

Compact urban forms have had limited success. They are unattractive to many suburban households with families. Mixed-use and high-density living developments have been primarily successful in inner city areas.

Self-contained decentralised eco-settlements have the advantage that they work with market forces. Their disadvantage is that self-containment is unrealistic.

A polycentric sustainable urban structure is dependent on successful take up of efficient public transport between centres.

Walkable neighbourhoods have been identified as a way of improving urban sustainability by ensuring residential, business and other land uses are within walking distance of public transport. The problem is accessibility to public transport does not necessarily reduce car use.

The real estate market and transport infrastructure are the key determinants of the spatial urban form and, hence, sustainable urban form.

A plan for making a city more sustainable needs to take into account local circumstances including the existing urban form and the operation of the local real estate markets.

Reshaping cities to achieve a more sustainable urban form should centre on transport infrastructure and redrawing real estate markets to ensure against urban sprawl.

The best opportunities to improve the sustainability of a city is through the upgrading of low-income neighbourhoods. This task will necessitate substantial public support and new public transport infrastructure.

The process of adjustment to a more sustainable urban form could take decades. Strategies will entail sustained public investment to support private real estate investment.

Bibliography

Downs M (2005) Smart growth: Why we discuss it more than we do it, *Journal of the American Planning Association*, 71, 4, 367–368.

Jenks M and Jones C (2010) *Dimensions of the Sustainable City*, Springer, Heidelberg.

Schmitt P (2013) Planning for polycentricity in European metropolitan areas—Challenges, expectations and practices, *Planning Practice & Research*, 28, 4, 400–419.

17 Neighbourhood and housing market dynamics

Residential neighbourhood change is bound up with the dynamics of the local housing market. The housing market of a community adapts over time to a range of external and internal influences. The housing stock gradually becomes older and may need repair or improvement to meet modern standards. In some cases, demolition and new building may occur.

At the same, residents who occupy the housing stock also get older. As their circumstances alter, through say a change of job, retirement or household composition, then their attitudes towards their housing also evolve. As a result, households may move home. Furthermore, new residents move into a neighbourhood with potentially different perspectives.

Chapters 5 and 7 examined some aspects of local housing markets including the spatial structure and housing submarkets. In this chapter, the focus is the neighbourhood. In particular, it considers the following:

- Neighbourhood decay and revitalisation.
- Neighbourhood succession including filtering and gentrification
- Policy questions.

Neighbourhood decay and revitalisation

The evolution of a neighbourhood can be tracked in terms of market indicators such as changing prices compared with other areas and vacancy rates or the speed of sales/letting. It can also be judged by the general upkeep of the housing and changes to the housing stock. The changing character of a neighbourhood may also be reflected in the socio-economic-demographic profiles of the residents.

The forces propelling this change can emanate within the neighbourhood but also can be external. Although the housing market is influenced by macroeconomic factors such as interest rates and real incomes, these are unlikely to have specific effects on a particular neighbourhood. External influences on a neighbourhood are likely to emanate from within the urban area.

As previous chapters have identified, the transport infrastructure of an urban area sets the framework for the spatial structure of the housing market. New/improved roads such as a bypass can have a positive or negative effect on real estate values in a neighbourhood. Changes to the public transport system including routes may have knock-on effects at the neighbourhood level.

DOI: 10.1201/9781003027515-19

Within a neighbourhood, internal influences on change relate to the interaction of the decisions of residents. It involves the shifting characteristics of the housing stock, the tenures of individual units and the behaviour of owners and occupiers. A particular mechanism for neighbourhood change is the role of externalities or spillover effects. One resident's or owner's actions can impinge on neighbouring properties.

The value of a house or flat is a function of its physical features, the characteristics of the neighbourhood in which it is found and its location. Location encompasses a range of accessibility relationships. Accessibility is not just in terms of its distance to the city centre (as in the basic access space model) but also to local schools, shops and other amenities.

The value of an individual dwelling is, therefore, partly dependent on the quality and type of the immediately surrounding housing. In turn, value can be determined by the level of investment (and expected future decisions) of the owners in these nearby properties. In some cases, these owners may be landlords.

Different types of owners have different motivations towards maintenance and repair of their buildings. They also have different financial capacities to undertake these activities. Low-income owner occupiers may have serious financial constraints to invest in improving their home. In addition, they may see their home more as a consumption good rather than as an investment that they should nurture to increase its value.

In similar fashion, some private residential landlords may see little incentive to investing large sums in upgrading their properties if they believe rents would not rise accordingly. The spillovers from negative investment decisions are likely to be exaggerated in high-density areas. Blocks of flats may require common agreement by individual owners to undertake major repairs such as roof maintenance.

The consequence of negative investment decisions can be seen by looking at neighbourhood decay. Neighbourhood decay can be recognised when widespread reductions on outlays for repair maintenance and repair renovation and modernisation become apparent. Expenditures on properties are less than necessary to maintain them in their original condition or more likely to ensure they meet modern standards.

The physical signs of effective disinvestment are manifested in the state of the housing stock in a neighbourhood. The incentives to invest are also deflated in inner-city neighbourhoods by cheap house prices with the predominance of owner occupiers on low incomes. The effect is exacerbated by the predominance of small properties.

Similarly, privately rented accommodation, especially if it is in multiple occupation, can contribute to lower values of neighbouring properties. Together, relatively cheap house prices and a concentration of low-income households and private tenants have negative spillover market consequences. They can also contribute to weak expectations of capital (and rental) growth.

Disinvestment and neighbourhood decay can also be impacted by public policy. Designation of neighbourhoods for redevelopment or rehabilitation sends out signals that may instil doubt about the future. Effectively, the future of the area can be blighted. The uncertainty provoked is likely to discourage residents from investing in the short term.

Potential home buyers will also not move in if the area is to be redeveloped, trapping the present population. As a result, the price of housing will fall, possibly becoming negligible. Even if the neighbourhood is designated for improvement, potential home buyers may defer purchase until evidence of progress (see later).

The problem of blight demonstrates the role of market forces. An active local market is an essential precondition for housing investment and neighbourhood stability. It enables individual households to move home to adjust their circumstances to passage through the family life cycle. It also validates through market prices the benefits of investing in and improving a property.

Inner-city residential markets can suffer from a number of restraints. Thin numbers of transactions may make valuations difficult and put off mortgage lenders. Wider constraints on mortgage finance in inner-city areas are discussed in Chapter 7. Loans may be difficult to find to purchase properties. Loans, if available, are at higher than standard interest rates or with shorter pay back periods. Overall, these restrictions not only reduce the marketability of properties but also deflate prices.

Although mortgage finance can be restricted for older housing in inner-city neighbourhoods, it is generally readily available for new housing. Much of this new housing is at the urban periphery. The ease of purchasing new and suburban housing contributes to a relative discouragement of investment potential of older housing in inner-city areas.

All these factors collectively can be factored into inner urban decay. Markets in these neighbourhoods experience the influence of low household incomes, the private rented sector and mortgage finance difficulties. These areas are also suffering population decline and falling demand. The implications include properties in a poor state of repair, high vacancy rates as properties are abandoned and house prices selling only at nominal values.

In a flourishing neighbourhood, market forces work in the opposite direction. There is likely to be a strong demand for housing, so properties coming on the market sell quickly. Demand to live in the neighbourhood could generate a house price premium compared with that in surrounding areas. Such a premium could be dissipated if there is new housing development.

If supply or planning constraints hinder or do not permit further new housing development, then real prices will rise. Excess demand could then be accommodated by using the existing housing stock more intensively. In the medium to longer term, supply restrictions lead to households adding an extension or a loft conversion to their existing homes. These decisions are stimulated both by the lack of suitable large properties to move to and the increase in the investment worth of their homes.

In these examples, housing turnover in a neighbourhood as an indicator of the state of the market cannot be interpreted in isolation. Low turnover of sales or lettings in a neighbourhood may be an indicator of weak demand, but it could also indicate a stable elderly population.

Similarly, a high rate of transactions does not necessarily indicate strong demand. A high rate of turnover can be associated with areas of predominantly rented properties from private landlords and poor-quality housing. The rate of mobility could reflect that people seek to move on if they can to upgrade to better housing. High turnover could also imply decline in a neighbourhood as owner occupiers sell out to private landlords.

Another type of neighbourhood with high turnover is one that comprises owned small houses and flats. These are often effectively 'starter' homes for households at the beginning of the family life cycle. These homes are typically occupied for periods of five years or less before the households move to larger properties. It is, therefore, important to interpret (changes in) turnover within the context of the neighbourhood type.

Neighbourhood change in terms of turnover patterns or tenure change or (dis)investment in the housing stock is subject to bandwagon effects. A bandwagon is normally expressed as people buying more of a product as the more other households' purchase. In a spatial setting, spillovers play a key role in influencing bandwagon effects within a neighbourhood housing market and land use change.

Anticipations about the future character of a neighbourhood and the reactions of households, investors and developers are crucial constituents in any bandwagon process. Neighbourhood changes can, therefore, induce bandwagon effects, with further consequences. They could involve the introduction of new house types or households or more houses being sold to landlords.

A bandwagon effect can emanate from both the demand and supply sides of the market. It could also originate from changes in local amenities such as a school closure or the environment, say by rerouting of road traffic. Inherent in this concept of a bandwagon effect is the notion of a threshold or tipping point. Beyond the tipping point, the process of neighbourhood change takes hold and accelerates.

This concept of a threshold effect on neighbourhood change has also been researched in the context of the levels of poverty. A local poverty threshold effect is likely on educational attainment, crime, school leaving and the duration of unemployment. The empirical evidence on these dynamics, including those of neighbourhood housing markets, is very limited.

An important conclusion from this analysis is that an active neighbourhood market is a crucial precondition for supporting housing investment and neighbourhood stability. A further lesson is that neighbourhood change tends to be incremental. However, there is a potential for negative and positive bandwagon effects, and these are subject to threshold or tipping points.

From an urban policy outlook, the message is that regeneration initiatives designed to address urban decay require to generate an active property market. In addition, regeneration needs a long-term commitment in terms of public support to reach a threshold point at which sustainable positive change can occur. This issue is taken up throughout Part 3.

This analysis of neighbourhood decay and revitalisation has ignored that the type of households living in an area are likely to change. The modification, and even possibly transformation, of the local socio-economic demographic profile of households is inherent in neighbourhood change. As part of this process, household movement into and out of a neighbourhood is fundamental to altering its character. The detail of how this occurs is now considered.

Neighbourhood succession

The replacement of one group of residents by another with different characteristics is known as succession. There are different ways in which this occurs. The 'sector theory' of urban growth and structure developed by Hoyt in 1939 is a useful starting point to consider this issue.

The theory centres on steadily rising incomes persistently generating demand for modern housing on the edge of the city. This modern housing is preferred by high-income households who can afford it. As a result of urban growth, high-income residential neighbourhoods gradually move outwards along the radii of a city.

This creates a process of succession as lower income groups then move into the properties vacated by higher income groups. In the process, they move out from the inner-city areas. Their housing is taken by immigrants into the city on even lower incomes. This model of succession offers some insights but is incomplete.

Filtering

An associated but broader approach to explaining succession is given by the concept of 'filtering'. It recognises that changes in occupancy of housing by different groups of households also involve revisions in housing quality and price. While it applies to individual housing, it can be seen as a perspective on neighbourhood change.

There are a number of different definitions of filtering but that developed by Ratcliff in 1949 is applied here. The filtering process is linked to a house changing occupancy and price. Specifically, it is the changing of occupancy as that housing that is occupied by one income group becomes available to the next lower income group, as a result of a decline in price.

The basis of this filtering process like the sector theory is that higher income groups move because they find their present housing 'obsolescent'. Their dissatisfaction could be as a result of a change in technology or tastes and, say, the attraction of modern housing. At the same time, lower income groups take the opportunity to upgrade their housing and move to improve their living conditions too.

The essential attributes of the filtering process are twofold. First, households move upwards in terms of quality to a 'better' home. Second, existing vacated dwellings move 'down' in terms of relative price and a deterioration in quality making them accessible to lower income bands.

The filtering process, as described, is a simplification as the following example illustrates. Large town houses were built in the nineteenth century in the UK and other western cities. Originally, these large houses were owned and occupied by wealthy businessmen and their families (and servants). These properties have now filtered down and rented to students and young people without families and are often subdivided into flats and bedsits.

These buildings are no longer in the best state of repair, whereas when they were built they represented the state of the art of modern housing. Once, they were fashionable houses occupied by the top echelons of urban society. Now, they are inhabited by relatively lower income groups. Filtering has happened, but the process involves not simply changes in occupancy, quality and price. It encompasses tenure change from owning to renting. The quantity of housing consumed per household has also fallen.

As noted above, although filtering is defined in terms of individual housing, it can be extended to examine the relationship between neighbourhoods. Given local housing commonalities, filtering can explain the process of neighbourhoods changing relative to others. At this neighbourhood level, patterns of residential mobility are linked to changes in the housing stock. Household migration patterns drive changes in prices and the quality of the housing stock across the neighbourhood.

Filtering (and the sector theory) highlight the role of income, via income groups and changing income, as the principal pressure for neighbourhood succession. It is consistent with the long- term driver of suburbanisation explained in Chapter 10. In the sector theory, in particular, lower income groups, perhaps immigrants, initially move into inner-city locations. As their income increases, they move outwards into surrounding neighbourhoods.

The filtering process as set out above does not incorporate the influence of band-wagon effects and thresholds or tipping points. Yet, there is undoubtedly a dimension to the process that incorporates these elements along the lines of Schelling's theory of neighbourhood succession. Schelling applied his ideas to the changing racial composition of neighbourhoods, but it could be applied to other household groups.

Schelling argues that as black residents move into a predominantly white neighbourhood, there are spillover effects. If a certain threshold percentage (or tipping point) of black households is achieved, then it will eventually become all black. This transformation process can also be envisaged as a filtering type process including price and potentially quality adjustments to the neighbourhood's housing. The increasing incomes of black households are an important push factor.

Gentrification

Filtering involves low-income households moving into neighbourhoods formerly occupied by higher incomes. Gentrification is the reverse process whereby higher income groups take over neighbourhoods where low-income groups live. The term was coined by the sociologist, Ruth Glass, in the early 1960s.

She illustrated it as follows:

> One by one, many of the working class quarters of London have been invaded by the middle-classes—upper and lower. Shabby, modest mews and cottages—two rooms up and two down—have been taken over, when their leases have expired, and have become elegant, expensive residences...Once this process of 'gentrification' starts in a district it goes on rapidly until all or most of the original working-class occupiers are displaced and the whole social character of the district is changed.
>
> (Glass, 1964)

Although first identified in London, gentrification has occurred all round the world. Essentially, gentrification is a neighbourhood displacement process in which not only the 'rich' replace the 'poor' but also individual properties are upgraded. As a result, the neighbourhood is transformed into a middle-class area. The gentrification process includes elements of neighbourhood change that have already been discussed threshold/tipping points, price and quality changes to the housing.

Gentrification can be seen as a cycle of change. First, investment in high-quality housing for the rich in what are now central locations during the nineteenth century. Second, disinvestment as the original rich occupants move out to outer urban locations to be replaced by low-income households through a filtering process. The example above highlights these changes. Third, the houses are recolonised decades later by higher income groups who reinvest and improve the housing stock.

As described, gentrification occurs in inner-city areas, but it can occur elsewhere. Gentrification may happen anywhere that is cheap housing stock with historic character. Villages on the edge of a city with reasonable access to the centre may see the agricultural workers displaced by middle-income groups. These households buy up the properties and renovate, improve and extend the housing. Such households are attracted by the semi-rural lifestyle. The nature of the process is mirrored by extensive second home purchases in deep rural areas.

The search for lifestyle also seems to drive inner-city gentrifiers although the reasons can be complex. It is discussed in Chapter 10. Undoubtedly, motivations include the opportunity to purchase housing with character at a reasonable cost. The initial phase of gentrifiers, including white and non-white households, may be attracted to a more 'ethnic' neighbourhood. Nevertheless, there is displacement of existing white and non-white residents.

The displacement occurs through the increased real estate rental and capital values. The increase in households with higher incomes seeking to buy in a neighbourhood and the upgrading of buildings bring these higher values. It also leads to housing tenure changes towards owner occupation. The rise in values stimulates landlords to sell up, and in the process, their typically lower income working class tenants need to move.

As described so far, the gentrification process is simply a housing market adjustment. But it is useful to delve deeper into how it occurs. It partly depends on the difference between the value of an unimproved house and an improved house. This difference is crucial to the financial viability. It is sometimes called the 'rent gap', and it will change as gentrification becomes more established in a neighbourhood.

At the starting point of the process, the gap is likely to be small so creating a barrier to upgrading. In financial terms, it is simply not worth it to spend large sums upgrading a property if there is no increase in value. Not all pioneer gentrifiers see the activity in purely financial terms, but the state can be an unwitting supporter. In the early stages of gentrification in London, it was backed by housing improvement grants from the central government. These grants were available nationally as part of the policy to modernise the housing stock.

The widespread potential for gentrification is limited by the characteristics of the housing stock that is attractive to high income groups to improve. In many older inner-city areas, the housing is not attractive enough. The market conditions also do not make it financially viable to upgrade the housing stock. The improved value of a house is less than the original cost plus the improvement expenditure even with a grant.

In the UK, this is often referred to as the 'valuation gap' and is invariably a major constraint on the viability of urban renewal. The final value of an improved house is very much a function of the income capacity of those prepared to pay and the nature/state of the other housing in the neighbourhood.

Where neighbourhoods are transformed by gentrification, its permanent continuance is still dependent on the establishment of an active market as set out earlier. First wave pioneers have to be replaced by households who view the neighbourhood as a distinctive fashionable submarket. They are prepared to pay accordingly.

Meanwhile, there is some evidence that some of the original gentrifiers are at the beginning of the family life cycle. Ultimately, they move out to the suburbs and locations that meet their child rearing needs including access to good schooling, etc. Family life cycle may not be as overt an influence as incomes (and race) on neighbourhood succession, but it plays a part.

Policy questions

Gentrification cannot just be seen as the displacement of the poor by the rich or middle class as a result of market forces. There is a wider perspective on neighbourhood gentrification that stems more explicitly from public policy. In other words, equivalent displacement can also occur as a consequence of public policy. National urban policy in the UK, for example, has arguably endorsed gentrification in its widest sense as a solution to inner-city problems.

Since the middle of the 1970s, in the UK, urban policy has encouraged private housing development in inner-city areas. This is achieved through a combination of land and financial subsidies to promote owner occupation. Mechanisms encompass building new houses on land owned by a local authority or renovating/improving council (public sector) housing for sale.

Urban policy has taken the view that to transform neighbourhood decay revitalisation requires an input of new house types and households. At one level, this is promoting social mix or tenure diversification (see Chapters 20 and 22). But it can also lead to a form of gentrification if imported middle classes ultimately lead to the displacement of the existing community. At the very least, subsequent changes to the neighbourhood, for example, shop types, may not necessarily benefit the original community.

Regeneration and public policy initiatives such as, say, a new public transit line can also lead to gentrification of a neighbourhood. Physical improvement of an area may make it more attractive to higher incomes to move into it. The increased demand pushes up house prices in the neighbourhood. Similarly, the greater accessibility of an area as a result of a new tram link could lead to a rise in local house prices. In these cases, increasing house prices could contribute to the displacement of the original low-income residents. Gentrification may be an unintended consequence of public policy.

Summary

The changing characteristics of a neighbourhood can be judged by a combination of pointers. These include local market indicators, the physical state of the built environment and the socio-economic-demographic profile of the residents.

Neighbourhoods evolve as they are influenced by internal and external factors. External influences on a neighbourhood are likely to emanate from within the urban area such as those linked to transport infrastructure. Within a neighbourhood, the actions of an individual owner can create spillover effects on nearby properties.

The value of a specific house can be influenced by the low level of investment of its neighbours. Neighbouring low-income owner occupiers may not be able to invest in improving their home. Local private residential landlords too may not be prepared to invest if rents do not justify it. The spillovers from such negative investment decisions are exaggerated in high-density areas.

Neighbourhood decay can be attributed to insufficient investment, and it is discernible by the state of the housing stock in a neighbourhood. Incentives to invest are depressed in inner-city neighbourhoods. The reasons are not only a concentration of low incomes and private rented accommodation but also cheap house prices and the predominance of small properties.

Public policy can impact on disinvestment by designating neighbourhoods for redevelopment or rehabilitation. The uncertainty engendered may blight or discourage investment and deter people from moving into the area until its future is clear. Blight shows that an active local market is essential for housing investment and neighbourhood stability.

Markets in inner-city areas can also suffer from a number of other restraints such as the small number of transactions and the unavailability of ready mortgage finance. With these neighbourhoods experiencing population loss, all the market forces combine to create an engine for urban decay. It is exemplified by properties in a poor state of repair, properties difficult to sell or abandoned and house prices selling only at nominal values.

The reverse is true in a flourishing neighbourhood where there is a strong demand for housing, so properties coming on the market sell quickly. It could be reflected in a house price premium compared with surrounding areas. Where these areas have constraints on new development, it is likely to see an increase in extensions and loft conversions.

Neighbourhood change is subject to bandwagon effects through the spillover effects of the decisions and activities of local households and owners. Expectations about the future nature of a neighbourhood can induce these bandwagon effects. These effects could be stimulated, for example, by the introduction of new house types or households or more houses being sold to landlords.

Intrinsic to the notion of a bandwagon effect is the existence of a threshold or tipping point. Once the tipping point is reached, the process of neighbourhood change quickens. The implications for urban regeneration initiatives are that they require a commitment to reach a threshold point to achieve fundamental positive neighbourhood change.

Neighbourhood change brings modifications not only to the housing stock but also to the local socio-economic demographic profile of households. As neighbourhoods evolve, household movement in and out of the area can see the characteristics of the population be transformed. Hoyt's sector theory of urban growth, for example, encompasses a process of neighbourhood succession. Lower income groups move out from inner-city areas into properties vacated by higher income groups who move to more modern housing on the periphery.

Filtering is one way to explain neighbourhood succession. As part of this process, not only are specific house types occupied by different income groups but also the quality and price of the housing changes. In particular, low-income groups can move into housing occupied by a higher income group because of a fall in price. Filtering, in practice, is more complex involving not just changes in occupancy, quality and price but tenure change from owning to renting and the subdivision of housing.

The basic filtering process logically should incorporate bandwagon effects and thresholds or tipping points as proposed by Schelling. The changing racial composition of neighbourhoods can be envisaged as a filtering type process. Although the increasing incomes of black households are an important push factor in the process, there are also likely to be price and quality adjustments to the neighbourhood's housing.

Gentrification is the reverse process to filtering whereby higher income groups take over neighbourhoods where low-income groups live. It is a neighbourhood process in which the 'rich or middle class' upgrade old properties leading to the 'poor' being displaced. Again, there are threshold/tipping points to the process that is not necessarily confined to inner-city locations.

The displacement occurs because low-income groups are priced out of neighbourhoods through the increased real estate rental and capital values. Often, rented accommodation is sold off to owner occupiers by landlords attracted by the high values on offer. Gentrification depends on owners prepared to bridge/fund the gap between the value of an unimproved and an improved house. It is probable that pioneer gentrifiers do not see the challenge simply in these financial terms.

In some cases, gentrification has been supported by grants from the government. The widespread potential for gentrification is limited by the lack of distinctive characteristics of much of the inner-city housing stock. This housing stock is often small and unattractive to high-income groups to improve. The improvement of this 'more standard' stock is hindered by the financial costs. For much of this housing, the improved value of a house is less than the original cost plus the improvement expenditure even with a grant.

While the pioneer gentrifiers may not have been motivated by financial gain the permanent continuance of the upgrading is dependent on the establishment of an active market. House buyers need to be prepared to move into the area and pay a high price to live in a distinctive neighbourhood.

Gentrification in the wider sense of the displacement of the 'poor' by the 'rich or middle class' can also occur as a consequence of public policy. Urban regeneration can support the movement of higher income groups into an area that can lead to gentrification. In particular, policies that promote social mix or tenure diversification may ultimately lead to the displacement of the original community. Gentrification can also be an unintended consequence of public policy by pushing housing prices up in a neighbourhood.

To summarise, the analysis of neighbourhood dynamics highlights the significance of an active property market to ensure, or as a sign of, the continuing prosperity of a locality. Change tends to be incremental. Nevertheless, there is the potential for negative and positive bandwagon effects that are subject to threshold or tipping points.

The major driver of substantive neighbourhood change is the succession of one grouping of households by another. It is most evidently seen through changes in racial groups or via gentrification. More typically, the process can be seen as the spatial restructuring of different income groups within a city. It is associated with adjustments in the price, tenure and quality of housing within a neighbourhood.

Learning outcomes

The changing characteristics of a neighbourhood can be judged by local market indicators, its physical and its residential socio-economic-demographic profile.

Within a neighbourhood, actions of individual owners may create spillover effects on investment in nearby properties.

The inability/unwillingness of low-income owner occupiers and private landlords to invest may have negative impacts on other owners and instil decay in an inner-city neighbourhood,

Public policy may blight or discourage investment and deter people from moving into an area.

An active local market is essential for housing investment and neighbourhood stability.

House prices in inner-city areas can suffer from a small number of transactions and the unavailability of ready mortgage finance.

In a flourishing neighbourhood, constraints on new developments are likely to see an increase in extensions and loft conversions.

Neighbourhood change is subject to bandwagon effects as spillover effects of the decisions and activities of local households and owners influence nearby residential investors.

There are thresholds or tipping points beyond which the process of neighbourhood change quickens.

Neighbourhood change results not only in adaptions to the housing stock but also to the local socio-economic demographic profile of households.

Filtering is a neighbourhood process of change that involves low-income groups moving to houses once occupied by higher income groups as the quality and price of this housing fall. Often the process includes tenure change from owning to renting and the subdivision of housing.

The changing racial composition of neighbourhoods can be envisaged as a filtering type process.

Gentrification occurs when higher income groups upgrade old properties and take over neighbourhoods where low-income groups live. The displacement occurs because low-income groups are priced out of neighbourhoods through the increased real estate rental and capital values.

The major barrier to improvement of the standard housing stock is hindered by the finances. The improved value of a house is invariably lesser than the original cost plus the improvement expenditure even with a state grant.

The long-term continuance of gentrified neighbourhoods is dependent on house buyers prepared to move into the area and pay a high price to live there.

Urban regeneration can support the movement of higher income groups into an area that can lead to gentrification. It occurs partly by pushing housing prices up in a neighbourhood.

Bibliography

Glass R (1964) *London: Aspects of Change*, MacGibbon & Kee, London.

Hoyt H (1939) *The Structure and Growth of Residential Neighbourhoods in American Cities*, Government Printing Office, Washington.

Jones C and Watkins C (2009) *Housing Markets and Planning Policy*, Wiley-Blackwell, Chichester.

Ratcliff R (1949) *Urban Land Economics*, McGraw Hill, New York.

Part III

Regeneration and urban growth policies

18 Urban regeneration policies

Objectives

This chapter provides the contextual base to Part 3 of the book. The assessments of individual policy initiative assessments follow in subsequent chapters. The chapter also offers an historical perspective on urban policy, especially in the UK and the United States. However, many of the initiatives discussed have parallels in other countries. It focuses on real estate–led urban policies and seeks to place them in a wider historical policy framework.

The chapter is in the form of a commentary. It does not ignore urban social policies but provides limited detail. The story begins in the 1960s and is not only one of riots, changing governments and perspectives but also one of unexpected consistencies. This overview does not attempt to be comprehensive, as the specifics of initiatives are dealt with in individual later chapters. The chronology of urban policy begins with the 1960s and highlights key initiatives and their underpinning logic, together with policy debates. The review is structured as follows:

- Historical context
- Traditional approaches to urban renewal
- Urban social policies of 1960s and 1970s
- Economic regeneration of 1970s and 1980s
- Broadening out of urban policy
- Urban policy under a Labour Government 1997–2010
- Urban regeneration post 2010

Historical context

Previous chapters have narrated how western core cities shifted from a manufacturing-based economy to one centring on services. The anatomy of this urban change encompassed deindustrialisation and industrial decline in inner-city areas. At the same time, there were parallel processes of manufacturing decentralisation as well as residential suburbanisation. These processes reinforced the problems and the pains of change. From the 1960s, as set out in Chapter 9, inner-city decline encompassed not only lost jobs but also lost skills and work opportunities. It led to high inner-city unemployment as the skills of many workers in traditional industries became obsolescent.

DOI: 10.1201/9781003027515-21

Inner-city areas also suffered population decline. The young and economically active moved away to employment in thriving urban areas such as new and small towns or simply to the suburbs. The inner core became predominantly occupied by the poor and ethnic minorities. There were also high crime levels and occasional riots and protests about poverty and prospects and the general state of life in these areas. The concentration of these travails in the inner-city area led to the term urban 'donut' to be commonly used in the United States to describe a city's spatial structure.

Much of the housing in these inner-city areas was in a poor state, partly because of its age. Invariably, it was small and not up to quality or space standards expected by the norms of present society. Some of it lay empty and abandoned with no one knowing who owned it. There was a widespread need for slum clearance and redevelopment.

The nature of the central urban decline was epitomised physically by large tracts of land lying unused, and derelict docks, industrial units or warehouses, vacated by closures. Very often, this former industrial land and docks were contaminated by the historic industrial processes once applied on their sites.

The urban regeneration challenge to address this legacy is, therefore, multifaceted. The overall impact is the result of a range of linked processes. It is not simply about rebuilding the physical fabric of the city. There is a spatial concentration of poverty. The inner-city urban economy needs to be reformulated with new land uses that generate incomes for the local residents. Better housing needs to be provided by improving the stock or replacing it.

It is a social, physical and economic urban renewal planning challenge. Left to market forces, the task is unlikely to make much headway. In a sense the inner-city localities are suffering functional obsolescence as the previous land uses are no longer viable. There needs to be mechanisms to promote land use succession and create a new revitalised inner-city core or donut.

The problem is that the economics of redevelopment is constrained by the scale of the dereliction and negative spillovers or externalities. It means widespread low or minimal land and property values. Any contamination needs to be treated, and derelict buildings be replaced before new development can occur. Although land values are low, the capital expenditure required to redevelop an individual site/building is prohibitive. The result is that without public financial support, land use succession is invariably unviable, at least in the initial phase of regeneration.

Traditional approaches to urban renewal

The long-established approach of urban regeneration was to focus on the physical problems of cities. It was a response to the high densities and poor sanitary conditions of urban housing built after the industrial revolution. Policies took the form of replacing the slums by new modern housing through public programmes. Slum clearance in UK cities began in the mid-1800s on a low scale by individual local authorities under public health laws, without rebuilding. The first national slum clearance programme in the UK began in 1933 with replacement public housing and was interrupted by World War II. With national housing shortages, it was not until the late 1950s it began again in earnest.

UK slum clearance programmes accelerated through the 1960s before stalling in the mid-1970s. It was not planned to rehouse all the displaced households within the city boundaries. The city authorities created overspill public housing estates in neighbouring areas or formal rehousing agreements with surrounding new towns. Slum clearance evolved into comprehensive redevelopment of neighbourhoods. The slum clearance bulldozer demolished not only residential slums but also adjacent shops and workshops.

City retail centres in the UK were also subject to physical redevelopment from 1960 initiated and supported by the local planning authorities. Many major city centres underwent extensive redevelopment schemes replacing traditional high streets with planned pedestrianised shopping malls. Initially, these took the form of open-air precincts, but in the 1970s these were developed to encompass covered malls and multi-level shopping.

In the United States, there was a parallel residential slum clearance programme initiated by an Act of Congress in 1949 that continued until funding ended in 1974. Federal grants of two-thirds of the net project cost to the city were made available, where the net cost was defined as the difference between the total cost of acquiring and clearing properties and the income received from selling the cleared land.

These policy initiatives were initiated before the predominant era of deindustrialisation. The logic of the policies was on physical change. Critics argued that there was economic pain even before the onset of deindustrialisation. The urban renewal programmes removed cheap industrial premises that restricted opportunities for start-up businesses and local employment. Small shopkeepers lost their livelihoods and never reopened elsewhere. Displaced residents sometimes found their new locations had little nearby employment. In some cases, low-income workers had to travel long expensive commutes to their place of employment.

In many ways, the break-up of traditional communities also saw rising crime rates and increased dissatisfaction about life in inner-city areas. This dissatisfaction was also exacerbated by the standard of some of the replacement multi-storey housing that proved to be unpopular, and much of it eventually had a short shelf life. In addition, subsequent reduced employment opportunities with deindustrialisation contributed to increased poverty.

An alternative physical solution to the problems of cities is the building of new towns. A national new towns' building programme began in the UK in the early 1950s. The concept is derived from the ideas of the garden cities developed by Howard at the turn of the twentieth century. There were similar proposals by Wright in the United States in the 1930s. These schemes for ideal 'Utopian' cities incorporate detailed decentralised physical layout designs that contrast with the overcrowded cities at the time.

Each new town had its own public agency, a development corporation, charged with building it. Funded by central government (plus borrowing), these agencies created the new towns on green fields. They were responsible for planning the town and building most of the housing that was initially owned by a corporation. Each corporation also ensured the development of industrial units and the main town centre and local shopping centres.

UK new towns are broadly seen as a success in terms of providing low-density good quality housing and employment opportunities. They were also a success initially in decanting people from city slums. Today, they are established urban areas in their own right, even if many residents commute to employment elsewhere (as indeed occurs in all towns). However, the utopian plans of the original proponents to revolutionise society in terms of social justice and cooperation did not materialise.

New towns as a form of planned decentralisation of overcrowded cities with housing shortages have been taken up as policy around the world. In achieving this goal, they are essentially drawing people and economic activity away from cities. From a national perspective, arguably as discussed in previous chapters, a decentralised spatial economy is more productive. However, while new towns have a role in addressing the problems of cities, they can ultimately contribute to the demise of cities. The role of new towns in urban renewal, therefore, has limitations.

Urban social policies of 1960s and 1970s

The focus of urban policy initiatives shifted from the mid-1960s to addressing the problems associated with concentrations of poverty in cities. In the United States, legislation was passed in 1964 establishing a five-year federal 'Model Cities' experiment in more than 150 cities. It was seen as part of a 'war on poverty' strategy including community action programmes. They were designed to develop new antipoverty initiatives and alternative forms of the municipal government. The emphasis was on social impacts.

The Model Cities programme was designed to provide greater understanding of the lives of the impoverished and improve ways for dealing with their problems that could be ultimately repeated on a wider scale in other cities. The vision was the elimination of urban poverty. In fact, the Model Cities programme did not start on the ground until 1969. It incorporated a wide range of activities aimed at reforming local government procedures, new and improved community development practices.

The original legislation also sought to establish fully 'self-sufficient' neighbourhoods. It also had ambitions to enhance infrastructure and transportation systems, provide better housing, employment and educational opportunities; reduce numbers on welfare; lower crime rates; and encourage greater participatory democracy. In practice, the Model Cities programme created social amenities such as day-care centres, legal aid centres, as well as job training and summer youth schemes in impoverished neighbourhoods.

The Model Cities programme inspired equivalent policies in the UK. These policies were introduced under the banner of tackling urban deprivation. The term brought together a range of urban problems linked to unequal access to resources. It encompassed material poverty, poor housing and inequality in the provision of services by the central and local government. It also implied that this urban deprivation was to be found in small pockets within cities of an otherwise affluent society.

The UK policies had strong parallels with those in the United States. The Urban Programme was initiated in 1968 to provide central government resources to support social service schemes in areas of severe deprivation. These schemes included play schemes, adventure playgrounds and neighbourhood legal aid centres.

The Educational Priority Area (EPA) project from 1968 to 1971 attempted to apply positive discrimination strategies to areas suffering from deprivation. Their task was also to identify lessons that could impact on the causes of educational underachievement in deprived neighbourhoods. In an EPA, there was extra money for school buildings and teachers received additional pay to work there.

Shortly after, a national Community Development Project (CDP) was created with 12 experimental local projects in deprived areas in different regions of the country. The emphasis was on localised community issues, and the teams based in each area were tasked with action research. The basic idea was that community workers in the CDP team would work with local residents' groups and local government officers. In this, they would identify and implement neighbourhood solutions.

The underpinning philosophy of the CDP initiative was that the causes of urban deprivation lie in part within individuals and families and also the lack of access to neighbourhood services. The solutions, therefore, lay in self-help and joined-up services provision. Services provision included tackling educational disadvantage and the reform of social work. The local projects each ran for up to five years, but there were staggered starting dates with the first ones established in 1970. The final CDP experiment finished in 1978.

A variation on this theme brought about the Inner Area Studies instigated in 1972 by a Conservative Government. In these two-year studies, management consultants worked with local authorities in parts of three cities to find local corporate management solutions. Again, they were action research studies that sought ways in which a local government could interact better with inner-city residents and provide improved services.

The ethos of both the CDP and Inner Area Studies projects was that urban problems are local and specific to an area. However, it is arguable that the research achieved much towards their original goals. On completion of the projects, there was little evidence of positive change, and the final reports from the different initiatives generally accepted this finding.

In one sense, the projects were a product of their time and policy development. During the period of the projects, the scars of deindustrialisation on the inner city began to accumulate. Perhaps, not surprisingly, the concluding reports in different ways rejected the presumptions at the beginning of the projects.

Their conclusions were that problems of poverty were not deemed to be restricted to isolated urban pockets. Poverty, to different degrees, was experienced by much wider layers of inner-city residents, especially in manual occupations. The reasons were more 'structural' and linked to the national economy and, especially, the urban economy. The problems of living in declining industrial cities were far beyond the capacity of either self-help or better aligned local public services to resolve.

Economic regeneration of 1970s and 1980s

In 1977, there was the first formal recognition by the UK Government of the role of urban economic decline and an 'inner-city problem'. In that year, the Labour Government published a white policy paper entitled, 'The Inner Cities', that set out its perspective on the way forward to address the issues.

It clearly acknowledges the problem as economic and identified the following constituents:

- Population/ physical decline
- Loss of jobs caused by deindustrialisation and decentralisation
- Concentrations of the elderly and unskilled in inner cores as the skilled economically active leave cities for suburbs and small towns
- High levels of unemployment in inner-city areas partly caused by characteristics of the population, e.g., unskilled.

The white paper's prognosis is that cities are in decline but argues that there should be policies to target the revival of inner cities. This is based on a broad cost/benefit analysis that asserts the benefits outweigh the costs. The argument was that saving the inner cities will use existing social/ physical infrastructure and avoid the necessity and cost to build on greenfield sites.

Perhaps, strangely, the policy response was not economic, but a set of physical improvement policies for cities. The designation of new towns designed to rehouse people from 'crowded' cities was stopped (for the reasons discussed earlier). The government created seven 'Inner Area Partnerships' in 1978 — three in London and one each in the provincial cities of Birmingham, Liverpool, Manchester and Newcastle in England. There was an equivalent one in Glasgow called Glasgow Eastern Area Renewal (GEAR) although this, strictly speaking, was established in 1976.

These inner-area partnerships comprised the central government and a range of public agencies including the relevant local authorities. Planned policies focused on social infrastructure, but by 1979, when Labour loses power, these initiatives had not got very far. The most successful of these initiatives was the GEAR project that continued until 1986. This project is considered in detail in Chapter 20.

Mrs Thatcher was elected prime minister in 1979 with a commitment to a monetarist strategy for the macroeconomy. This involved freeing of markets from state intervention together with a commitment to reducing public expenditure and a shift of priorities away from social issues. At the heart of these policies is a belief in market forces. These views under President Reagan signaled a retreat from urban policy in the United States that continued into the mid-1990s.

Market forces were signalling a move to optimum decentralised locations a major cause of urban decline. So how does UK urban policy fit it to this policy paradigm, if at all? Part of the answer lies in an alternative formulation of the 'inner-city' problem.

The Conservative Government continued to see the characteristics of the problem in economic terms. However, the processes that caused it are redefined as a lack of confidence in the inner areas by private sector investment. In other words, the problem is a form of market failure. The policy solution is then one to make markets work by restoring the confidence of the private sector. Based on this analysis, there are three principal Conservative flagship initiatives:

* Enterprise zones
* Urban development corporations (UDCs)
* Grants to developers

Enterprise zones

Enterprise zones (EZs) were the first initiative. They were the political brainchild of Sir Geoffrey Howe in 1977 to free constraints on industry as part of the wider monetarist critique of the economic problems of the country. At broadly, the same Professor Sir Peter Hall coming from a very different political perspective also promoted the idea of EZs. He proposed them after visiting Hong Kong where he was impressed with the enterprise of the residents as a last desperate attempt to revive entrepreneurship in cities.

EZs as originally conceived by Howe were aimed at freeing constraints on private industry. The constraints in question took the form (in late 1970s) of the following:

* Planning legislation
* Land taxation
* Price controls
* Pay policies

Their removal he argued would let the free market work and bring a return of private confidence.

These ideas were combined with the regeneration of urban areas to create the EZs experiment in 1981 as a laboratory for new ideas to promote enterprise. The final version was watered down from Howe's original concept. They became small areas where there was a combination of tax incentives as a carrot to locate with reduced state controls on firms, notably less planning development controls. Each zone was to last for ten years. A series of these zones were established between 1981 and 1996. The use of EZs was revisited in the 2010s with a slightly different formulation (see below).

The concept of EZs created interest around the world. The idea was taken up enthusiastically in individual states of the United States (not the federal government). States designated a panoply of EZs with different sizes and types of locations and tax incentives. By 1995, 34 states still had active zones although four states had not renewed their programmes. In 2017, there were still 21 states with zones. In some cases, the EZs were aimed at residential neighbourhood revitalisation. The EZ concept as a regeneration tool was also translated to other European countries, such as Ireland and France. Chapter 21 gives an assessment of these zones.

UDCs

This policy initiative was also developed while the Conservatives were in opposition. It is based on the New Town Development Corporations (DCs) that were given special powers to build new towns without any role for a local council/democracy. They are outlined earlier and were seen as very successful and efficient. The policy translated the application of these corporations to the inner-city context as agencies with powers to physically redevelop areas.

UDCs were established under a 1981 Act of Parliament. They covered areas within town and cities and had similar powers to New Town DCs except they did not have the power to build houses. In particular, UDCs had the power to do the following:

- (compulsory) purchase land,
- assemble sites,
- reclaim and service large areas of derelict land,
- build infrastructure and improve the environment,
- provide land for private sector development,
- give financial assistance to developers where necessary and
- ensure quick planning decisions.

There were in fact four 'generations' of UDCs through the 1980s, with each new generation having a modified type of area and, hence, remit. These generations in chronological order were as follows:

- Dockland areas of London and Liverpool set up in 1981,
- Four English UDCs designated in 1987 in large areas of dereliction in the peripheral regions plus ones in Cardiff in Wales and Belfast in Northern Ireland,
- 1988 – mini UDCs announced in (off) city centres in Bristol, Manchester, Leeds and Sheffield,
- In early 1990s – Plymouth, Birmingham Heartlands.

UDCs have an antecedent, the New York State UDC in the United States. It has operated from 1968 but is now subsumed in the Empire State Development Corporation. The agency has planning and compulsory purchase powers to support development in inner-city areas, and in some cases where necessary with financial support. Initially, the UDC's main output was affordable housing, but since 1975, it has focused on commercial/industrial and civic developments.

A full assessment of UDCs is given in Chapter 22. The use of UDCs was rediscovered in the 2010s (see below). EZs and UDCs have common themes in the sense that they relate to small areas, were time limited, and were ultimately about the promotion of the private sector. However, both approached the task with very different mechanisms.

Grants to developers

There were riots in the summer of 1981 at the deepest point of a recession; first in an inner-city neighbourhood of Liverpool, followed by ones in areas of London and Manchester. The government wanted to be seen to do something. Given its perception of a lack of private investment as the problem, it organised a coach tour of Liverpool with 26 chairman or chief executives of pension funds, insurance companies and building societies. These financial institutions were challenged to find ways of investing in inner-city areas.

As a result of this trip, a number of initiatives were set up. First, the Merseyside Task Force was set up. This taskforce was a team of equal number of seconded civil servants and managers from the private sector. It looked for development projects that could attract private finance in Liverpool with public support.

At the national level, each of the financial institutions on the Liverpool bus trip seconded a member of staff to a think tank with civil servants to develop new regeneration approaches. Two ideas were developed from these discussions – a company called Inner City Enterprise (ICE) and grants to developers.

The purpose of ICE was to identify investment opportunities in inner-city areas which would not be financed by an individual institution because of the risk involved. ICE was initially owned by 40 financial institutions, and it acted as a way of unitising or securitising the risk of development projects in secondary locations of cities. It eventually became a property developer specialising in the inner city, but it was sold by the financial institutions. Later, it was bought again and no longer exists in its own right.

In contrast, grants to developers was effectively adapted from Urban Development Action Grants (UDAGs) in the United States. They were aimed at supporting development in inner-city areas that otherwise would not happen. These grants to private developers are subject to a series of conditions and are designed to be the minimum grant necessary to make a project viable. Their issue was conditional on the scale of private/ jobs investment generated by the individual project. These grants have become a continuing approach to urban regeneration although it is now called gap funding. The logic and impact of these grants are considered in detail in Chapter 23.

The UDAG programme began in 1977, with the first grant awarded in 1979. Just as in the UK, there was a swing from urban social policy to a new focus on economic regeneration in the United States. It also brought a move towards bricks and mortar hardware projects. The grants were very similar to their British cousins as they were designed to attract private development to distressed parts of cities. The UDAG pro-gramme continued until 1988.

Broadening out of the urban policy

The headline initiatives of the 1980s – EZs, UDCs and grants to developers – all fall under a banner of real estate–led urban regeneration. This title reflects the objectives of the policies that were to directly support private property development and invest-ment. By the late 1980s, there was a growing awareness that real estate–led initiatives in isolation were too narrowly focused. A fuller account of the arguments is given in Chapter 22.

As a response to these criticisms, urban policy was extended to incorporate a wider economic agenda and social goals. Various labour market initiatives such as train-ing schemes for the unemployed were introduced under the banner of the Urban Pro-gramme. There were also longer-term education policies including inner-city City Technology Colleges and Compacts. Compacts involve setting (school children) tar-gets and local employers guaranteeing a job if these targets are achieved.

The initiatives of the 1980s did not involve local authorities as they were deliberately bypassed in the interests of efficiency. The return of local authorities to national ur-ban regeneration initiatives in the early 1990s occurred with the introduction of City Challenge. In this scheme, local authorities competed for funds that were 'created' by reducing the Urban Programme and housing budgets. Local authorities had to submit bids for community projects that needed to demonstrate support from business and residential communities.

Another criticism of the urban policy in the 1980s was the lack of a national strategy as the location of many of the initiatives appeared ad hoc. In 1994 'English Partner-ships', a national urban regeneration agency was set up and took over administration of grants to developers (except Scotland). It was portrayed as a roving UDC. It took on unfinished developments of the UDCs when they were wound up (all by 1998). Eventu-ally, it was recast as the Home and Communities Agency under the subsequent Labour Government and charged with developing inner-city land for housing.

A further initiative that emphasised a strategic perspective was 'City Pride' initia-tive was launched in November 1993. The essential logic of the scheme was to take a long-term overview of urban regeneration in a particular city. Civic and business lead-ers were required to prepare a prospectus, detailing a vision of a city's development taking a ten-year perspective. Three cities – Birmingham, London and Manchester – participated in the scheme and these prospectuses were a bid for resources from the central government. This idea of vision statements was taken up subsequently for Scottish cities in the 2000s.

Urban policy in Scotland from the 1980s was organised in a distinctively different way from the rest of the UK. There were EZs and grants to developers were available, but there were no UDCs. The equivalent work to UDCs was undertaken first by a national economic agency, the Scottish Development Agency, which was eventually reformulated as Scottish Enterprise.

In 1989, four 'Urban Partnerships' were set up in 1989 with a similar approach to City Challenge in England. These were partnerships of local public agencies, business and community groups in large social housing estates on the periphery of urban areas. These were some of the most deprived communities in Scotland. The aims of these partnerships were partly about the physical development/ improvement/transforma-tion of these estates. They also included tenure diversification/new housing for sale and the training and education of the local population.

Following a Scottish white paper, 'Progress in Partnership' in 1994, the approach was extended across the country. The funding was subject to competitive bids between local authorities along the lines of City Challenge. Priority was given to the most de-prived neighbourhoods as identified in the 1991 Census.

Urban policy under a Labour Government 1997–2010

With the 1997 election of a Labour Government in the UK, there is no immediate change to urban policy. In November 2000, a white policy paper was published enti-tled 'Our Towns and Cities: The Future – Delivering an Urban Renaissance'. It reaf-firmed a commitment to urban cores. The goal of the urban policy becomes 'making our towns and cities places for people'. The document is high on rhetoric, for example, 'A new vision of urban living', but subsequent policies were very diffuse.

There is a new emphasis on social exclusion and health inequality within cities. So-cial Inclusion Partnerships are set up in small deprived areas across urban Scotland. There are extra resources for education, health, transport, housing, criminal justice and leisure plus new initiatives such as education action, employment and health zones and town centre improvement schemes. These town centre schemes were called 'Busi-ness Improvement Districts' (BIDs).

There are now over 200 BIDs in the UK that are funded via a levy on business/ commercial rate payers in the town centre. BIDs are a worldwide phenomenon (under different titles/funding models) that originated in Canada. In the UK, they are focused on town retail centre management. In contrast, in the United States, they are linked to a much wider range of mixed-use businesses in suburban and central locations.

Employment Zones were areas designated local authorities with some of the highest concentrations of long-term unemployed – disadvantaged local labour market areas. There were 15 areas selected as zones with the aim to help long-term unemployed peo-ple aged 25 years or older to find work. Within these zones, those eligible receive help from an employment agency to develop personal action plans over time to increase their employability. These zones were wound up in 2004. More general schemes took their place, and these were not spatially focused.

Urban regeneration companies

An important urban policy initiative was the establishment of Urban Regeneration Companies (URCs) and Regional Development Agencies. The URCs were public/pri-vate partnerships that promoted private property development in parts of cities as part of a wider urban strategy. They had similar objectives to UDCs except they had no block funding and no planning powers. Essentially, URCS would be responsible for implementing individual initiatives within a planning strategy. This strategy was agreed by the relevant Regional Development Agency (or a regional government) and the local authority. The Regional Development Agency would then provide the neces-sary public funding.

The first three were set up in 1999 in provincial cities – 'Liverpool Vision', 'New East Manchester' and 'Sheffield One'. In all, 13 URCs were set up in England, with others in the other nations of the UK. Of particular interest is the Clyde Gateway URC established in December 2007 in eastern Glasgow. The area encompasses the building of a new motorway through it and the main site for the Commonwealth Games in 2014. This initiative is discussed in more detail in Chapter 20.

URCs were superceded in England in 2007 when the government promoted the concept of City Development Companies which were later rebranded as Economic Development Companies (EDCs). These agencies have the same model as a URC – 'independent'-led private sector approach but are seen as specialist economic development bodies and not limited to urban regeneration. These companies were designed to have a limited lifespan of 7–15 years to regenerate their areas. The plans were part of a broader strategy to increase confidence in private sector investment.

Pathfinder agencies

In parallel to these URCs, there was also a programme of nine Pathfinder agencies established in 2002 to address areas of low housing demand in neighbourhoods of northern English cities. Typically, these were areas of terraced housing suffering from low demand/weak capital values and many were unwanted/abandoned.

Pathfinder agencies were partnerships of central, regional and local 'stake holders' with a 10–15 years' timescale to complete their task. They were funded primarily by the central government. Interestingly, for the first time, their goal was not set in terms of physical improvement of an area but related to regenerating the local housing market (see Chapter 17). In particular, they were set the target of reducing the gap in average house prices between their area and its respective region

The approach of these pathfinder agencies was a combination of demolition and refurbishment of the existing stock plus new build housing. There was clearance of surplus/obsolete buildings followed by a reshaping of the area. Usually, there was a masterplan for the neighbourhood drawn up involving the community.

The agencies bought up land and property where necessary undertook site preparation and reclamation if required and any desired environmental improvement. New building was supported through 'gap funding to developers', and housing renovation grants could be offered to residents. It was also recognised that there was a need to undertake parallel economic regeneration.

Enterprise areas

The most deprived 2,000 areas in the UK were declared Enterprise Areas (EAs) in 2005. The government heralded EAs as providing a flexible range of policies which local authorities can promote to tackle the barriers to enterprise that exist in deprived areas. In England and Scotland, they were the 15% most deprived wards/areas whereas in Northern Ireland and Wales, it was 42% to reflect higher deprivation in these countries. They were located in many of the urban areas of the UK and encompassed some of the former EZ areas.

Within these areas, there was exemption from stamp duty for residential transactions below £150,000 and a Business Premises Renovation Allowance (BPRA). The BPRA provided 100% first year capital allowances for the capital costs of renovating or converting premises that have been unused for a year in these areas. The scheme ran from April 2007 until April 2012. EAs were arguably EZs remoulded as their logic is also resolving property market failure. The focus on established neighbourhoods has parallels with some EZs in the United States.

The Labour Government, therefore, introduced a myriad of urban policy initiatives, and the list above is not exhaustive. The real estate−led urban regeneration initiatives can be seen as having a direct hereditary line to those of the 1980s. A key difference is that the schemes are more embedded within an urban planning framework. Wider social and economic regeneration policies also follow similar themes developed in the 1990s.

The urban policies and priorities of the UK in the 1990s and 2000s also have similarities with those of the Clinton administration in the United States. American policies were quite diverse and included investment in training and education, access to jobs, leveraging private investment and expanding home ownership. Distinct projects included the 'Bridges to Work' project that sought employment opportunities in the suburbs for inner-city residents. Community development banks were established whose primary mission was to lend, invest and provide basic banking services for low- or moderate-income communities.

Perhaps the most important distinctive initiative was Urban Empowerment Zones initiated in 1994. The first cities chosen for zones were Baltimore, Chicago, Detroit, New York and Philadelphia. There are 40 zones located across urban America. They were set up in competitive bidding process. As part of the process, communities had to present strategic plans for revitalization. Just like the City Challenge in the UK, input was required from local residents, public agencies and the private and non-profit sectors. The process also led to financial commitments from private firms to invest in new factories.

Each zone received a block grant to be applied to a broad range of activities, including social services and physical improvements. Businesses in these zones received a tax credit for each employee resident in the zone. Zone businesses also received other tax benefits for investments in properties (similar to EZs) and access to finance. These zones have had a continuing life. There are no explicit new urban policies under the subsequent Bush administration from 2001 to 2009. Its emphasis was on supporting affordable housing.

Urban regeneration in the UK post 2010

Urban policy since the coalition government was elected in 2010 has been dominated by the strategy of fiscal austerity. The government prioritised cutting public expenditure to address the national financial deficit that had grown after the global financial crisis. Soon after its election, all the major publicly funded initiatives of the 2000s were closed down, as funding was withdrawn. It brought the end of the Pathfinders, RDAs, and URCs/EDCs, and two years later, EAs. The Clyde Gateway URC was an exception.

Fiscal stringency effectively limited urban policy, and new initiatives focused on economic regeneration. First, there was a further reawakening of EZs. Twenty-four EZs in England were announced in 2011 – active from April 2012. These EZs were defined differently from the original model as the incentives focused primarily on a holiday from property tax. This saving applies to individual firms for five years.

In 2015, the government increased the number of zones to 43 including extending the area of eight of the original 24. In the following year, further four zones were announced. In parallel, the Scottish Government announced equivalent EZs in January 2012 with similar incentives, but branded as EAs. The Welsh Government has also established eight zones in total.

These EZs are no longer experiments in free enterprise but embedded in local economic strategy. In England, they are integrated into the activities of local economic development agencies called local economic partnerships (LEPs). These agencies are partnership organisations between private industry and local government with funding from the central government.

Since 2011, the UK Government has operated a national Local Growth Fund to which in England LEPs bid for investment funding for local projects. These projects, known as Growth Deals, are designed to boost local economic growth and bring in private sector funding. These deals are often linked to the local EZ. Funding encompasses finance for infrastructure such as access roads, supporting new technology industries, improving labour skills and unlocking housing sites for development.

In Scotland, the equivalent funding is known as City Deals. It is useful to focus on the Edinburgh City Deal to illustrate the nature of the initiative. As part of a formal agreement, the UK and Scottish Governments are each investing £300m. The agreement covers the city region encompassing six local authority areas. Contributions from these councils and three local universities are expected to take the total investment through the deal to about £1bn.

The funds are designed to be used to support and develop the strengths of the city region's economy. It is to support the finance of new innovation 'hubs' in association with the universities, including in robotics and space technologies. Under the project banner, the Scottish Government plans at the time of the announcement to invest up to £25m in the region's workforce to reduce skills' shortages and gaps. The Scottish Government also plans to invest £65m in new housing and £10m towards a new concert venue. It has committed £140m to transport projects, including £120m for improvements including the Edinburgh city bypass. In the south of Edinburgh, a business park is to be developed.

The final economic initiative of this period is the opportunity for English cities to establish Mayoral Development Corporations to regenerate areas. The activities of these corporations cover small areas of cities and have planning and compulsory purchase powers. They are funded by a city mayor who has oversight of a city region. The mayor's office can provide loans for capital projects to be paid back from the proceeds of development returns.

The corporations prepare business plans that are approved by the mayor, so these plans are effectively bids for capital funds. They are, therefore, a cross between the original UDCs and URCs, having more powers than the latter but not the money of the former. So far, there have been only four set up, and each has a different type of regeneration task.

The first one was set up in 2012 to rework the site of the London Olympics and to deliver regeneration for the east of the city. The second one, also in London, centres on regenerating an area around a new station that will provide an interface with a high-speed train network. In 2017, a further one was announced to regenerate the site of a former steelworks. The latest aims to repurpose a declining town shopping centre. The work practices and remit to attract private investment of these development corporations are very similar to the UDCs of the 1980s.

Across the Atlantic, urban policy under the Obama Government during this period also suffered from financial constraints after the global financial crisis. As a result, the signature policy, the Strong Cities, Strong Communities programme was limited in scope to demonstration projects in selected 'distressed' areas. It had a number of different elements focused on economic regeneration. The first part of the programme involved employing outside federal expertise to support six pilot local authorities to promote economic development.

A second element was the Economic Visioning Challenge in which teams from three cities competed for prizes for the best idea to stimulate the local economy. Prize entries varied from ways to promote successful industrial clusters to repurposing shopping centres to a lifelong learning initiative to a plan to use shipping containers as inexpensive commercial outlets in underused areas such as city parking lots.

Summary

Many western cities have been economically and physically disfigured by the long-term decline of manufacturing, but they have also suffered from decentralisation of industry and the suburbanisation of the population. The consequences were concentrations in the inner city of poverty and high unemployment. Much of the housing in these inner-city areas was also in a poor state. Even from the 1950s, there was a widespread imperative for slum clearance and redevelopment.

By the 1970s, there were substantial areas of vacant land and buildings awaiting forlornly for redevelopment. New uses were partly constrained by the scale of surrounding dereliction and the cost of resolving historic contamination. The spatial concentration of poverty in these areas further hindered the viability of regeneration. Market forces in isolation could not resolve this cocktail of functional obsolescence, dereliction and deprivation.

A regeneration policy challenge was to promote new land uses and create a reinvigorated inner-city core. The starting point was pervasive low land and property values. As a result, individual private development was constrained by the inability to create sufficient uplift in capital value to enable viability. The catalyst for stimulating extensive land use succession is almost certainly public financial support.

The traditional approach to urban regeneration was to physically adapt cities by reducing densities and improving housing conditions. Slums were demolished and new housing built. The first national programme in the UK began in the 1930s and resumed again in the late 1950s. In the latter phase, many slum residents were rehoused beyond a city's boundaries. Redevelopment of cities also included retail town centres. These urban renewal programmes often had negative economic consequences for residents, even before the arrival of deindustrialisation.

New towns are a further physical option to resolving the overcrowding and decay in cities. An extensive programme of new towns was initiated in the UK over 30 years from the 1950s. The policy has been subsequently adopted around the world. New towns can be very successful in addressing the problems of cities. However, by drawing away population and economic activity from cities, the policy does not have universal application.

The nucleus of urban policy initiatives from the mid-1960s swung towards social issues linked to the elimination of concentrations of poverty in cities. They took the form of policy experiments. In the United States, there was the Model Cities programme that created social amenities such as day-care centres, legal aid centres, as well as job training and summer youth schemes in impoverished neighbourhoods. In the UK, these ideas translated to a number of parallel initiatives under the umbrella of tackling urban deprivation.

These initiatives included the Urban Programme, Educational Priority Areas, Community Development Projects and Inner Area Studies. All had a common theme of supporting specific deprived neighbourhoods. The implicit underpinning assumption was that solutions to urban poverty lay in self-help or improving services provision in these neighbourhoods. The final reports of the CDPs and Inner Area Studies rejected this original diagnosis as the evidence of deindustrialisation took hold in inner-city areas. Improving local public services was no panacea for the consequences of urban economic structural change.

1977 brought the first formal acknowledgement by the UK Government that urban economic decline was source of the inner-city problem. Its policy response was to promote the revival of inner-city areas. The premise was that the use of existing social/physical infrastructure in urban area avoids equivalent expenditure of development on the periphery.

The policy solution unexpectedly took the form of providing social infrastructure in the urban cores. There were a series of partnerships with local authorities and other public agencies set up to undertake this task in the largest English cities. The approach followed that of the GEAR project in Glasgow that outlived them and continued until the mid-1980s.

In the 1980s, macroeconomic policy was dominated by monetarist thinking with an obligation of reducing public expenditure. At the same time, there is a conviction that market forces should be freed to operate unfettered by state intervention. In the UK, urban policy continued by redefining the inner-city problem as a lack of confidence by private sector investment. From this perspective, the problem is one of market failure. Following this paradigm, there are three principal Conservative flagship initiatives – EZs, UDCs and grants to developers.

EZs were located in small areas and offered a combination of tax incentives as a carrot to locate there with reduced state controls on firms, notably less planning development controls. Each zone existed for ten years. A series of these zones were established between 1981 and 1996. The potential of EZs gained the attention of governments around the world and the concept was taken up in many countries.

UDCs covered areas within towns and cities and had substantial legal and financial powers to intervene and reshape the land market, provide infrastructure and promote private development. There were in fact four 'generations' of UDCs through the 1980s, with each new generation applied to a different type of urban problem area, from docklands to run down edges of city centres.

Grants to developers were effectively derived from UDAGs in the United States. The basic idea was to make development possible in inner-city areas through a government grant. The criteria for the funding of projects were on the basis of the scale of jobs and private investment generated. These grants have become a mainstay of urban regeneration although it is now referred to as gap funding.

These headline initiatives of the 1980s were all forms of real estate–led urban regeneration, supporting private property development. By the end of the decade, there was a growing chorus of criticism that they were too narrowly focused. As a result, urban policies were expanded to encompass labour market/training and education initiatives. New initiatives also incorporated partnerships between local authorities, the private sector and local communities. A more strategic approach was to some extent also adopted, for example, through long-term vision statements for cities.

Under the Labour Government, from 1997, urban policy becomes quite diffuse although set within an overall aim of 'making our towns and cities places for people'. There are many new spatially focused initiatives covering different aspects of urban life. Perhaps, the most important urban policy initiative was the establishment of URCs with finance allocated by Regional Development Agencies. These agencies continued the theme of property-led urban regeneration begun in the 1980s. A key difference from previous such initiatives was the incorporation of local planning strategies into the choice of developments to promote.

In parallel to these URCs, there was also a programme of nine Pathfinder areas set up to revitalise low housing demand in neighbourhoods of northern English cities. Pathfinders were partnerships of public and private 'stakeholders'. They had a time scale of 10–15 years to complete their task, set out in a masterplan. A plan typically involved clearance of surplus/obsolete buildings, rehabilitation and new house building as part of a comprehensive redesign of the area.

The most deprived neighbourhoods in the UK were also declared EAs in 2005. Within these areas, there was a combination of tax benefits for house purchase and for renovating or converting industrial premises that have been unused for a year. The scheme can be seen as EZs restyled to established neighbourhoods as occurred in some states within the United States.

Many of the urban initiatives of the UK from 1990 through the next two decades corresponded closely with those of the Clinton administration in the United States. An important distinctive programme was Urban Empowerment Zones initiated in 1994. As part of the process, communities in selected American cities had to present strategic plans for revitalization. A grant was allocated to each zone to support social services and physical improvements. Businesses in these zones also received tax benefits including subsidies for each employee resident in the zone.

The arrival of the coalition government in the UK in 2010 saw urban policy in retreat. Its priority of fiscal probity led to the speedy cancellation of, virtually, all the initiatives of the previous government. Regeneration initiatives that have occurred have prioritised local economic development. EZs and UDCs were revived although in both cases, the powers and incentives for development have been diluted compared with the originals.

A key difference is that this new generation of initiatives is embedded in local economic development and planning policies. These policies are implemented in partnership with the private sector and local government with funding from the central government. Within a similar if broader perspective, the central government has also introduced 'growth' or 'city' deals that are bespoke arrangements for individual cities. They are designed as a holistic strategic approach to boost local economic growth and leverage in private sector funding, including from universities. Projects funded within these frameworks range from new transport infrastructure, investment in high technology sectors, improving labour skills to promoting housing development.

In the United States, during this period, urban policy was also constrained by public finances. The Strong Cities, Strong Communities programme had the highest profile. It comprised a series of demonstration economic regeneration projects in selected 'distressed' areas.

To summarise, urban policy has evolved over the decades from a primary social/managerial standpoint in the 1960s. By the end of the 1970s, the diagnosis of urban

problems had become economic, influenced particularly by the rise of deindustrialisation. While the dominant ethos since then has been to stimulate economic change,
policies demonstrate a degree of cyclicity. Individual approaches fall out of favour
only to return sometime later.

Overall, there has been a shift to more holistic approaches to urban policy and local
economic development. Policies have also moved very much from the remit of local
authorities and the central government, and a range of public agencies, to partnerships
between the public and private sectors.

The policies described are what has primarily happened in the UK and, to a lesser
extent, in the United States. Many of these initiatives (moulded to local circumstances)
are found around the world. There is a considerable international urban policy transfer with, for example, the UK copying the United States policies on grants to property
developers and Canadian business improvement districts.

Learning outcomes

Many western cities experienced economic and physical decline, leading to concentrations of poverty in the inner areas. By the 1950s, much of the older housing in these
inner-city areas was in a poor state.

By the 1970s, substantial areas of inner-city land and buildings were vacant and
functionally obsolete.

The ubiquitous low land and property values in inner-city areas represented a significant barrier to redevelopment by the private sector.

Urban regeneration was traditionally undertaken by demolition and rebuilding.
There were extensive slum clearance programmes from the 1930s on.

An extensive successful new towns programme was begun in the 1950s to address
the overcrowding and decay in UK cities. They decanted people and economic activity
away from cities.

From the mid-1960s, the emphasis of the urban policy moved to policy experiments
to address social issues. The initiatives were designed to remove pockets of poverty in
cities. The underlying rationale was that urban poverty could be resolved by self-help
or improving neighbourhood services.

By the mid-1970s, it was recognised that improving local public services was no cure
for urban economic decline. The UK Government's policy response was to seek the
economic revival of the urban core rather than to build new social infrastructure in
decentralised locations.

The immediate policy solution unexpectedly took the shape of providing social infrastructure in inner urban areas through initiatives in the old core cities.

In the 1980s in the UK the inner-city problem was reconceptualised as a lack of
confidence by private sector investment. Policies were then ostensibly designed as addressing market failure. The three primary Conservative flagship initiatives from this
decade are EZs, UDCs and grants to developers.

EZs were small areas in which firms received a combination of tax incentives, and
there were reduced planning controls for ten years.

UDCs had substantial legal and financial powers to intervene and reshape the
land market, provide infrastructure and promote private development in areas of
dereliction.

The aim of grants to developers was to make development viable in inner-city areas
through a government grant.

By the end of the 1980s, these real estate−led regeneration initiatives were subject to a significant critique of being too narrowly focused. The scope of regeneration was extended to encompass labour market/training and education initiatives.

The election of a Labour Government in 1997 brought a myriad of urban policy zones relating to many aspects of urban issues.

URCs were an important urban policy initiative of the early 2000s that pursued the property-led urban regeneration approach of the 1980s, but set within local planning strategies.

Nine Pathfinder agencies were established too to address low housing demand in neighbourhoods of northern English cities. These projects sought to address their tasks by a combination of demolition, rehabilitation and new house building, as part of a comprehensive redesign of the area.

EAs were launched in 2005 in the most deprived neighbourhoods in the UK. Within these areas, tax benefits for house purchase and for renovating or converting industrial premises were available.

An important distinctive programme in the United States was Urban Empowerment Zones started in 1994. In selected American cities, zones received a grant to support social services and physical improvements. At the same time, firms received tax advantages encompassing subsidies for each employee resident in the zone.

The coalition government in the UK cancelled virtually all the initiatives of the previous government as part of an austerity programme. However, this government revived EZs and UDCs as part of local economic development strategies.

Growth or City Deals have also been implemented to support the economic development potential of individual cities. Public and private funds are brought together to support a range of activities from new transport infrastructure, investment in high technology sectors, improving labour skills and promoting housing development.

In the United States, during this period, urban policy was also constrained by public finances. The Strong Cities, Strong Communities programme had the highest profile. It comprised a series of demonstration economic development projects in selected 'distressed' areas.

Since its inception in the 1960s, urban policy has evolved from a primarily social/managerial outlook to a focus on local economic development.

Bibliography

Banks S and Carpenter M (2017) Researching the local politics and practices of radical Community Development Projects in 1970s Britain, *Community Development Journal*, 52, 2, 226–246.

Jones C (1996) Property-led local economic development policies: From advance factory to English Partnerships and strategic property investment, *Regional Studies*, 30, 2, 200–206.

US Department of Housing and Urban Development (1995) The Clinton Administration's National Urban Policy Report, Department of Urban Housing and Urban Development, Washington.

Weber B A and Wallace A (2013) Revealing the empowerment revolution: A literature review of the Model Cities program, *Journal of Urban History*, 38, 1, 173–192.

19 Urban competitiveness and the real estate market

Objectives

Around the world, there is a great belief in the application of real estate–led policies to promote urban economic growth. At its basic, yet in some ways arguably, its highest profile form is the building of landmark tall skyscrapers as a demonstration of a city's economic worth. It is also seen as a catalyst for growth in cities around the world. Despite the worldwide prevalence of such initiatives, the building of offices (or other land uses) is not a simple route to the economic development of cities or to urban revival.

The specific goal of this chapter is to consider the wider role of real estate–led policies to address urban competitiveness, especially in the context of offices. This chapter initially examines the role of building tall landmark buildings as demonstration projects. From this base, the relationship between real estate and the economic competitiveness of cities is appraised. The starting point is the economics of cities, including a recap on agglomeration economies and clusters of successful businesses.

It identifies, in particular, the role of real estate market constraints on office development and local economic development. A number of examples where cities have explicitly sought to remove these constraints are considered. Office urban dispersal policies are then reviewed. Finally, the chapter develops an overarching logic to real estate–led local economic strategies.

The chapter is structured as follows:

- Building of landmark buildings
- Agglomeration economies, cities and industrial clusters
- Competitiveness and real estate constraints
- Office dispersal policies
- Logic of real estate–led local economic development strategies.

Building of landmark buildings

The building of state-of-the-art landmark offices in developing countries today is increasingly viewed as a catalyst for the promotion of a city as a business centre. During the 1990s, Dubai set out to attract worldwide attention through innovative real estate developments and thereby establish the city as an international business hub. This strategy is exemplified by the building of the Burj Khalifa, the highest (mixed use) office block in the world.

DOI: 10.1201/9781003027515-22

Its strategy had a number of dimensions – a series of global landmark or signature high-rise buildings. Development was undertaken by state-supported companies, and the demand was attracted by a series of free zone 'cities' or specialist zones. The return on investment was seem not just in real estate rentals but in the wider economic benefits to the macroeconomy. This impact would stem from the spending of visitors and highly paid staff of international companies attracted to the Emirate.

Dubai has seen a rapid physical development of a central business district and satellite centres. The first of these buildings in this strategy was the 'seven-star' Burj Al Arab hotel, the tallest free-standing hotel in the world when it was completed in 1999. It was quickly followed by the Emirates Office Tower, the tallest commercial building in Europe and the Middle East at the time, the next year.

This was just a beginning of a development boom encompassing offices, hotels and residential properties. It now has some of the tallest buildings in the world flanking the main thoroughfare/freeway, Sheikh Zayed Road with the central business district (CBD) area stretching from the World Trade Centre to 'Downtown' Burj Khalifa. It has established a critical mass in terms of an office centre. The Dubai strategy of the development of tall office blocks has a long-standing history. They have been used as marketing symbols in the competition between cities for more than a century.

At the beginning of the 1900s, New York sought to establish itself as the world's financial centre (overtaking London). From 1890 to 1973, the world's tallest building was in the central business district of Manhattan. This period culminated in the 102-storey Empire State Building opened in 1931. It remained the world's tallest building from 1931 to 1972. Later in the century, there was competition between New York and Chicago in the 1970s. The World Trade Center, in New York, became the world's tallest building in 1972 with 110 storeys. It was overtaken by the construction of the Sears Tower in Chicago that opened in 1974.

At the turn of the millennium, the Mayor of London advocated the building of more high office blocks for the city. His argument centred on competition between cities for multi-national companies. High-rise office blocks would ensure the future of London as a world city through the attraction of international companies. He argued that large international companies do not wish to be housed in slab-like 'groundscrapers' but are seeking signature buildings that can give them a sense of identity.

As Table 19.1 notes most of the highest office towers built in the twenty-first century have been built in the fastest growing cities of Asia. While tall office blocks can be seen as a way of attracting international businesses, they are also symbolic of the success of a city's economy. However, the promotion of tall buildings is only a partial view of the role of real estate in supporting a city's competitiveness. Before considering this issue further, the chapter looks again at agglomeration economies and its relationship with competitiveness.

Agglomeration economies, cities and competitiveness

Services now normally account for the vast majority of jobs, certainly in western cities, rather than the traditional manufacturing economic base. This phenomenon underscores the importance of office employment and, hence, the office market within a modern urban economy. Business services in a discernible sense are now the economic base activities of many cities. It is from these activities that consumption expenditure is generated and, hence, the demand for shops and housing.

Table 19.1 World's Tallest Office Buildings in Order of Height in June 2021

Building	City	Country	Floors	Date Completed
Burj Khalifa Tower	Dubai	United Arab Emirates	163	2010
Shanghai Tower	Shanghai	China	121	2015
Ping An Int. Finance Centre	Shenzhen	China	115	2017
Lotte World Tower	Seoul	South Korea	123	2017
One World Trade Centre	New York	United States	94	2014
Tianjin Finance Centre	Tianjin	China	97	2020
Guangzhou Finance Centre	Guangzhou	China	111	2016
China Zun	Beijing	China	108	2019
Taipei 101	Taipei	Taiwan	101	2004
Shanghai World Financial Center	Shanghai	China	101	2008
Petronas Towers	Kuala Lumpur	Malaysia	88	1998
International Commerce Centre	Hong Kong	China	198	2010
Nanjing Greenland Financial Complex	Nanjing	China	89	2010
Willis (Sears) Tower	Chicago	United States	108	1974
Kingkey 100	Shenzhen	China	100	2012
Guangzhou International Finance Centre	Guangzhou	China	103	2010
Jin Mao Tower	Shanghai	China	93	1998
Al Hamra Tower	Kuwait City	Kuwait	77	2011
2 International Finance Centre Finance Centre	Hong Kong	China	90	2003
CITIC Plaza	Guangzhou	China	80	1997
Shun Hing Square	Shenzhen	China	69	1996
Empire State Building	New York	United States	102	1931
Central Plaza	Hong Kong	China	78	1992
Bank of China Tower	Hong Kong	China	72	1990
Bank of America Tower	New York	United States	54	2009
Almas Tower	Dubai	United Arab Emirates	74	2009
The Pinnacle	Guangzhou	China	60	2012
SEG Plaza	Shenzhen	China	72	2000
Emirates Office Tower	Dubai	United Arab Emirates	56	2000
Tuntex 85 Sky Tower	Kaohsiung	Taiwan	85	1997
Aon Center	Chicago	United States	83	1973
The Center	Hong Kong	China	73	1998
John Hancock Center	Chicago	United States	76	2011
Tianjin World Financial Centre	Tianjin	China	100	1969
Keangnam Hanoi Landmark Tower	Hanoi	Vietnam	70	2011
Shanghai Shimao International Plaza	Shanghai	China	60	2005

Source: Skycraperpage.com

The right type of service employment can also stimulate economic growth. The location of top management jobs and head quarter office functions in a city contribute to the success of its local economy by generating activities that feed off them. New services linked to the knowledge economy are the drivers of future economic growth through creating new production processes and jobs. These two examples illustrate the role of agglomeration economies in cities.

The trademark of cities is agglomeration economies that derive from the clustering of economic activities. It is useful to recap on Chapter 3 that these economies can be decomposed into localisation, activity-complex and urbanisation economies. *Localisation economies* are cost savings that emanate from when firms from one industry, such as pharmaceutical services, locate close by in one place. The advantages are derived from the viability of specialist services at lower costs because of scale economies or the availability of nearby skilled labours. The proximity of like firms also generates knowledge spillovers through access to information on innovations, marketing, research and development.

Activity-complex economies are financial benefits that stem from firms being part of a sequential production process supported by adjacent locations. As part of this process, there are linkages between firms that ensure the rapid transfer of information through face-to-face contact. *Urbanisation economies* arise from the wide range of business services available in a city. These economies encompass the good communications, municipal services and a large labour pool within an urban area.

From a household's perspective, cities also offer *social agglomeration economies* in the form of a range of services, encompassing shops, cultural utilities and medical facilities. They make cities attractive places to live (and work). Overall, agglomeration economies are a major driver of the location choices made by business and households. In particular, urbanisation economies are linked to the size of a city, potentially supporting the attractiveness of established centres.

The competitiveness of a city is related to its characteristics and these agglomeration economies. The individuality of a city can be seen in terms of the scale of economic activity, its industrial structure and the physical environment. However, Porter in 1998 developed the theory of competitive advantage that stresses only the importance of the role of local industrial 'clusters' on urban economic development. In accordance with this argument, the key to success of a city is entirely down to the nature of the clusters of economic activities within it.

Porter's arguments have been very influential on urban policy thinking as demonstrated by the UK Growth or City Deals noted in Chapter 18. These local economic development initiatives invariably include the promotion of industrial clusters with growth potential, building on existing local expertise. For example, the Edinburgh City Deal is financing innovation hubs in robotics and space technologies specifically for this reason.

The argument is that localisation economies bring competitive tension, local rivalry through peer pressure. In this way, clusters promote forward innovation and efficiency as well as cooperation. It also breeds the formation of new linked businesses. These new firms support and reinforce the cluster, and together, they stimulate local economic growth. It is envisaged that a large dynamic cluster also attracts a pool of talented workers to the city.

Yet, the cluster concept as a tool of local economic development is too simplistic. It is not easy for governments to directly foster the enterprise of a cluster of firms in a given private industry. Setting aside the practicalities, the nurturing of 'growth' clusters can also create over-specialisation, especially if the specific industries suffer a downturn. Many cities have declined because their primary industrial clusters became uncompetitive (deindustrialisation) partly because they did not innovate.

This argument suggests that competitiveness and urban growth are not simply about the productivity of industrial clusters, i.e., localisation economies. It is more broadly related to the nature and scale of economic activity and the attractiveness of the city to people and investors. In other words, competitiveness is linked to the specific urbanisation economies available in a city.

This conclusion needs to be tempered by competitiveness in the context of a city is very different than that for a firm. Cities do not compete like firms who are directly trying to sell their goods, usually at another's expense in lost sales. But cities do compete to a degree in attracting footloose firms and workers as places to live.

This wide-ranging competition is reflected in a league table that is recalculated and published every six months by Z/Yen for financial centres around the world. This Global Financial Centres Index rates cities on the basis of five elements – business environment, human capital, infrastructure, their financial sector and reputation. These can be subdivided as follows:

- **Business environment** covers political stability, institutional and regulatory environment, macroeconomic environment and tax and cost competitiveness.
- **Human capital** translates to the availability of skilled local people, flexibility of the labour market, education and development and the quality of life.
- **Infrastructure** relates to the city's built infrastructure, information, communication and technology (ICT) and transport infrastructure.
- **Financial sector** equates to the scale of the city's financial services market in terms of volume of business, clusters of different segments, availability of capital and liquidity.
- **Reputation** includes the nature of the city brand, level of innovation and cultural diversity.

These criteria are designed for assessing the competitiveness of financial centres around the world. Clearly, the criteria are biased towards judging financial centres and relate not necessarily to individual cities but to the national financial environment in which they are located. Nevertheless, the criteria reinforce the argument that a city's competitiveness is based on a broad range of economic characteristics.

From this perspective, real estate is an influence on competitiveness through the built environment of a city and, hence, on the quality of life. It could therefore be argued then that real estate has a minor role in the economic development of cities. However, it is first important to note that the factors that contribute to urban competitiveness should not be seen in individual terms. They are part of the package a city has to offer to attract business. A positive package could be a combination of a highly qualified workforce, modern office space and relatively low business costs, coupled with a good quality of life on offer.

If one of these ingredients is absent, then the overall attractiveness may not simply be diluted, it may be negligible. In particular, it can be argued that for services-based economies, the availability of offices is crucial. There need to be offices available of the right quality in the right location and at the right price to attract and keep businesses in a city. Similarly, good housing and residential environments are necessary to attract highly skilled in-migrants. This argument stresses the requirement of a city to have attractive real estate opportunities to support the economic growth of a city.

Competitiveness and real estate constraints

The imperfections of the real estate market mean that it may not be able to respond quickly to support and enhance the competitiveness of a city. Chapter 8 notes the restrictions on new development as a result of urban land and planning/conservation constraints. New supply also takes time to come to fruition. This occurs because of the time-consuming processes of land assembly, meeting planning obligations and potential conflicts with neighbouring uses and finally, construction time lags.

There is therefore an argument for public policy to intervene to promote competitiveness by addressing supply constraints. Given the complexities of land assembly in city centres, in many instances, large-scale redevelopment also needs public intervention and support. There are numerous instances around the world where public policy has recognised the importance that property constraints have on a city's competitiveness. This section now looks at a number of international case studies of office centres that have sought to remove stock constraints in different ways.

Edinburgh

Edinburgh is a good example where the local authority in the 1980s recognised the limitations of the city's office stock for business. The traditional office centre of the city is the 'New Town' purposefully designed and built in the late eighteenth century as town houses. This new town area was once referred to as the 'golden rectangle' of the office market. It encompasses some of the most architecturally prestigious squares of the city, and it had been designated a United Nations UN World Heritage Site.

Edinburgh is the second most important financial centre after London in the UK. Banks and insurance companies were located in the New Town area. It was therefore at the heart of this business sector of the city. By the 1980s, many of these houses now converted into offices were no longer suitable for the demands of commerce with the ICT revolution. Following pressure from business leaders, it led to a rethink of conservation policies and a new development strategy by the city council.

There was a clear conflict between conservation policies and the future of the city as a financial centre. The city council chose to resolve this dilemma by the creation of two alternative modern office hubs, one in a central location and the other on its periphery.

The 1985 Structure (strategic) Plan introduced a more relaxed approach. This signalled the creation of a new financial and commercial centre in what is now known as the Exchange District. It is close to the new town. The new office area was created based on a 'master plan' with a new civic square. The scheme involved the redevelopment of an old obsolescent rail goods yard/terminal and other nearby derelict areas.

The city's council was instrumental in supporting this process from the beginning by the sale of substantial land holdings and the building of a new civic square. The council and other public agencies initially combined to build an international conference centre in the area. New bespoke head quarter offices were developed privately over the next decade. Private speculative developments then extended the new business district. The outcome today is a collection of plate glass office blocks including most of the headquarters of the city's largest companies.

A peripheral office centre was also promoted by the city. It took the form of a large business park development on the fringe of the city. It was primarily developed by a public agency set up by the city. A master plan was drawn up by a celebrity architect, appointed by the public agency. Edinburgh Park is five miles from the city centre, two miles from the airport and close to an out-of-town shopping centre.

The first buildings were completed eventually in 1995 and included bespoke buildings for major relocations from the city centre, although some of the jobs are 'back office' functions. Its public transport links have improved over time: a railway station opened in 2003. In 2014, the tramway link from the city to the airport opened with stops in the locality. The park has established itself as a major office centre in its own right, and private development continues to expand. Together with the adjacent shopping centre, it can be described as an edge city.

City of London

London in the 1980s also saw the relaxation of planning constraints on the city's financial centre after many years of constraint. The historical analysis below highlights the underlying processes that first emphasised constraint on development and then the forces that ultimately led to a U-turn and the promotion of redevelopment.

The City of London is the traditional financial services centre of the UK, and the area originally comprised one square mile. Extensive bombing during World War II had left vast areas of dereliction, and the area was rebuilt to meet rising demand in the post-war boom. This rebuilding was undertaken under the planning guidelines of the City of London Corporation, the elected body for the area. The City of London Corporation acquired, over the years, large tracts of land partly through compulsory purchase.

Through planning powers and ownership of land, the corporation was able to shape new developments and ensure that the new office building followed the characteristics of the existing office stock. This rebuilding of the City that began in the 1950s in a piecemeal way was to broadly standard accepted norms and height restrictions. The first high-rise office blocks began to be built in the 1970s.

The rebuilding of the City was designed to continue the status quo and maintain its framework of long-standing prestigious addresses for the banks and financial institutions. Such well known addresses included Lloyds in Lime Street and the Stock Exchange and the Bank of England on Threadneedle Street. Location in one of the signature addresses of the City was seen as essential to the credibility of firms in the financial sector. Banks were required by the Bank of England to have their head office in the area.

This cosy financial world was disrupted by dramatic changes to the financial services industry. First, there was a revolution in the way stocks and shares were transacted with the end of face-to-face trading. Instead, shares were bought and sold electronically, and the use of ICT brought the demand for large undivided floor areas with ubiquitous computers.

Second, traditional stockbroking and jobbing firms were absorbed into integrated financial services offered by banks. At the same time, deregulation of the financial sector at the time brought a surge in global business, and many international companies sought a presence in the City. It was not possible for all these changes to be accommodated within the existing office stock of the City's square mile.

The City was subject to 22 conservation orders covering more than a quarter of the square mile in 1981. The existing office stock built for a bygone era was a major constraint on the economic future of the City. The exclusivity of the city cultivated over many decades by minimising new development was under threat.

The local property constraints meant that these firms would be forced to look elsewhere for office accommodation, unless new space was built. At this time, office space meeting the new technological requirements was being developed close to the City. There was also an alternative centre planned for the docklands nearby. The Corporation decided to reverse its conservationist policies and actually embrace an expansionary phase with relaxed planning constraints.

Despite the removal of supply constraints, the rapid increase in the demand for offices from the financial services sector could not be accommodated in the square mile. It necessitated a rethink in the locational requirements of the financial services sector and a dilution of the inelastic demand/insistence on a City address. Office developments have been built on the fringe of the original City, and it is now accepted by the sector.

Paris

Paris has a long history as a leading financial centre in Europe since the eighteenth century. The present internal central business district is known as the 'Golden Triangle' and bounded by the Arc de Triomphe, Place de la Concorde and the Avenue de l'ena. The office buildings in this area are only 6–8 storeys built in the second half of the nineteenth century. The buildings are part of a major urban planning/comprehensive redevelopment initiative that created wide boulevards. Buildings were subject to height limits that are still applicable, and there are regulations governing the facades of buildings.

These buildings were subject to extensive renovation in the late 1980s to adapt them to modern ICT business requirements. The area is home to many administrative offices of large companies, but the stock is highly constrained. There is the continuing long-standing height constraint of office development in the centre of 37 metres that has been rigorously upheld. There is only one exception, the Tour Montparnasse, a 59-storey office block completed in 1973, and built on the site of a railway station.

Paris has addressed the office supply problem by developing a modern office centre at La Défense on the western edge of the city and link it with rail and metro links to the centre. La Défense is now the financial centre of the city and the principal business district over an area of 1.6 sq km on the western extremity, 10 km from the centre. It now comprises 3m sq m of office space and the same size as the Golden Triangle and represents 9–10% of the city market.

The site of La Défense was designated in 1951, and in 1958, the public agency, Public Establishment for Installation of La Défense, was set up to plan it. Today, it has 1,500 businesses including the headquarters of leading French banks and its largest companies, together with many international companies. The employment population is more than 140,000 people.

New York

New York has also adapted its built environment in its own way to meet the demands of modern offices. In Manhattan, planning regulations or ordinances covering permitted floor area to plot size, height, type of use, etc. have been varied in the interests of promoting economic development. At various points in time, New York has further offered tax subsidies not only for new office construction in Manhattan but also for luxury housing and hotels. The World Trade Center was in part publicly supported and developed by the Port Authority. The goal was to reinvigorate the Wall Street area as the financial hub of the city.

Offices dispersal policies

Public policy designed to promote the economic development of city centres by relaxing planning constraints is relatively recent. Decades earlier, in a different era before globalisation, policy concerns were quite different. The boom of office development in certain cities as part of the economic recovery after World War II brought calls in the 1960s to halt, even reverse the trends. In some countries, it brought a deliberate office dispersal policy. This section examines the detail of the arguments and impact of the property-led policies that were applied. It focuses in detail on London and then widens the discussion to encompass Paris and New York.

Dispersal from London was instigated by the UK Government in the 1960s because of fears of over centralisation of decision-making in the capital city. The arguments in favour centred on the role of headquarters as growth poles in local economic development (see Chapter 10). At this time, the concentration of firm headquarters in London was seen as increasing traffic and population congestion in the city.

There were fears that the removal of high-level decision-making in provincial cities to London would have negative consequences for these urban areas. The impact would be seen directly through loss of high-status jobs and earnings and local competitiveness. In addition, there could be indirect effects through the loss of purchases in the wider local economy. The solution in the UK was to restrain the growth of the capital city and redistribute employment opportunities by imposing office development constraints.

In November 1964, the UK Labour Government introduced a ban on all further office building in central London. Initially, offices of more than 2,500 sq ft required an 'Office Development Permit' before they could be built besides the normal planning permission. The policy was known as the 'Brown Ban', named after George Brown, the minister responsible. The logic was not simply to relieve congestion but also to encourage decentralisation of offices to the rest of the UK.

The government also set about decentralising its own offices and sponsored the Location of Offices Bureau to assist the private sector to disperse from the capital. The ban was rigorously applied in the City of London until 1968, but the constraints were substantially relaxed from 1970. The office size requiring permission was raised gradually although the permits were not formally abolished until 1979.

The precise impact of the policy is difficult to quantify partly because of insufficient records. But also, as discussed in Chapter 8, decentralisation was occurring anyway. Many of the dispersed jobs went to satellite centres around London rather than to the peripheral regions/provincial cities. Many of these jobs, whether in the private or public sector, involved low-level or back-office services. Head offices remained in London.

Other countries have also attempted similar policies. In 1969, France introduced not only office development permits but also office occupation permits as a means to reduce the dominance of Paris. The French policy package included development taxes which increased towards the centre of Paris, but it had limited success.

An alternative, more comprehensive dispersal strategy was implemented at broadly the same time in New York. The city's dispersal strategy was a sub-centre strategy. This approach actually did not apply office development constraints but incentives. It aimed to promote sub-centres through transport improvements, zoning incentives and city participation in land development.

The focus of these incentives was urban renewal areas designated in the sub-centres of Brooklyn, Queens and the Bronx surrounding, New York. As part of the strategy, authorities were prepared to provide land at below acquisition cost. But as the strategy produced only one individual development, it was a clear policy failure.

It is interesting to note that none of these office dispersal policies based on restricting real estate development were particularly successful in achieving their goals. Restraining supply in a city centre does not necessarily mean that office demand can be redirected to those locations desired by the policy. Decentralisation did occur in London and New York during the life of the dispersal policies, but to some extent, it was independent of the initiatives. Office dispersal also did not take the desired form as it was mainly back offices and did not go to the expected locations.

These conclusions suggest limitations to the policies designed to shape local economies through intervening in real estate markets. On the other hand, the promotion of landmark buildings and the relaxation of constraints discussed earlier point to a complex set of potential outcomes reflecting the interaction of policy and market forces. The next section seeks to shed light on these issues by discussing a broader theoretical underpinning of real estate–led local economic development strategies.

Logic of real estate–led local economic development strategies

The removal of office development constraints or dispersal policies or landmark buildings can be seen as subsets of real estate–led economic development policies. All these policies are aimed more generally at manipulating the spatial location of land use. These policies have a long pedigree without possibly any fundamental assessment of their logic.

Regional real estate–led local economic development policies can be traced back to the 1930s in the UK when the original concept was relatively simple. Firms will locate in areas partly where they can find suitable premises easily. Based on these arguments, regional policy set out to build estates of factories in the peripheral regions of the UK. The goal was to attract manufacturing firms into areas of high unemployment. This policy of the state building 'advance' factories ahead of private demand continued for many decades and was applied not just in urban but also in rural areas.

Beyond this simple belief, the application of real estate–led policies can be justified on the inefficiency inherent in the property market, not least because of development time lags (see Chapter 8). The imperfections of the real estate market can mean that it does not meet the demands of a local community. There could be a lack of or under provision in certain areas or property of an inadequate size and standard to enable the expansion of output by individual firms.

The property market, therefore, may create serious supply constraints to both national economic growth and local economic development. From this premise that the property market is imperfect, it is possible for the public sector to intervene to resolve issues that arise. On the one hand, public policy can address what can be called real estate market failure by not meeting demand. On the other hand, the policy can seek to divert employment/property demand to a substitute location or away from a given area as illustrated by the dispersal policies earlier.

The key point of property diversion policies is that they do not occur in a vacuum. Their efficacy is partly dependent on the competitor or substitute locations and partly reflects a prerequisite of the existence of (footloose) demand. A public agency can influence a firm considering a range of potential locations by offering an attractive financial package. This could occur through direct provision of premises/serviced sites or through a range of property-related subsidies.

Looking just at offices, cities are generally competing to attract firms from others of a similar status or with the equivalent set of functions within the (global) urban hierarchy. Major office property developments in global financial centres are competing internationally. Provincial cities are in a very different 'market', usually domestically, but not necessarily, within the same country. The rise of the Dubai office market supported by state-backed development companies has arguably been achieved by harnessing and focusing demand from occupiers that could also seek substitute centres around the Middle East.

If the publicly supported office or industrial development is successful in generating employment, it improves the competitiveness of the area. In addition, it may have a demonstration effect on promoting further private investment and development. Dubai is also a good example of how the demonstration effect can lead to further private sector investment, in this case, supported by agglomeration economies.

The success of real estate–led policies can be measured by increased local economic activity, but there is clearly a cost/subsidy involved. Where different cities are directly competing for the same footloose companies, it can lead to an auction between public agencies in the areas to offer the best incentives. It could feed a rise in subsidy and, hence, an increase in the cost per local job generated. Logically, this cost should be factored into the cost effectiveness of policy strategies.

Summary

The construction of high-rise landmark offices is increasingly seen as a symbol of a city success as a business centre. The development of Dubai is the epitome of such an approach as it sought to establish itself as an international business centre. This strategy encompassed a series of signature skyscrapers developed by state-supported companies. Dubai has seen a rapid physical development of a central business district that has established a critical mass in terms of an office centre.

The development and exploitation of high multi-storey office blocks have an enduring history enabled by innovations in construction technology. They have been used as marketing icons in the competition between cities for more than a century. New York was the original advocate of the strategy with the world's highest buildings to be found in Manhattan for most of the twentieth century.

The case for the promotion of more high buildings in London was made by its mayor at the beginning of the millennium. His argument was that the future of London as a world city depended on having signature skyscrapers available to attract multinational companies. These views appear to be shared by the fastest growing cities of Asia where the most recent highest buildings have been built.

The role of real estate local economic development strategies inevitably focuses on their impact on business services. Expansion of employment in services, particularly linked to the knowledge economy, is the driver of a city's economy. With most business employment to be found in services, the office market is crucial to facilitating growth. But, so is the role of agglomeration economies in cities.

Agglomeration economies derive from the clustering of economic activities and provide the underpinning of the existence of cities. Such economies can be decomposed into localisation, activity-complex and urbanisation economies. The individual competitiveness of a city is defined by its characteristics and the agglomeration economies that are generated. The specifics of a city can be seen in terms of the scale of economic activity, its industrial structure and the physical environment.

The recent public policy has focused only on the benefits of localisation economies based on Porter's theory of competitive advantage. It emphasises the importance of successful local industrial 'clusters' in stimulating innovation and urban economic growth. However, competitiveness of a city can be seen more broadly to relate to urbanisation economies available within it. These economies are the basis by which cities compete in attracting footloose firms and workers as places to live.

Nevertheless, these factors that underwrite urban competitiveness should not be seen purely in individual terms. They are part of the bundle of characteristics that a city has to offer to attract firms seeking a new business location. If some of these constituents are missing from the portfolio then the overall attractiveness may not simply be weakened it may be insignificant. This argument stresses the requirement of a city to have attractive real estate opportunities to support the economic growth of a city.

These opportunities may not be provided because of the imperfections of the real estate market. There is therefore be a strong case for public policy to intervene to promote competitiveness by addressing supply constraints. In fact, there are many examples around the world where this has occurred, and the chapter considers a number of examples of cities removing office market constraints.

A driving force for public intervention is often to ensure that a city can be in a position to respond to technological change. One of the examples is Edinburgh where the city council set about creating not only a new central business district but also a large business park on the periphery. Planning restrictions were rethought, and a public agency was established to work with the private sector to achieve these goals.

The City business centre in London in the 1980s faced the same dilemma. Conservation policies were constraining the city's growth as a financial centre with the rise of information technology. There followed a policy about turn and the active encouragement of redevelopment. It enabled the construction of offices with large undivided floor areas suitable for the computer age. The new developments also encroached into neighbouring areas beyond the City's original square mile.

Paris' solution to meet the rising demand for modern offices was to maintain its conservation policies in the centre but build a new centre, La Défense. The new office centre was planned by a public agency and has become the financial hub. La Défense is on the western edge of the city. It is connected to the city centre by rail and metro links.

In New York, the city adjusted its city centre's environs to ensure it could address the demands of new offices and the wider requirements of a world finance centre. Planning regulations were relaxed and subsidies provided for key developments. In particular, financial support was provided for the World Trade Center developed by the Port Authority.

In the 1960s, the public policy concerns were very different. Office development booms in certain cities after World War II in the 1960s brought concerns of over centralisation. As a result, in some countries, there were office dispersal policies. In the

UK, a ban on all further office building in central London was introduced by requiring and refusing development permits. The ban was rigorously applied in the City of London until 1968 but gradually relaxed from 1970 and abolished in 1979.

France introduced an equivalent strategy including development taxes to disperse offices from Paris. In New York, office dispersal policies focused on incentives to develop in surrounding sub-centres. None of these dispersal policies achieved their goals. In general, office dispersal was mainly in the form of back offices and not to the desired locations.

Overall, real estate–led economic development policies are aimed at engineering change in the spatial location of land use. They can be justified by reference to the imperfections of the property market. Left to its own devices property market outcomes could result in serious supply constraints to the economy. Public policy may intervene in the real estate market to address these failures.

Property diversion policies are partly dependent on the existence of (footloose) demand. Besides making available suitable premises, a public agency can offer an attractive financial package. These activities can be applied at different spatial scales. Global financial centres are competing internationally whereas provincial cities are vying with domestic locations. The prize is potentially a long-term demonstration effect on promoting further private investment and development.

Learning outcomes

The construction of high-rise landmark offices has a long history as symbols of a city success as a business centre.

The development of Dubai is the epitome of such an approach as it sought to establish itself as an international business centre.

The case for the promotion of high buildings is that signature skyscrapers are attractive to multi-national companies.

With the business employment of most cities dominated by services, the office market may be critical to facilitating growth.

The individual competitiveness of a city is defined by its characteristics and the agglomeration economies that are generated.

Recent public policy has focused on promoting industrial clusters and localisation economies, as a means of stimulating innovation and urban economic growth.

Competitiveness of a city can be seen to relate to more widely to its urbanisation economies.

Urbanisation economies are the basis by which cities compete in attracting footloose firms and workers as places to live.

If some elements of the desired compendium of urban characteristics are missing, then the overall attractiveness may be severely deflated. It is a prerequisite of a city to have attractive real estate opportunities to ensure the economic growth of a city.

The imperfect nature of the real estate market may not produce the necessary range of properties that business requires. Public policy may intervene to promote competitiveness by addressing these supply constraints.

Cities around the world have intervened to remove office market constraints and promote new development. Often planning restrictions have been redrawn and a public agency established to work in partnership with the private sector to achieve these goals.

In the 1960s, the public policy concerns about over centralisation in large financial centres led to office dispersal policies. These dispersal policies were not very successful.

Property diversion policies are essentially reliant on footloose businesses. Global financial centres are competing internationally whereas provincial cities are vying with domestic locations.

The prize is potentially a long-term demonstration effect on promoting further private investment and development.

Bibliography

Jones C (2013) *Office Markets and Public Policy*, Wiley-Blackwell, Chichester.

Porter M (1998) Clusters and the new economics of competition, *Harvard Business Review*, 76, 6, November-December, 77–90.

X/Zen (2020) *Global Financial Centres Index*, https://www.longfinance.net/programmes/financial-centre-futures/global-financial-centres-index/

20 Physical and housing-led urban regeneration

Objectives

The nature of decline in many cities has been revealed as including concentrations of low incomes/high unemployment and population decline in inner urban areas. Housing is also often in a poor state of repair, and there were large tracts of unused land and derelict buildings in these areas. The urban policy task is generally to address these social and physical features of these areas as well as bring economic regeneration. In particular, economic revival involves finding new uses for land left vacant by the functional obsolescence of its previous use.

This chapter assesses different ways to undertake physical urban regeneration including housing-led approaches. Under the banner of physical regeneration is new social infrastructure, the promotion of tourism or cultural-driven strategies and the location of new retail centres. These policies are described as physical because they involve new buildings often for old, but they also have economic consequences. There is in practice a blurring between physical and real estate−led regeneration as the former can stimulate private investment. In fact, its role in economic regeneration is just, if not more, important.

The chapter is structured as follows:

- New social infrastructure and physical renewal
- Tourism-led regeneration
- Retail-led regeneration
- Housing-led regeneration
- Barcelona experience
- Glasgow experience
- The Glasgow Eastern Area Renewal (GEAR) project and the regeneration of the east of Glasgow

New social infrastructure and physical renewal

The original physical solution to inner-city problems was slum clearance. By the mid-1970s, there was too much resistance to this approach from local communities in western cities. It was replaced by the provision of new social infrastructure and the physical improvement of the housing stock. This answer involved capital expenditure

DOI: 10.1201/9781003027515-23

by the state on inner-city areas. New social amenities provided could range from new schools/colleges to health centres to sports facilities.

The improvement of the physical standards of housing directly upgrades the quality of life of inner-city residents by the removal of damp and unsatisfactory living conditions. State subsidies for such physical improvement may be channelled in a number of ways, including through a specialist agency such as a housing association. This improvement process could also involve a transfer of tenure to renting.

Physically improving the smaller housing stock can include combining the original units to make larger ones. Creating larger housing in this way can address overcrowding. New social or affordable housing for rent could also be built on vacant land to increase the supply.

Capital expenditure by the public sector in this way may be seen as addressing elements of urban deprivation or concentrations of poverty in these areas. It can be justified directly through a clear link between housing and health. More broadly, the policy is aimed at tackling deprivation as seen as encompassing poor health, lack of education and inadequate living standards.

Tourist-led regeneration

Tourism-led regeneration extends the notion of physical-led regeneration by building often landmark buildings that are designed to generate urban economic growth. It also has the dual impact of physically transforming streetscapes ravaged by urban decline. Sometimes, it is referred to as culture-driven regeneration and can be seen as part of place marketing, promoting the image of the city.

It can incorporate building-based flagship projects on symbolic sites usually in the city centre, together with public realm schemes, and the reuse of strategic buildings. Typical examples are new museums/art galleries or concert halls or other visitor attractions to improve the quality of life that are viewed as contributing to an urban renaissance. Historic buildings can be redesigned and recast often as locations for gift shops and restaurants. There are public good benefits in the form of quality architecture and a revived built environment.

Cultural activity is then the catalyst and engine of regeneration with a high-public profile and frequently to be cited as the sign or symbol of regeneration. Often, there is an associated continuing programme of events such as arts festivals within the new facilities partly to ensure usage. The publicity associated with these events is used to rebrand a city as a place to visit and live and an enhanced sense of place. In some cases, the reclamation of an open space can be celebrated in its own right through its use as a garden festival or an EXPO trade venue site.

There are many recent examples around the world of such schemes designed as part of a regeneration strategy. They include festival shopping in Boston and Baltimore in the United States or Covent Garden in London. Museums of contemporary art were built in Barcelona and Glasgow as part of plans to promote the cities as tourist destinations. It has become a ubiquitous approach to urban regeneration. It can also be seen as part of a spirit of urban entrepreneurialism.

A key argument of cultural regeneration is to attract tourists to a city, but it can also be seen in a wider perspective of competition between cities for tourists, sometimes on a global or continental scale. Furthermore, following Porter's notion of promoting successful industrial clusters set out in the previous chapter then creative industries can be seen as a key growth sector in urban economies. At the same time, expanding cultural opportunities can be seen as improving the quality of life in a city, and so attracting investment and well-qualified workers.

Like investment in cultural infrastructure, the building of sports facilities has similar goals. It is also aimed at attracting tourists, encouraging inward business investment and redrawing the image of the city. The drivers of such a policy are the desire for new employment opportunities caused by the loss of the manufacturing industrial base. The policy started in the United States from the 1970s as cities such as Indianapolis sought to attract domestic sports franchises. Elsewhere in the world, it is associated with hosting international sports competitions.

Whether tourist-led regeneration is about the provision of cultural or sports facilities, it is the public sector that usually provides the initial capital investment. In some cases, the funding for facilities comes from the central government so that the city is also a beneficiary of inward investment. The facilities also provide a platform for the future. The private sector is the main financial beneficiary in terms of the expenditure from tourists.

In the longer term, if the regeneration is successful in improving the image of the city, there could also be benefits through increased wider economic activity and employment. The economic base of the city is broadened and strengthened. However, it may need to be supported by continuing events or newly constructed attractions. In addition, given that many cities are trying the same strategy to attract tourists, not all such strategies are likely to have such significant impacts.

The benefits to the city may not be translated to inner-city residents as the physical regeneration in the shape of new buildings may be confined to city centres, rather than the surrounding nearby areas. Sporting areas and associated infrastructure are more likely to be built-in inner-city areas rather than cultural facilities. In either case, they may be little used by low-income/inner-city residents.

In addition, the jobs created in tourist-related activities such as hospitality are likely to be low paid. At the same time, expanded accommodation for tourists could potentially diminish housing opportunities especially for low-income households. There are a number of other dangers that could mean that inner-city residents could lose out from a tourism-based strategy. At the extreme, the urban core may be 'gentrified' by tourists prepared to pay high rents or buying second homes. Traditional shops could be transformed into selling goods aimed at the tourists. Much of the city's infrastructure expenditure could be aimed at meeting the demands of tourists rather than locals.

To summarise, tourism-led regeneration embraces improving the physical fabric of a city, together with more positive images of a city and attracting (global) tourists. It also seeks to improve the quality of life through enhanced cultural opportunities. The longer-term goal is to raise the attractiveness of the city as a place in which to invest and live and restructure the economy. There are some very high-profile successful cities in this regard, but there are issues about the benefits to low-income households. The sameness of the strategies queries whether the approach can be applied successfully on a universal basis.

Retail-led regeneration

The role of retailing in regeneration is complex. Market trends in retailing outlined in Chapter 12 have contributed to the decline of small-/medium-sized towns as well as traditional suburban centres in the UK. It is a major cause and physical demonstration of urban decline with large numbers of vacant boarded-up shops. This demise, especially of small/medium town retail centres, has led to a process of reappraisal of their long-term future. Many are 'repurposing' to a mixed-use future, whereby shops are replaced with housing and leisure.

Nevertheless, retail investment can be used as a means of adapting and improving localities. There is no single retail regeneration model. The traditional route is the building of indoor shopping malls in city centres, with the first wave occurring in the 1970s. These shopping malls are often adjacent to the high streets often on land formerly used for alternative uses.

This approach grew in policy prominence from the 1980s as a way of boosting the retail attractiveness of city centres and the local economy. These shopping malls were built by private enterprise, but the driving force was city governments. The momentum for change was partly linked to tourism-led urban regeneration and the place-making agenda noted in the previous section. The policies went much further than simply the provision of new shopping malls to embrace the whole streetscape of a city centre.

With rising incomes, the logic of the policies was based on retail centres as places of consumption that appeal to the 'leisure tourist shopper' rather than just the functional/utilitarian shopper. The urban design of the central area, following the ideas, for example, of Gehl, saw streets as social spaces. Shopping malls and historic streets with attractive building facades were integrated often with pedestrianisation to physically improve the streetscape and promote the shopping/leisure experience.

By attracting tourists and day trip shoppers, centres could support a greater range of shops than the city's population on its own could afford. The strategy is an extension of the competition for tourists through cultural attractions. In effect, the retail element is an integral component. The attractiveness of the retail morphology to leisure shoppers also enables the city to successfully compete with the threats from surrounding out-of-town shopping malls. In addition, the establishment of a strong regional centre contributes to drawing demand from nearby small towns and their relative loss of attractiveness noted above.

Outside the city centre, Chapter 11 records the growth of out-of-town retailing from the 1970s in the UK. Shopping centres, including retail parks and supermarkets, have been built in decentralised locations. Many have taken advantage of vacant sites abandoned because of deindustrialisation. This land may be functionally obsolete for its original use, but retailing is a physical replacement and a means to regenerate the locality. It was also seen by policy makers as an easy first step in a wider regeneration strategy.

The industrial decline of inner urban areas and associated unemployment also brought the disappearance of local shops catering for these poor communities. The lack of inner-city shops spawned discussion about social exclusion (from shops and other services) and health inequalities. In particular, rundown district shopping centres built in the 1960s and 1970s catering for low-income neighbourhoods by the 1990s were in need of revitalisation.

The regeneration solution was often redevelopment of these centres, usually owned by a local council, by the private sector. The new replacement (or simply a new) out-of-town retail offering provided better and a wider range of shops for the local community, made viable by tapping into a wider car-borne catchment area. A further positive is that these new retail outlets were also a potential creator of jobs in disadvantaged areas where there were few employment opportunities. The retailers might also offer to undertake to provide training opportunities for local residents, particularly the long-term unemployed. On the other hand, the nature of this work is generally low paid.

In conclusion, retail-led regeneration takes a number of different forms. From one perspective, it is a dimension of tourism-led regeneration with cities expanding their shops and improving their centres to attract leisure shoppers. Retail-led regeneration can also be harnessed to fill vacant and derelict sites forsaken by manufacturing industry or to revitalise shopping facilities in poor inner-city neighbourhoods. Yet, at the regional level, the retail-led regeneration of some areas by expanding shops also divert demand from existing shopping centres and can result in substantial even terminal decline.

Housing-led regeneration

Earlier in the chapter, the physical improvement of the housing stock was considered so here housing-led regeneration is taken to mean the attraction of new households into an (inner city) area. This task would be achieved by the building of new housing or the improvement of the housing stock for sale. In both cases, there is a physical improvement of the neighbourhood and wider regeneration goals. There is a range of potential overlapping objectives for promoting housing for sale in a locality. The proponents of the policy list the benefits as follows:

- *Attracting new households into the area, thereby increasing or attempting to stabilising the local population*
  Many inner-city neighbourhoods are losing population, so building new housing in an area can contribute to reversing this trend by encouraging in migration.
- *Introducing more attractive housing and increasing local tenure mix*
  Through building new housing for sale, it is likely to change the profile of housing in two ways. New housing balances the older housing that dominates inner-city areas. Owner-occupied housing dilutes the predominant concentration of rented properties. Together, these changes improve housing choices for households and, hence, the scope for neighbourhood stability.
- *Reducing the concentration of old and low-income groups means less strain on public services per capita*
  Purchasers of new housing are likely to be younger than the existing inner-city residents. Although younger adults could be in the child-bearing phase of the family life cycle, they are likely to put less demands on public services.
- *Increasing population means a rising demand for private services such as shops which is to the benefit of existing residents*
  An increase in the local population means that there is increased demand for neighbourhood shops. The viability of local shops and services improves, reducing the likelihood of closures and even encouraging openings.
- *'Ontological security' – there is increasing self-identity/confidence as an owner occupier, thereby improving labour market opportunities for households*
  It is possible that owner occupiers have a greater sense of self-worth that translates to confidence in the labour market. There is no real hard evidence on this phenomenon, but intermingling could transmute to those of working age in the original population

- *Owner occupiers who buy are likely to be younger with more skills/education than the original population, reducing stigmatisation/ improving social mix*

    The arrival of new home buyers who are young, economically active and probably better qualified than the original population revises the neighbourhood dynamic. It changes the social mix of the area even if some gentrification occurs. The changed social character of the area could bring positive consequences to its external image and particularly for employers. It could reduce any stigmatisation from living in the area

- *Improving social mix may increase the educational attainment of school children of original population*

    There is a belief that social mixing of children from different backgrounds overall enhances learning and educational achievements.

- *Improving tenants' attitude to the maintenance of the housing stock following the example of the new owner occupiers*

    There is a conviction that owner occupiers' positive approach to maintaining their property acts as a demonstration to tenants nearby.

- *Meeting a real latent demand for home ownership*

    The low proportion of owner-occupied housing in many inner cities belies the fact that many households wish to buy there.

The bases of some of the arguments above have not been subject to critical empirical review, so some are more acts of faith. Besides these arguments, there is an important independent policy driver. Housing may be the only practical use for much of inner-city vacant land. As noted earlier, the vacant land has arisen from industry closing down following deindustrialisation, because the optimum industrial location is now elsewhere.

Perspective on this policy approach can be seen through some examples. In the UK, these housing-led policies focusing on promoting inner-city owner occupation began in the mid-1970s. They arise from worries about housing opportunities to buy a home in cities. Many provincial cities, led by Liverpool and Glasgow, concerned with population loss offered subsidised sites for builders to construct houses for sale. The houses were aimed at low-income households, in some cases, giving priority to people on council (social) waiting lists.

Housing-led policies were then extended from the early 1980s to the improvement of unpopular public sector stock that was then put up for sale on the open market. The local authorities of the inner London borough of Wandsworth and the cities of Liverpool, Edinburgh, Glasgow were in the vanguard of this strategy.

This housing-led approach moved on to building houses for sale in the areas adjacent to large public sector estates. The first example in 1982 was at Cantril Farm, a large overspill council estate on the edge of Liverpool. As part of this initiative, the community was also renamed as Stockbridge Village.

The Pathfinder programme established in 2002 and considered in Chapter 18 can be seen as a further extension to housing-led urban regeneration policies. While aimed specifically at addressing areas of low housing demand in northern English cities, they had a subtheme of attracting people back to the areas.

Overall, these housing-led regenerations are partly motivated by physical improvement goals. They also seek to undertake wider social restructuring of community and local economic development. Today, such policies are now a mainstream element of UK urban regeneration strategies.

Experience of Barcelona

Barcelona is one of the most outstanding examples of city physical regeneration in the world. Looking back to the mid-1970s, Barcelona was a grey, densely populated, industrial city in a deep economic crisis. Yet, by the beginning of this century, the city had become a major tourist destination with a reputation for high-quality urban design on a grand scale. The process was enhanced by the holding of the 1992 Olympics in the city.

Its achievements include a long list of new public buildings encompassing the Olympic park and village, conference centres, libraries, museums, offices, community centres, health centres, schools and markets. There have also been new public spaces – squares, beaches, promenades, parks. The old city and the port have also been renovated, creating tourist attractions.

The physical regeneration of the city extended to new roads of all types together with an overhaul of public transport. The new infrastructure has been extensive including the building of large storm drains, rail stations, airport, streetscapes and telecommunications networks.

The UK Urban Task Force report published in 1999 described Barcelona in its report 'Towards an Urban Renaissance' as a 'beacon of regeneration' and 'the most compact and vibrant European city'. It has received many awards and is seen as the symbol of successful long-term planning through the implementation of its 1976 strategic plan.

Planning for the reconstruction began in 1974, before the death of Franco, under whose rule Barcelona had declined and stagnated. The General Metropolitan Plan for Barcelona was put forward in 1976 and implementation began in 1979.

One of the original main goals was to provide public facilities for the local residents. Of the land that was initially acquired, some 39% was for parks and gardens, 22% was for forests, 32% for schools and other public facilities and only 7% for housing. The land was acquired when prices were low, and much of it was formerly used by the industry that had disappeared. In the early stages of the delivery, quick and inexpensive projects were undertaken. Visible success was seen in the provision of parks, developing public spaces and other public facilities.

In 1986, Barcelona won the nomination to host the 1992 Olympic Games, and this proved to be a step change in the regeneration process. It enabled the public funding from national and regional governments for city-wide large-scale public works projects. Not only were sports facilities provided for the event but also social and transport infrastructure was built connecting the city core with its surrounding hinterland. The city was also opened up to the sea with the construction of the Olympic Village.

After the Olympics, the city sought to build on the success by growing its reputation and attracting global investment and tourism. It remained a public-led regeneration process, but private sector investment was encouraged. The city began to take a more partnership-based approach with the private sector to support its large-scale programmes. It created an urban planning think tank to cultivate development proposals including major transformative projects in the city.

There was now an emphasis on culture in its widest sense through selective conservation of historic buildings, and flagship architecture projects in the city centre were developed. It was part of a conscious promotion of the city's image through urban design. At the same time, there was further redevelopment of the old industrial port transforming it into a leisure area.

The city's economy was metamorphosed by the physical transformation. The city is now a service-based economy. Barcelona's core has embraced the knowledge economy, cultural industries, retailing, leisure and tourism. There has been a dramatic increase in tourists from Spain, Europe and the rest of the world supported by a rise in hotel capacity.

High profile examples of regeneration areas in the city include Diagonal Mar (new name), part of Poblenou on the edge of the urban core. An American developer converted this old industrial area near the sea to a mixed-use area of luxury flats, shopping malls, hotels and offices. The Forum of Cultures festival, a four-month event was held in the area in 2004. This was an international festival held in a specially built conference centre surrounded by a new water park.

Also in Poblenou is 22@, designated in 2000 as the city's technology and innovation district on industrial land. The vision of the project is a focus on knowledge-based activities, together with the adoption of new technologies. The plan for the area includes subsidised housing, public spaces and green areas to create a mixed-use neighbourhood.

The regeneration of the city has brought an increase in its population with foreign residents now the order of a fifth of the total. The success of the regeneration has also led to a rise in the cost of housing in the city. The cultural refurbishing and the growth of tourism have encouraged gentrification in the old city. The example of the Diagonal Mar project noted above is indicative of the pattern of change. Once a low-income neighbourhood, it now has exclusive new housing. Many of the inner-city flats have been converted for short-stay and low-cost tourists.

As a result of the housing problem in 2019, the city council started providing one- and two-bed flats using shipping containers as temporary housing in the old city. High numbers were seeking emergency accommodation having being driven out by eviction/gentrification. The tide of gentrification and tourist accommodation has pushed up rents and capital values that has driven all but the rich out of the city. Nearly, all social housing is on the outskirts. The result is that low-income households have moved to the periphery and stimulated urban sprawl.

Overall, Barcelona has been transformed into a successful, attractive city with a thriving tourist industry. The city can be viewed as a great success for long-term strategic planning that actually delivered. The subsequent economy is based predominantly on tourism. However, the success of the global tourism ambitions has created tensions between segments of the city's population. The success has priced out many low-income households from the core. Similarly, the shops in the central area cater for the tastes of tourists who also crowd in the busy public spaces. In a sense, gentrification has extended beyond simply the housing stock.

In terms of the different approaches to physical regeneration outlined at the beginning of the chapter, Barcelona initially followed a strategy of developing social infrastructure. The designation of the Olympics began a tourist-led approach with first the focus on sports facilities and then on cultural tourism. The story also highlights the integration between retail-led and tourism-led regeneration.

Glasgow experience

There were strong parallels between Glasgow and Barcelona in the 1970s. Both cities were ports suffering industrial decline with deindustrialisation that was accelerating. The physical fabric of both cities was poor including slum housing. Much of Glasgow's housing tenements and public buildings were grey edifices stained by a century or more of soot from its industrial past. The city had some of the most deprived neighbourhoods in the UK.

In the early 1980s, the city began a process of reinventing itself, or at least its image to the outside world. There was a two-pronged strategy that took the form of place marketing and a series of cultural events matched with the physical improvement of the city centre and its environs. In addition, there was already in place a major urban regeneration initiative, GEAR, for an inner-city area in the east of the city. This project is considered separately below.

In 1983, the city launched its marketing slogan, 'Glasgow's miles better'. It was aimed not just at external negative perceptions but also at an internal audience to improve pride in the city. It was the beginning of an active strategy to promote the city as a place to work and live. In 1983, the Mayfest musical festival was established (until 1997), and a series of cultural festivals was developed throughout the year. An extensive citywide campaign was undertaken to stone clean the buildings of grime and return them to their original sandstone colours.

Glasgow also focused on the physical improvement of its city centre. The city had a strategy of developing attractions including reinvigorating existing museums and opening new art galleries and new music venues, etc. In 1988, the city hosted a summer garden festival over five months on reclaimed docks to showcase change. The festival attracted 4.3 m people. The site has now been redeveloped to incorporate offices and Glasgow Science Centre.

Two years later in 1990, Glasgow was the designated European City of Culture with a year-long calendar of special events. As part of the bid for the title, it built the Royal Concert Hall, owned by the council, in the city centre. The music venue then became the basis for the Celtic Connections music festival that runs every year in January. The three-week long festival now attracts visitors and artists from around the world.

The Gallery of Modern Art was opened in 1996 in a converted historic building in the city centre. Other existing museums were revamped including the transport museum. It was moved to a new building in 2011 on River Clyde, the main river that flows through the city. Two major auditoriums have been built along the Clyde, including a major music arena for popular concerts.

The main shopping centre was resculptured adapting the historic streetscapes to modern needs, providing accessibility for shoppers to car parks. The design of the retail centre importantly took close account of retail anchors such as major stores or in-town centres. Three new shopping malls were integrated into the existing main shopping streets. The city held an international design competition to develop the pedestrianisation layout of the centre.

At the beginning of the 1980s, a major physical problem for the city centre was the many warehouses lying empty adjacent to the city's shopping centre. As warehouses they were obsolescent as this economic activity had moved to decentralised purpose-built premises. The council decided to promote the area as a residential location and renamed the neighbourhood as the 'Merchant City'.

Over the next 20 years, the area acquired new housing through a combination of new building but mainly conversions. More detail of the public funding and the regeneration process is given in Chapter 23. The new housing was small flats aimed at young adults. Besides housing, the Merchant City also became the home for a wide range of pubs and restaurants.

The city council in conjunction with a public agency has also promoted an area adjacent to the city centre as an extension to the central business district. A motivation is also to physically regenerate an area of the city near the River Clyde comprising derelict land and dilapidated buildings. To support the regeneration of the area, in 2001, the council designated the area as the 'International Financial Services District'. Public funding has taken the form of improving the streetscape and a new bridge to support redevelopment.

The city has pursued a policy of pitching for international sporting events. The most important event held was the 2014 Commonwealth Games. It attracted athletes from 71 nations. As a result, new sports facilities and an athletes' village were built in the east of the city with financial support from the Scottish Government. After the Commonwealth Games, the athletes' village was adapted and expanded to create 700 new homes – 300 for sale and 400 for social and affordable renting. The majority of the homes were arranged in terraces with some flats alongside the River Clyde.

These policies have contributed to modernising the city's economy. There has been an emphasis on changing the image of the city and the quality of life it offers. The strategies have combined the provision of new amenities and a focus on festivals/events to promote tourism. As a tourist destination, the city is particularly popular for weekend 'city breaks'. Its attraction is enhanced by the shopping centre improvements that have made it one of the largest shopping centres in the UK outside London.

Individual policies have strong parallels with those in Barcelona, tailored to Glasgow's circumstances. A key difference between the outcomes is the existence of gentrification in Barcelona, in part reflecting the differences in tourism and the absence of luxury housing development. More generally, the city's strategy has similarities with the regeneration goals and policies of many other cities.

The GEAR project and the regeneration of the east of Glasgow

The application of physical regeneration to inner cities is more complex. The chapter now examines this approach by an assessment of the GEAR project in the east end of Glasgow. To recap from Chapter 18, GEAR stands for Glasgow Eastern Area Renewal. It was an initiative that began in 1976 and continued through to 1986. It is emblematic of the regeneration policy solutions applied at that time.

The area had suffered the ravages of industrial decline with a heritage of extensive derelict land. From being one of the most densely populated areas ringing the city centre at the end of the nineteenth century, it experienced dramatic population loss. The selective migration out meant that the population in the mid-1970s comprised a high percentage of low skilled and the elderly. In other words, it had the classic profile of the 'inner city' (problem) identified in Chapter 18. The area comprised only part of the east end of Glasgow. The River Clyde flows through the heart of the city from east to west. The southern boundary of the GEAR project is the (meandering) Clyde plus an area designated for an industrial estate, on the other side of the river. The western boundary is close to the city centre. The northern boundary is a railway, hence, taking the form of a straight line. There was an immediate problem about these boundaries – namely that they are quite arbitrary and narrow, and not reflecting economic functional or community realities.

The review of GEAR here begins with explaining the partnership structure of the project and then explains its goals, strategy and assesses the outcomes. All the partners in the project were from the public sector, namely the following:

- Glasgow City Council
- Strathclyde Regional Council
- Scottish Special Housing Association
- Housing Corporation
- Scottish Development Agency
- Greater Glasgow Health Board
- Manpower Services Commission

Only Glasgow City Council (GCC) is still in existence. Strathclyde Regional Council (SRC), as the name suggests, was the top level of a two-tier local authority structure in the city covering the wider subregion. Among its responsibilities were strategic planning, roads, social work and education provision whereas GCC's primary roles included local planning and provision and management of social (council) housing.

The other public agencies had specific narrowly defined roles. The Scottish Special Housing Association (SSHA) also built and managed social housing throughout Scotland where local authorities lacked the resources, normally in rural areas. The SSHA was incorporated in the partnership because it was seen as an efficient housing agency, and it could bring its expertise and resources to the task of improving the publicly owned housing stock. The Housing Corporation (HC) was a national conduit for funding housing associations. Housing associations were already operating in the area buying up and renovating the old tenement flats that lacked basic amenities and were in a poor state of repair.

The Scottish Development Agency (SDA) was a national economic development agency that also undertook environmental improvements. It was also responsible for chairing and coordinating the project. The SDA was eventually transformed into Scottish Enterprise with a wider range of powers including elements of the Manpower Services Commission (MSC). The MSC dealt with managing training and employment services particularly focusing on the unemployment aspects of economic development. The Greater Glasgow Health Board (GGHB) was responsible locally for the national health service.

The overall stated goal of the project was to bring new confidence in the area and the revival of industry. It had a long list of stated objectives:

- Increase residents' competitiveness in the labour market
- Arrest employment decline
- Overcome social disadvantage of residents
- Improve and maintain the environment
- Stem population decline and engender a better-balanced age and social structure
- Foster residents' commitment and confidence

The various partner agencies undertook the following responsibilities:

SRC: Enhancing transport infrastructure, education provision and social services
GCC: New council (social) housing, modernisation of council housing
SDA: Land assembly, site preparation, advance factory building (see Chapter 19), supporting business development and the physical environment
SSHA: Building new housing for sale, modernisation of council housing
HC: Supporting new housing and rehabilitation of tenements by housing associations
GGHB: Building of a health centre
MSC: Opening a job centre at the centre of the area to support the unemployed

The total expenditure attributed to this project by these public bodies in the period 1977–1986 was £315m.

A grouping of these activities and the expenditure on each reveals the following:

*   *Housing* (modernisation of council housing, rehabilitation of tenements, new housing) = *65%*
*   *Environment* (site preparation, physical improvement) = *12%*
*   *Factory Building and Business Development = 6%*
*   *Infrastructure* (transport, education, social services) = *9%*
*   *Health Centre = 2%*
*   *Job Centre = 2%*

It can be viewed as predominantly capital expenditure or physical development from housing and environmental improvement through to social infrastructure. Virtually, all was undertaken by different elements of the public sector. The only exception was the new housing for sale, albeit financially supported by public finance. Besides these capital expenditures, there were small amounts spent on promoting business development in conjunction with the new physical industrial units.

Assessment of outcomes

Once all this physical redevelopment was complete, an important question is how better off were the local residents? Houses had been improved, new housing constructed, industrial units built and jobs created. In terms of attracting private investment, some £184m had been attracted by 1986 in the form of private housing for sale (£87m), industrial plant/machinery (£48m) and commercial property development (£49m). These figures were inevitably an underestimate by taking too short term a time perspective for induced private investment. Subsequent private developments included the major shopping centre, Parkhead Forge.

A major failing of the project was the failure to address the level of local unemployment. After completion, unemployment rates in the area were still very high, among the highest in Glasgow and across the west of Scotland. The problem was that the local population had to compete in a city-wide labour market for the new local jobs, whereas there was little expenditure on training. Better qualified workers from elsewhere in the city were better positioned to take the jobs.

This failing raises questions about the strategy. Were there sufficient funds to tackle the scale of the problem and the restructuring of a large segment of a city? On the other hand, the £315m expenditure by 1986 was spent on only approximately 45,000 residents. Was it the right strategy? There is little link between the objectives that emphasise economic and social issues and the expenditure linked primarily to physical rebuilding. The difficulty in making these assessments is that the counterfactual, i.e., what would have happened without the project, is not known.

Despite these criticisms, GEAR was the most successful of the UK 1970s partnerships referred to in Chapter 18. The focus on the physical improvement of the area reflected the state of the area, but the underlying economic problems were not resolved although the residents were physically better housed. This is partly because a physical improvement strategy on its own has inevitably limited scope or potential. It can be seen as a demonstration of urban revival, but the evidence suggests that it is insufficient on its own. The GEAR project can be viewed as the last hurrah of post-war urban redevelopment policies that had been applied for the previous quarter of a century.

Clyde Gateway urban regeneration company

Twenty years after the completion of the GEAR project, the east end of Glasgow was designated as requiring further regeneration. In June 2006, the Clyde Gateway urban regeneration company (URC) was established. This time the initiative area covered a wider area of 840 hectares across the east of the city. Despite the efforts of the GEAR project the area 20 years on still suffered from the following:

- Much of the area was classified as in the 5% most deprived neighbourhoods in Scotland, including five of the 20 most deprived in the country.
- The population was only 20,000 people and declining, down from the 45,000 in the much smaller area covered by GEAR.
- Forty percent of the Clyde Gateway's area was vacant, derelict or polluted, with some of the land highly contaminated.
- Forty-six percent of adults had no formal qualifications in 2011, compared with 27% across Scotland.
- Educational achievements at school were much lower than the Scottish average.
- The proportion of residents with long-term health problems was 50% above that in Scotland as a whole.
- There were high levels of unemployment and worklessness, with 38% of the working age population claiming out-of-work benefits, almost three times the Scotland average of 13%.

The modus operandi of URCs is explained in Chapter 18, with a focus on real estate–led regeneration. In practical terms, its brief was broader encompassing physical and community regeneration. The URC was set up specifically to ensure the area received a beneficial legacy from the 2014 Commonwealth Games and the building of a new motorway through the area.

The overall goal of the URC over 20 years is stated as 'Tackling access to more and better jobs, better health services and a safer environment'. Its objectives are as follows:

- Improvement of the social and physical infrastructure ('sustainable place transformation') – housing choice/affordability with 10,000 new homes and the remediation of 350 hectares of derelict and contaminated land

- Increase economic activity – new employers, employability of locals – 21,000 new jobs over 25 years – building industrial units 400,000sqm of employment space, 46,000sqm retail and related development space
- Develop community capacity – social exclusion, stem population loss, etc.

These objectives parallel those for GEAR, and housing is still a key policy.

In the first ten years, it had received £200m funding from public sources, more than 70% from Scottish Government. There has been a strong emphasis on physical regeneration. A key priority was land remediation, and by the end of ten years, it had reached 68% of the final target. There were new roads, bridges and the refurbishment of railway stations. Modern sports and leisure facilities were provided including those from the Commonwealth Games. A woodland park with play areas, a nature reserve and a network of paths for walkers and cyclists have also been developed.

Unlike the GEAR project, the initiative has offered training courses to local residents. To promote these opportunities outreach officers have been appointed. In addition, there are community benefit clauses applied to contracts issued by the URC. There are also efforts to improve the health and wellbeing of the community.

New housing and commercial real estate development in the area has been relatively slow. After ten years, progress had yet to reach a quarter of any of the 20 years' targets. Retail development was primarily stand-alone fast-food restaurants and supermarkets. Offices/industrial units are generally in small-scale business parks.

The area is still in a weak economic state and is struggling to attract private investment. The low growth of the national economy over the life of the project to date has been a major constraint. Given the low starting point for the local economy to be successful the area needed a development that provided a step change. The Commonwealth Games and the new motorway did not deliver it. Urban regeneration has been mainly physically supported by public investment (but see Chapter 22 for more perspective).

Summary

Physical-led urban planning/regeneration policies were the traditional approach to reshaping cities. Slum clearance and rebuilding were ultimately replaced by the provision of new social infrastructure and the physical improvement of the housing stock. Improvements in housing standards upgraded the quality of life and health of inner-city residents but may involve a change of tenure.

Tourism-led or culture-led regeneration combines physical-led regeneration of cityscapes with the potential to stimulate local economic growth. The public realm is improved through good quality architecture, and obsolescent buildings can be revived with new functions. An enhanced range of visitor attractions also attract tourists to bolster the urban economy.

Cultural regeneration is also often paired with a series of (regular) events. These events reinforce positive messages about a city as a place to visit and live and give it an enriched sense of place. The attraction of tourists in this way has become a universal approach to urban regeneration. The result is that there is an increasing competition between cities for tourists. The expansion of cultural opportunities can be viewed as contributing to the standard of life in a city. It is also part of a wider competition between cities to attract investment and highly qualified workers.

The building of sports facilities by a city can have similar goals as a means of attracting tourists, encouraging inward business investment and redrawing the image of the city. For both the provision of cultural or sports facilities, it is the public sector that usually provides the initial capital investment. Private businesses in the city are main financial beneficiaries through increased income.

Tourism-led regeneration can also bring long-term benefits in terms of restructuring the local economy. However, as many cities are applying the same strategy, the sameness can dilute the benefits. What benefits there are do not necessarily filter to low-income households who may neither use any new facilities nor see their incomes rise. Gentrification too can have implications for the supply of housing for low-income households.

The role of retailing in regeneration is not straight forward. Small/medium town retail centres, certainly in the UK, are in decline. To be regenerated, these centres are 'repurposing' with shops being replaced with housing and leisure. Nevertheless, integrating new shopping malls together with the physical reshaping of city centres' streetscapes have created an attractive environment for leisure tourist shoppers. In the process large cities successfully compete for shoppers and draw demand from smaller centres.

In decentralised locations, out-of-town retailing in the UK has taken advantage of vacant sites abandoned because of deindustrialisation. Retailing can therefore be a reasonably easy first step to physically regenerate an area ravaged by industrial decline. Out-of-town shopping centres are also a vehicle for replacing local rundown shopping centres. The new centres are viable by drawing on a much wider car-borne catchment area. Besides physical improvement to these areas, these new retail outlets offer jobs in deprived areas.

Housing-led regeneration in the form of houses for sale can be a means of attracting new households into an area. In doing so, there are a range of potential benefits including arresting population decline, stimulating social mix and making local shops more viable. Housing may also be the only realistic use for much of inner-city vacant land.

There are a number of variants on housing-led regeneration, and their goals are only partly motivated by physical improvement goals. They are also seen as a mechanism for the social restructuring of communities and local economic development. Housing-led approaches are now a typical component of the UK urban regeneration strategies.

Barcelona is an exceptional example of the physical regeneration of a city. The city that was still in long-term decline in the 1970s has been transformed by high-quality urban design. The process has encompassed new public buildings and spaces, including the renovation of the old city and port, together with new transport infrastructure.

It can be viewed as the success of long-term planning. However, after the holding of the 1992 Olympics, it was able to draw on private investment to develop as a major tourist destination. The economic base of the city was extended to incorporate the knowledge economy, cultural industries, retailing, leisure and tourism. The population of the city has risen partly as a result of foreign in-migrants.

The success of the regeneration has led to gentrification and a housing problem for low-income households. Exclusive housing and flats for tourists have replaced homes once occupied by low-income households. The result is that low-income households have had to migrate from the core to the periphery and stimulated urban sprawl.

The challenge facing the regeneration of Glasgow in the 1970s was very similar. In the early 1980s, the city initiated a process of revival through place marketing and physical improvement of the city centre. The starting pistol was the marketing slogan, 'Glasgow's miles better'. The city's strategy was developing attractions including reinvigorating existing museums and opening new art galleries and new music venues, etc. Regular festivals were also a key component, including designation as European City of Culture in 1990.

In the city centre, the main shopping fare was remoulded, adapting the historic streetscapes with modern malls. Empty warehouses and other redundant public buildings were converted into flats. The council renamed the neighbourhood as the 'Merchant City', and it also became the location for a wide range of pubs and restaurants.

A series of international sporting events were held in the city, most notably the 2014 Commonwealth Games. It brought new sports facilities and an athletes' village located in the east of the city with financial support from the Scottish Government. These were used as a catalyst for the regeneration of this inner-city area.

Physical regeneration to the inner-city area of the east of Glasgow began in earnest with the GEAR project, from 1977 to 1986, and then again with the Clyde Gateway project from 2006. The GEAR project was a partnership of public agencies. It had the overall stated goal to bring new confidence to the area and the revival of local industry.

The project's spending was predominantly on infrastructure or physical development from housing and environmental improvement through to social infrastructure. Houses were improved, new housing constructed, industrial units built and jobs created. However, a major failing of the project was the inability to solve the high level of local unemployment. The new local jobs were taken mainly by outsiders. The local population were unable to compete in a city-wide labour market, while there was little expenditure on training. The GEAR project was the last UK inner-city initiative, that was focused on physical regeneration alone.

The subsequent Clyde Gateway URC covered a wider area of 840 hectares. Although it has a brief to develop real estate−led regeneration, its initial emphasis was physical and community regeneration. The URC sought to use the 2014 Commonwealth Games and the building of a new motorway through the area as a catalyst for regeneration. Many of its objectives parallel those of GEAR, and housing-led regeneration is still a central policy. Unlike GEAR, the initiative has offered training courses to local residents.

The scale of new housing and commercial real estate development over the life of the project to date has been modest. The project has suffered from the slow macroeconomic recovery from the global financial crisis. The Commonwealth Games and the new motorway have not proved to be silver bullets to revive the local economy, at least as yet.

Learning outcomes

Physical-led urban planning/regeneration policies were the traditional approach to reshaping cities until the 1980s.

Physical-led regeneration of cityscapes can support tourism and remodel the economic base of a city.

Cultural- or tourist-led regeneration is also often combined with a programme of festivals that underpin positive messages about a city as a place to visit, live and work.

There is an increasing competition between cities for tourists.

New sports facilities in a city are also a means of attracting tourists, encouraging inward business investment and redrawing the image of the city.

The provision of cultural or sports facilities is usually provided by the public sector. Private businesses are the main direct financial beneficiaries through increased income.

The benefits of tourism-led regeneration do not necessarily permeate down to low-income households, and they may suffer a loss of housing from gentrification.

Declining small and medium town retail centres in the UK need to be regenerated through shops being replaced by housing and leisure.

Many city centres have created an attractive environment for leisure-tourist shoppers by improving the physical central streetscape and integrating shopping malls.

Out-of-town retailing in the UK has taken advantage of vacant sites abandoned because of deindustrialisation and offers a way to physically regenerate derelict areas.

Out-of-town shopping centres can viably replace local shopping centres in inner-city areas by drawing on car-borne shoppers.

Building houses for sale can be a means of attracting new households into an inner-city area.

Housing-led regeneration is aimed at stopping population decline and stimulating social mix.

Housing may also be the only realistic use for much of inner-city vacant land.

Barcelona stemmed long-term economic decline by physically transforming the city, introducing new amenities and incorporating high-quality urban design. The result was a major tourist destination.

The economic base of Barcelona was expanded by the strategy to incorporate the knowledge economy, cultural industries, retailing, leisure and tourism.

The success of the regeneration in Barcelona has resulted in gentrification and the housing for low-income households displaced out of the urban core.

The city of Glasgow's strategy for regeneration has incorporated developing visitor attractions, regular festivals and sports events to attract tourists.

In Glasgow city centre, the main shopping area was redesigned to maximise the historic streetscapes and the interface with modern malls.

Empty warehouses and other redundant public buildings in Glasgow were converted into flats.

Physical regeneration to the inner-city area of the east of Glasgow began in earnest with the GEAR project, from 1977 to 1986. The project's spending was predominantly on infrastructure or physical development from housing and environmental improvement through to social infrastructure.

A major failing of the GEAR project was that the new local jobs went mainly to outsiders.

The subsequent Clyde Gateway URC in the east end of Glasgow began in 2006 in what was still a very deprived area. It hoped to harness the 2014 Commonwealth Games and the building of a new motorway through the area as a stimulus for regeneration.

Many of the Clyde Gateway's objectives are similar to those of GEAR, and housing-led regeneration is still a central policy. Unlike GEAR, the initiative has offered training courses to local residents.

The progress of the Clyde Gateway project has been slowed by a weak macroeconomy, and there has been no substantial revival of the local economy.

Bibliography

Degen M and Garcia M (2012) The transformation of the 'Barcelona Model': An analysis of culture, urban regeneration and governance, *International Journal of Urban and Regional Research*, 36, 5, 1022–1038.

National Audit Office (2007) *How European Cities Achieve Renaissance*, NAO, London.

Wannop U (1990) The Glasgow Eastern Area Renewal (GEAR) project, *Town Planning Review*, 61, 455–474.

What Works Scotland (2018) *Clyde Gateway Urban Regeneration Company Case Study*, University of Glasgow, Glasgow.

21 Enterprise Zones
Real estate-led regeneration

Objectives

This chapter is the first of three that consider issues raised by real estate-led regeneration, and focuses on enterprise zones (EZs). To recap, this title of real estate-led regeneration relates to the objectives of the policies to directly intervene in the market. Their purpose is to shape the market to support and attract private property development and investment into an area.

The key aim here is the direct intervention in the market. This is because, as Chapter 20 shows, urban regeneration policies can bring new land uses without reference to real estate markets. Private investment and development is a secondary or indirect consequence.

EZs were one of the headline urban regeneration policies of the UK Conservative Government of the 1980s, with the first phase established in 1981. At the time, they were seen as a controversial attack on planning and an experiment in free enterprise. In this chapter, we look at how they worked, what they achieved, and the underlying questions raised about their effectiveness.

The chapter begins by setting the development of EZs in an international context before detailing the precise measures of those in the UK. The next section focuses on the nature of the zones, choice of locations and the evolution of the policy until the designation of the last zone in 1996. The chapter then evaluates the policy beginning with its goals but also assessing the nature of its impact and its cost. This leads into a discussion of who actually benefitted from the policy. The final part of the chapter considers the legacy of EZs and their recent revival in the UK.

The structure of the chapter is as follows:

- International context
- Details of UK EZs
- Evaluation
- Limiting factors
- Legacy
- Revival of EZs

International context

The concept of an EZ has its modern-day origins in the free trade zone or export processing zone that was first established in 1959 at Shannon Airport in the Republic of

DOI: 10.1201/9781003027515-24

Ireland. This was followed by zones in Taiwan and South Korea in the mid-1960s and early 1970s, and a subsequent proliferation around the world. In the late 1970s and early 1980s, China evolved the concept into special economic zones. Similar ones were established in the former soviet republics of east/central Europe in the late 1980s/early 1990s.

These zones generally covered large areas and have a range of objectives. All normally provided a host of financial incentives to foreign and domestic manufacturing and commercial firms. Some offered cheap land and factory buildings for lease, plus tax reductions and holidays. For foreign firms, there could be unrestricted remittance of profits, together with duty-free entry of materials and machinery for export production.

Free trade zones (or special economic zones) continue to be established around the world and have increased in popularity since the 1990s. Within the European Union, zones are to be found in countries such as Hungary, Latvia, Lithuania, Poland, Portugal and Italy. The incentives and types of locations vary. In some cases, there are exemptions from property taxes.

Lithuania's six Free Economic Zones are located in the country's main cities except the capital, Vilnius. They provide ready-to-build industrial sites with physical and other infrastructure, support services and tax incentives. The main tax incentives are zero tax on business profits for the first ten years and no tax on dividends for international firms.

In the Middle East, Dubai is a prolific applier of such zones. Since 2000, there have been more than 30 free zone 'cities' established in Dubai beginning with Dubai Internet City and including Dubai International Finance Center (near the CBD) in 2004. Each of these zones is focused on an industrial theme. They offer a range of financial incentives to international companies including 100% exemption from personal income tax for 50 years, 100% exemption from corporate taxes for 50 years and 100% repatriation of profits.

These free trade zones are very different animals to the UK EZs initially designated in the 1980s. The UK model focused on urban regeneration, and combining real estate tax incentives and reducing planning constraints. As Chapter 18 notes, the UK EZ concept stimulated the establishment of EZs in the United States mainly in the form of tax incentives to resuscitate poor neighbourhoods. Different states apply different approaches in terms of size of zone and the incentives available. Tax benefits typically extend beyond real estate incentives to directly stimulate new employment or the employment of local residents (as in Empowerment Zones).

There were real estate tax incentives applied in renewal areas of Ireland from 1986. These incentives covered parts of Dublin as well as other major urban areas. Within designated areas, incentives were aimed at developers, investors and occupiers, and lasted with some amendment through to 2008. The original package of incentives including similar measures to those in UK EZs is discussed below. The package also included a double rent tax allowance, which occupiers could set-rent off against trading income for new leases on commercial buildings for ten years. There were also tax incentives aimed at residential investors unlike in the UK EZs.

In 2015, Ireland reintroduced real estate tax incentives with the 'Living City Initiative'. It gives tax relief for refurbishing or converting existing properties within a Special Regeneration Area. These are areas within Dublin and other provincial cities. It applies to residential and commercial premises and is due to run until 2022.

Between 1997 and 2014, France designated 100 EZs in distressed urban areas. Incentives applied only to firms with less than 50 employees. The real estate incentives took the primary form of exemption for five years from local business rates, corporate income taxes and property taxes. There were also potential employers' exemptions from tax contributions to social security dependent on the location of employees.

These international examples show the range of area-based tax incentives that can be applied. Many of the zones are aimed at inward foreign investment, and the incentives are fashioned accordingly. EZs in the UK were the first to apply the approach to urban regeneration, and other countries have followed. However, there is no standardised set of incentives applied and the characteristics and size of zones vary across states/countries. The next section looks in detail at the nature of these zones in the UK.

Details of UK EZs

EZs were first established in 1981 on primarily derelict sites needing regeneration. There were 33 zones designated in the period up to October 1996. There were 11 EZs announced in the first round during 1981/1982, followed by a second round of 13 in 1983/1984. Another nine ad hoc zones were periodically designated between 1987 and 1996. The zones were relatively small areas with the largest 454 hectares, and most a third of that.

The spatial pattern of zones is given in Figure 21.1. Their locations have the common theme that they are mainly sites of a major manufacturing industrial plant closure. Most of these are in the major city regions but their outliers such as Corby, a free-standing small town where a steelworks closed down. Figure 21.1 also shows that these locations were primarily in the north of Britain with only two in the south-east of England. The map in a sense represents a broad geography of deindustrialisation in the UK, certainly at the end of the 1970s.

Each EZ was not one defined area, but usually a collection of sub-zones or individual sites spread around the local community. A zone would typically contain the site of the major plant closure and other nearby derelict areas. In some cases, greenfield sites on the edge of the built-up area were included in the zone designation. Zones were generally devoid of a residential population, and boundaries were drawn to exclude existing businesses wherever possible.

Each zone was designated to last for ten years and was eligible for the following main measures:

- 100% allowances for corporation (business profits) and income tax purposes for capital expenditure on industrial and commercial buildings,
- Exemption from rates (occupation tax) on industrial and commercial buildings,
- A simplified planning scheme for each zone that gave automatic right of development for specified land uses.

Firms in EZs were also exempt from industrial training levies and some bureaucratic form filling.

The 100% tax allowances mean that developers and investors can set the whole cost of the construction of a building against their annual tax bill (provided their tax bill is sufficient). In turn, it means that if their tax rate is 45%, then the actual cost of construction to investors is only 55% of the nominal cost. The higher the tax rate, the

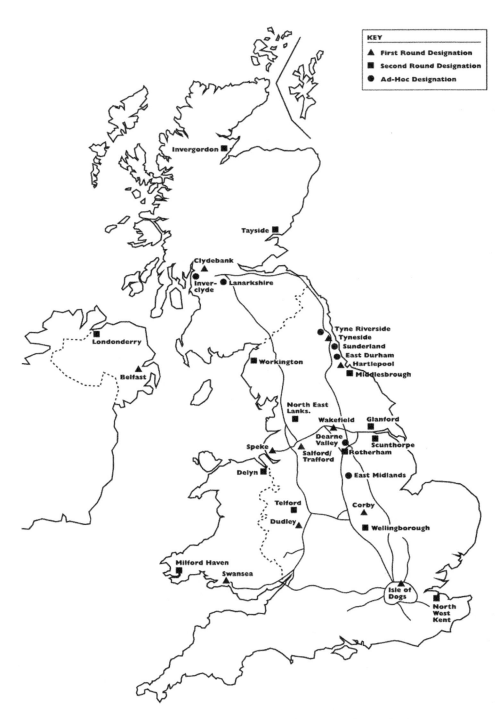

Figure 21.1 The Geography of Enterprise Zones in the UK Designated 1981–1996.
Source: Jones (2006).

lower the cost. If the funds are available, then the low development cost may outweigh the perceived risks of an EZ location.

These tax allowances were a way of promoting new building supply in an EZ. The simplified planning element was aimed at also making it easier to build by removing planning constraints. For each zone, there is a list of acceptable uses and types of development. A development could be built if it is simply on the list. The exemption from rates for ten years related to property tax that is paid by occupiers in the UK. It was therefore a way of attracting occupier demand to a zone. The measures together supported both the supply of and demand for properties on zones.

The stated aim of the policy was to stimulate free enterprise as a solution to blighted/derelict urban areas. A sub-plot was the promotion of new firms. The policy was framed as a way to address market failure, in that development would not otherwise take place on these derelict sites. The ten years of life of each EZ was seen as the period necessary to demonstrate to the private sector the potential of investing in the location. Many ad hoc zones designated after 1987 also had an explicit policy to attract inward investors and were located on greenfield sites.

Evaluation

In this section, the EZ programme is evaluated by reference to its stated goals. The first is the promotion of free enterprise. In many ways, this was the most high-profile reason for establishing EZs. A subsidiary linked goal was the stimulation of new firm formation on EZs. In the later ad hoc zones, the policy emphasis switched specifically to attracting inward investment. Finally, EZs were to act as demonstration projects to address market failure in areas, disadvantaged by major industrial closure.

Promoting free enterprise

To achieve this goal, EZs were strong on financial incentives. On the other hand, they were weak on the reduction of regulations, limited primarily to a relaxed planning regime. The subtext to the EZ policy of freeing private industry from government constraints referred to in Chapter 18 is very much a minor role.

The key attraction of EZs was the tax benefits that can be seen as a subsidy for business and development in these localities. From this perspective, EZs are arguably a form of spatial economic planning despite the rhetoric of Conservative Government policy. For this reason, many opposition-controlled Labour local authorities queued up to bid for a zone.

In addition, a number of EZs were actively promoted by local economic development agencies as places to set up in business. These agencies often arranged for public expenditure on land assembly, site preparation and the building of infrastructure such as feeder roads to open up the zones for transport. The result of these substantial public sector expenditures plus the taxation incentives is that the net costs per job generated were very high. For this reason, in December 1987 the government announced that further EZs would only be designated in exceptional circumstances.

Incubators of new firms

The promotion of private enterprise implicitly included EZs acting as incubators for new firms by providing modern cheap premises. The de-designation surveys after ten years of firms in the zones of the first two rounds found that 29% of firms were new starts. Most of these firms employed less than ten employees.

These statistics are consistent with the incubator role. However, the same surveys found that 38% of firms were relocations, while 23% were branches or subsidiaries of large companies. Further, and perhaps surprisingly, 10% were already in an EZ prior to designation. Even so, since 96% of firms employed less than 100 employees, it can be argued that these EZs in the first two rounds were true to the incubator ideals.

Vehicles for attracting inward investment

Originally, EZs were seen as attracting new firms, but as the statistics in the previous paragraph note, there were a high proportion of established firms that moved into them. For the ad hoc zones, the government reappraised the role of EZs. Ad hoc zones had an explicit policy to attract inward investors, and many of them included more greenfield sites to reflect this goal. Even so, in 1999 a survey found that just under half of all establishments on the ad hoc zones employed less than ten people. Overall, there was only a marginal shift to larger firms −7% of firms on ad hoc zones employing 100 people.

Resolving market failure

Although the government saw EZs as addressing market failure, its monitoring did not assess this goal. Instead, government-sponsored research primarily focused on the employment generated and types of firms attracted. Research into resolving market failure requires an assessment of local property markets. It can be tested by the following criterion: after ten years of life of an EZ, the local demonstration effect should ensure unsubsidised development is viable.

Two studies have looked at this issue. One published in 2004 observed that property markets in some of the zones studied were likely to collapse following the end of incentives. Another was based on zones in central Scotland. It included Clydebank, a town on the western edge of the Glasgow city region (see Figure 21.1) and once a world centre of shipbuilding. This study found that

> *...it would appear that EZs on a ten year time horizon (and on a spatial scale applied in the UK) that instigate a market demonstration effect based on public subsidies/ tax incentives will not ultimately generate sustainable local industrial property markets in areas fundamentally drained by spatial economic change.*
>
> (Jones et al, 2003)

There is therefore a question mark over whether EZs, as constituted in the UK, can turn around the real estate markets and economies of urban areas suffering from severe industrial decline. The key point comes once the life of the EZ is over with the loss of incentives for new investment. There needs to be a critical spatial mass of firms to build future growth on.

Areas subject to deindustrialisation, combined with a location that is not easily accessible to the modern motorway network, face substantial barriers to attracting new investment. This leads to a question about where EZs should be located. There is a case that EZs should be selected to accelerate and control urban change, not stem decline. In other words, EZs should be located on greenfield sites that can be described as potential 'winners'.

Limiting factors

The discussion above shows that the UK EZs devised in the early 1980s at best only partially achieved their goals. Many EZs did not rely entirely on tax incentives to attract firms, and they were enhanced by local physical infrastructure improvements. There were also fundamental queries about the long-term impacts on the local economy, at least for those EZs in the most disadvantaged locations. The lack of success was derived partly from the limitations inherent in the nature of EZs. These are now considered.

Displacement

There are a number of limitations to the efficacy of EZs that stem from market forces. A major problem is jobs transferring from elsewhere attracted by the incentives. It was noted above that in de-designation surveys, 38% of firms on EZs were relocations, while 23% were branches or subsidiaries of large companies. The problem is especially important when firms were moving short distances on to EZs – it became known as 'zone hopping'. In these cases, there is arguably no or little increase in economic activity as a result of the EZ. It is just being moved around.

The most high-profile example of zone hopping is firms moving from the traditional City of London financial centre to new office developments in the Isle of Dogs EZ (see Figure 21.1). This former dockland area centring on Canary Wharf area is now arguably larger than the City as many banks have moved their headquarters to this new location. While this is a clear case of displacement, it is also true, as discussed in Chapter 19, that the creation of an office centre in the docklands has reduced the space constraints on the financial sector in London.

There is a more subtle influence in the form of displacement when firms on an EZ take business from neighbouring firms who are in the same industry. This essentially relates to local firms offering the same services. Local services therefore provided by new firms on an EZ can displace similar existing firms off the zone, all competing in the local economy.

The most obvious example is retailing. If retailers are allowed to locate on EZs, then they can draw expenditure by customers from competing shopping centres. There were only a few UK EZs that had retail uses in their list of permissible developments. Notable exceptions were the out-of-town shopping centres of the Metrocentre on the Gateshead EZ and the Merry Hill Centre on the Dudley EZ (see Figure 21.1). These out-of-town centres attract car-borne shoppers from a wide catchment area resulting in extensive displacement effects. This in turn switched employment in shops from nearby centres.

Given these displacement effects, a true evaluation of the employment generated by EZs must be on a net rather than a gross basis. This is not necessarily easy and it depends on the nature of employment attracted to the zone. The analysis also depends on the boundary assigned for where new jobs on a zone are attracted from. Employment moving within a city region is clearly no net gain, but if it is transferring from a prosperous region to a poor one, then it could be viewed differently. One study estimated that only 46% of jobs established on EZs were new.

This issue of displacement does not just apply to UK EZs but also applies to any policies that offer differential fiscal incentives to locate in a zone. It therefore applies, for example, to Empowerment Zones in the United States. Similarly, local authorities with differential tax rates can generate displacement economic activity. Area-based policies may be a zero-sum game in terms of overall economic activity.

Incentives and the market

A key question for the success of EZs is 'Are incentives bid away via the real estate market?' Logically, incentives to locate in an area should push up rents/increase capital values. Tenants may be prepared to pay higher rents as there are no property taxes for occupiers for ten years. As discussed in Chapter 4, tenants calculate how much they can afford to pay in rent at a given location based on expected revenue generated and costs. The removal of property taxes increases the rent they can afford to pay.

There are a number of valuation problems in assessing this effect because often the only new properties in an area are on the zone. Nevertheless, differentials should exist between on- and off-zone properties. Logically, these differentials should narrow/disappear towards the end of the ten years of life of the zone as the benefit is phased out. Research has found that there is evidence of such differences, but it is not definitive partly because the real estate market is not efficient. The differential is also dependent on the degree of competition for units in a zone.

The result is that occupiers may lose all or some of the benefits of the ten years of property tax holiday. In that sense, the financial benefits of locating on an EZ may be illusory. However, the modern premises available on an EZ, compared to elsewhere in the area, may still be a sufficient attraction.

The benefits to investors too were not as straightforward as they seem. Investors received the capital allowances to support the construction but still need future rent revenues to justify the investment. Often developers stimulated investment by providing rental guarantees. But there can be uncertainties about the long-term future rental income. As noted above, once the EZ status ends, in some cases there are doubts about continuing demand for premises. Given this effect of market processes, the main financial beneficiaries are potentially not the occupiers or investors, but arguably private landowners/developers.

Legacy

The original UK EZ model emphasised the promotion of free enterprise, the relaxation of planning and localised tax incentives. The immediate policy transfer was in demonstrating the use of tax incentives at the local level, an approach followed by other countries. In the UK, the introduction of Enterprise Areas (EAs) in 2005 with their use of tax allowances can be directly linked back to the EZ model. However, as Chapter 18 explains, tax incentives in EAs were applied to refurbishing existing buildings, not new development.

The UK Government billed the introduction of EZs as an attack on planning by removing such constraints. In practice, the application of EZs was more nuanced. As noted earlier, many individual EZs benefitted from the support of a local economic development agency, underpinned with public infrastructure. Right from the start, then some EZs were integrated by a public agency into a local economic planning strategy. By the arrival of the later ad hoc zones, all had master plans developed by a public agency.

At the same time, UK planning was changing, switching from a passive model to a more positive paradigm. EZs as operated fitted into the new more positive approach to spatial planning that intervened in the land development process. The use of tax incentives and grants now naturally was seen as part of area initiatives. In other words, while in 1981 the rhetoric of EZs and planning were poles apart, over the decades they have been integrated under the banner of local economic development strategies.

Revival of UK EZs

EZs were revived in a UK policy announcement in 2011. From 2012, there were 24 EZs established in England from April 2012 with parallel initiatives in the rest of the UK. There are many similarities with the original EZs. Businesses that move into an EZ before April 2018 are entitled to a property tax (rates) discount of up to 100% over a five-year period (later eight years) rather than the original ten years. There is also an individual cap for a firm on this benefit. Simplified planning regulations are again part of the package, but this time subject to local approval. Zones still have little or no existing businesses on them at designation.

Notable differences exist. The 100% capital tax allowances are only available for large plant and machinery and not buildings, and in only the eight zones in defined deprived areas. There is a subtle difference this time in the overall logic of the programme. The stated focus is now on reducing the burdens on the private sector so that it could deliver economic growth and job creation. This translates to choosing zones based on economic potential or opportunity.

Unlike the originals, these EZs are now clearly linked to planning and a LEP local economic development agency (Local Economic Partnership) and its strategy for its area. Many EZs are linked to promoting firms in a particular industrial cluster. A LEP also has a crucial role in managing its EZ including promoting it as a place for a firm to locate. As part of its management role, a LEP is also able to bid for capital funds to improve the surrounding infrastructure of the EZ.

These new zones have some similarities with the original ones. The essential attraction of the zones is still real estate tax benefits. They are small areas around the country with some zones comprising a range of local sites. Some are offering retail opportunities. As with the original EZs, some are located in areas where factories have closed and in areas that need to be regenerated, such as docklands.

Most of the zones are very different in terms of the type of location and goals. Unlike the original EZs, most are in the southern part of England, and are to found on the edge of small towns. Many locations have good accessibility to the road/motorway network and/or an airport. In other words, the locations of the revived EZs are what might be described as 'winners' rather than casualties of urban decline.

The locations are also chosen to enhance growth potential following the logic of improving competitiveness by successful clusters, as discussed in Chapter 19. Most EZs are matched to a particular high-tech theme. There are invariably firms in that sector close by, and even adjacent to the sites of the EZs. In promotional material, firms in the same industrial sector are urged to co-locate to benefit from synergies. Some of the zones have acquired university campuses or research institutes.

Despite the new emphasis on 'winners', a study of the first 24 of these zones found that gross job creation between 2012 and 2017 was limited. Some of these jobs were in the public sector rather than the private. The problem of displacement is still a real issue. At least one-third of the jobs resulted from businesses that had moved from elsewhere, half of these from within the local area. Other employment on the zones linked to local services would have increased the scale of displacement.

The reasons for the relative lack of success can partly be attributed to the slow macroeconomic growth context. The new EZ model has diluted incentives – the business rates holiday applies only for a shorter period than the original model. Whereas the original model had incentives to construct new buildings and to occupy the completed buildings, the latest EZs only have the latter incentives, and for a shorter period.

Overall, the new EZs are no longer urban regeneration vehicles but aimed at local economic development. In the main, the new zones are located in very different types of areas, aiming to attract high-tech and logistics industries on to new premises in, for example, business/science parks. However, like the original model, displacement is a major detractor with many of the businesses transferring from elsewhere.

While EZs in the UK may no longer be centring on regeneration in 2021 the government announced the establishment of free ports that partly have this role. At the time of writing, the exact financial benefits of location in a free port are yet to be spelt out. They are promised to include real estate tax reliefs on purchasing land and constructing or renovating buildings. The new policy potentially represents a partial return to the original EZ model.

Summary

EZs developed from free trade zones or export processing zones that provided financial incentives including duty-free entry of materials and machinery for export production. UK EZs initially designated in the 1980s applied real estate tax incentives and the reduction of planning constraints to support urban regeneration. The UK EZ model was translated to the United States where many states experimented with variations. Other countries such as Ireland and France have applied area-based real estate tax incentives too.

In the UK, 33 EZs were designated between 1981 and 1996 with the aim of regenerating primarily derelict sites linked to a major manufacturing industrial plant closure. Each EZ was normally composed of sub-zones or individual sites distributed around the local urban area, including in some cases greenfield sites. Zones contained no residential population and usually also no existing businesses.

The life of a zone was for ten years and offered tax incentives. Over the life of the zone, there were tax incentives to reduce the cost of industrial and commercial buildings and exemption for occupiers from paying property tax. A simplified planning scheme was applied in each zone to make development quicker and easier. These combined measures promoted both the supply of and demand for properties on zones. The ten years of life was deemed a sufficient period to demonstrate to the private sector the attractiveness of investing in the location.

The main attractions of EZs to businesses were the tax benefits rather than the relaxation of government regulations. EZs were also supported by public agencies and new infrastructure. Together with the tax incentives and significant public sector expenditures, the net costs per job generated were very high.

EZs acted as incubators for new firms by providing modern cheap premises. The later ad hoc zones also had an explicit policy to attract inward investors. As a result, many of them incorporated more greenfield sites in their offering. However, the result was only a minimal shift to larger firms.

Two studies have examined whether after ten years of life of an EZ the local demonstration effect ensured unsubsidised development was viable. Both studies concluded that property markets in some of the zones in areas of industrial decline had not generated a sufficient step change in attractiveness before the incentives ended.

A particular problem about the EZ model is displacement. This can take a number of forms including firms moving on to a zone to take advantage of the perceived financial benefits. Displacement can also occur if firms offering local services located on a zone take business from neighbouring firms who are in the same industry. One study estimated that more than half of employment established on EZs had been displaced from elsewhere.

The incentives that attract firms to EZs may be illusory. Logically, the existence of tax incentives is ultimately embedded in higher rental and capital values. In practice, the differentials are clouded by the different types of properties on and off zones, the inefficiency of real estate markets and the extent of competition for units in a zone. More expensive modern premises on a zone may still be more attractive than old units off a zone.

EZs were successful in demonstrating the attractiveness of applying tax incentives to small areas. In a sense, EZs were an international demonstration project. However, right from the beginning some EZs rather than being beacons of free enterprise were integrated into local economic development policies. Ultimately, with ad hoc zones EZs became closely aligned with spatial planning.

The establishment of UK EZs was revived in 2011, and in 2021 there are 44 zones in England. The main similarity with the original EZs is that businesses are entitled to a property tax holiday, although it now applies only for a fixed number of years from the date of occupation. There are also simplified planning regulations applicable.

There are also significant differences. The capital tax allowances are generally unavailable except in deprived areas and then only for large plant and machinery and not buildings. The logic of the programme has also switched to delivering local economic growth potential. Many EZs are linked to firms in a particular industrial cluster. The new EZs are embedded in planning and managed by a local economic development agency (LEP) for their area.

Real estate tax benefits are still the core financial attraction of the revived zones. While some are designated in areas where there have been industrial plant closures or in areas of cities to be regenerated, most are found in small towns. Many of the locations of these latter EZs are situated near the inter-urban road network and can be portrayed as well placed to benefit from the changing patterns of urban growth.

Most EZs aim to foster clusters of high-tech economic activity by linking to nearby existing firms in a particular industrial sector. The success to date has been limited by displacement, the relatively weak macroeconomy and the low level of incentives compared to the original model.

Learning outcomes

EZs developed from free trade zones.

The UK EZ model of the 1980s applied real estate tax incentives with a relaxation of planning constraints to support urban regeneration.

There were 33 EZs designated between 1981 and 1996 to regenerate small areas of primarily derelict sites linked to a major manufacturing industrial plant closure.

Each zone lasted for ten years. For these ten years, tax incentives reduced the cost of industrial and commercial buildings, occupiers paid no property tax, and planning was relaxed.

A decade was seen as sufficient time to demonstrate to the private sector the attractiveness of investing in the area.

EZs were also supported by public expenditure to provide new infrastructure such as access roads. The net costs per job generated was very high.

A function of EZs was to be incubators for new firms by providing modern cheap premises.

Real estate markets in some of the zones in areas of industrial decline were not sustainable once the incentives ended.

The net impact of EZs suffers from the displacement of economic activity from elsewhere.

Tax incentives on EZs may be embedded in higher rental and capital values. Nevertheless, paying higher rents for more expensive modern premises may still be more attractive than old units off a zone.

EZs in practice were not promoters of free enterprise as most were integrated into local economic development policies.

The revival of UK EZs in 2011 was based on diluted tax incentives, just a property tax holiday, for five then eight years from the date of occupation.

The new EZ programme now aims to promote local economic growth potential, and each zone is managed by a local economic development agency.

The primary locations of the new zones are not areas needing regeneration but greenfield sites on the edge of small towns with good road connections.

Most EZs aim to encourage clusters of local high-tech economic activity by linking to existing nearby firms in a particular industrial sector.

Bibliography

Jones C (2006) Verdict on British enterprise zones, *International Planning Studies*, 11, 2, 109–123.
Jones C, Dunse N and Martin D (2003) The property market impact of British enterprise zones, *Journal of Property Research*, 20, 4, 343–369.
Ward M (2020) Enterprise Zones, Briefing Paper No 5942, House of Commons Library, London.

22 Urban Development Corporations

A costless solution?

Objectives

Urban Development Corporations (UDCs) are specialist public agencies that usually work outside the direct responsibility of local authorities. The case for UDCs is that such a specialist agency with the requisite powers can undertake its task efficiently and without delays caused by local planning processes. It has also been seen as a way that central or a state government can impose its development plans on a local community.

The chapter is about the urban regeneration processes undertaken by UK UDCs and their outcomes. Many of the methods – retail, tourism and housing-led regeneration – were discussed in Chapter 20. The chapter therefore focuses on the types of areas covered by UDCs, their powers and time span, and their activities and goals. It examines their historical development and criticisms of the approach: from physical vs social regeneration to issues raised by timescales and the role of the macroeconomy. It begins by setting them in an international context.

The structure of the chapter is as follows:

- International context
- Details of UK UDCs
- Review of first-generation UDCs
- Review of second-generation UDCs
- Review of third-generation UDCs
- Dependency on macroeconomy
- Regeneration of places or people

International context

The use of the term UDC has a number of different interpretations across the world. Perhaps the first reference to the term is found in the United States in the 1960s. At that time, UDCs administered central government grant programmes linked to anti-poverty social programmes in defined urban areas. The nearest American forerunner to the UDC's model in the UK was the New York State UDC that has operated from 1968, but is now under the banner of the Empire State Development Corporation.

The UDC has planning and compulsory purchase powers to sponsor private development in inner-city areas, and in some cases offer financial support. One of its most profile projects was the regeneration of the 42nd Street area of New York including Broadway and Times Square. The agency developed a plan in 1984 that included four

DOI: 10.1201/9781003027515-25

new high-rise office towers, a retail centre, a new hotel, refurbishment of Times Square subway station and the restoration of the theatres. The grand scheme did not come to fruition, but it is indicative of the work of the UDC.

UDCs around the world are often state agencies that undertake development in the public interest. In Malaysia, for example, the Sabah UDC was formed in 1972 as the main agency of the Malaysian government responsible for real estate development. Its development activities include residential suburbs, commercial buildings, industrial estates and tourist resorts.

Caribbean governments have addressed the challenges of urbanisation through the formation of UDCs with powers to facilitate regeneration in specific areas. UDCs can be found in Jamaica, Antigua and Trinidad and Tobago. Jamaica's UDC was created in 1968 with the mandate to improve the urban fabric of metropolitan regions across the island. The UDC was conceived as a developer in the public interest, combining public resources with the expertise and dynamism of the private sector.

The UDC has planning powers in designated areas. The UDC acts as a developer in its own right with construction, for example, along Kingston's waterfront including new offices, hotels and conference facilities. A recent high-profile development is Harmony Beach Park in Montego Bay. It also manages financial incentives for private investors to support regeneration of distressed areas.

These examples are not exhaustive, but they illustrate the role of UDCs across the world. In all cases, they have planning powers often to develop master plans for neighbourhoods. Some can act as direct developers in the public interest to support economic development. There are clearly variants in their powers and activities. In the UK, the UDC model is designed to foster private development in urban regeneration areas, and this is the focus of the rest of the chapter.

Details of UK UDCs

The overarching role of UK UDCs is to free up the local land market by removing supply constraints in closely defined parts of distressed areas. Essentially, their role is to address market failure to bring private investment into areas of dereliction. The logic is that without public intervention, areas of urban wasteland represent too much of a risky challenge for private investment. Even the development viability of plots of land adjacent to abandoned individual buildings could be scuppered by the value of the surrounding real estate.

The specific goals of UDCs within their designated areas can be divided into the following:

- To assemble sites, and reclaim and service large areas of derelict land to bring it into effective use,
- To revive private sector confidence by the provision of infrastructure and improvement of the environment,
- To provide land and an attractive environment for private sector development,
- To give financial assistance to developers where necessary,
- To ensure that housing and social facilities were available to encourage people to live and work in the area.

Each UDC was able to develop its own strategy to address the local circumstances.

To do so, they were equipped with wide-ranging statutory powers. Inside the boundaries of their designated areas, they had powers to:

- Compulsorily purchase land,
- Buy land to assemble sites,
- Reclaim and service areas of derelict and contaminated land,
- Build infrastructure and improve the physical environment including the refurbishment of existing building,
- Provide land for private sector development,
- Give financial assistance to developers where necessary,
- Provide health, training, educational and community facilities and
- Plan the local area.

Planning powers were seen as supporting confidence in the area and ensuring an efficient development process without a role for local democracy to slow decisions. UDCs were governed by boards of directors appointed by central government to whom they also reported. Funding was also derived primarily from central government, but there was the potential to raise additional resources through capital receipts generated from land and real estate sales.

Most of their budgets were spent on land development via purchase, site assembly and preparation. There were also significant amounts spent on marketing, promotion and advertising as part of improving the image of their places. Developing promotional campaigns and publicising development opportunities was part of the UDCs' remit to attract private investment. Only a very small percentage was spent on community projects and social infrastructure.

There were four generations or waves of UDCs designated during the 1980s and early 1990s with the details shown in Table 22.1. The first UDCs, the London Docklands Development Corporation (LDDC) and the Merseyside Development Corporation (MDC), were set up in 1981 to address the regeneration of derelict docklands. While superficially the same task, the LDDC covered a much bigger area and included a large population. In many ways too, the location of the LDDC in the national capital was much more attractive to private investment than that of MDC in the northern provincial city of Liverpool. The time spans for these UDCs were 17 years.

Some six years later, a second generation of five were initiated in 1987 – the Black Country, Teesside, Trafford Park, Tyne and Wear and Cardiff Bay – in England and Wales. Most of these UDCs covered large swathes of deindustrialisation and contamination in northern cities. The exception was that in Cardiff, that continued the theme of redundant dockland, and this was the only UDC that did not have planning powers. These UDCs were permitted 11 years to complete their tasks.

This was quickly followed by a third generation of 'mini-UDCs' in 1988 in the provincial cities of Leeds, Central Manchester, Bristol and Sheffield. These third-generation UDCs are much smaller, with all but one less than 500 hectares. By this generation, the regeneration timescale was reduced, and was ultimately to be six to eight years. This partly reflected the size and type of area that the UDC was charged with regenerating.

Finally, there were the two more ad hoc designations in 1992 of Birmingham Heartlands and Plymouth. These designated areas did not have the common theme of previous generations and can be seen as an afterthought. The Plymouth area is very small, and was announced after the Royal Navy closed its docks there. Birmingham

Table 22.1 Characteristics of UDC Areas in England and Wales designated 1981–1992

Name	Date of Designation	Size Acres	Original Use Characteristics of Area	Population at Designation
London	1981	5313	Derelict dockland	40,400
Merseyside	1981	865	Derelict dockland	450
Cardiff Bay	1987	2701	Derelict dockland	5,000
Black Country	1987	6420	Former manufacturing	35,405
Teesside	1987	12004	Former steel/chemical sites	400
Trafford Park	1987	3131	Rundown industrial estate	40
Tyne and Wear	1987	5869	Former shipyards and industrial land	4,500
Bristol	1988	1038	Off city centre	1,000
Manchester	1988	462	Off city centre	500
Leeds	1988	1334	Off city centre	800
Sheffield	1988	2224	Off city centre	300
Birmingham	1992	2471	Former manufacturing	12,500
Plymouth	1992	173	Former dockland	–

Heartlands is a former manufacturing area with similarities to the second-generation areas. It was added after some years of lobbying by the city council.

The life of the fourth-generation UDCs was only six years, on a par with the third generation. All UDCs in England were dissolved by 31 March 1998. Cardiff Bay in Wales continued for another two years. The de-designation of UDCs did not mean that their tasks were complete, but simply that they had been transferred on to a new-regeneration vehicle (see below).

The UDC's approach to regeneration was continued by Urban Regeneration Companies (URCs) as outlined in Chapter 18. In the 2010s, there was a revival of UDCs in the form of Mayoral Development Corporations to regenerate areas. The activities of these agencies mirror UDCs as they cover small areas of cities and have planning and compulsory purchase powers. A key difference is that they are funded by a city mayor. As Chapter 18 notes, these mayoral UDCs are a cross between a URC and the original UDC model.

While the funding of mayoral UDCs is different from the UDCs of the 1980s, their modus operandi is the same for attracting private investment. In terms of their specific area tasks, the four mayoral UDCs to date are all very different. There is one in a northern region that is regenerating the site of a former steelworks. It has a pedigree that can be traced to the UDCs of the 1980s. A major difference between the 1980s is that the policy framework is not one of market failure. There is also an emphasis on using master plans, community involvement and longer time spans.

These latest mayoral UDCs are still too early in their lives to be evaluated, so the chapter focuses on the original UDCs to assess the effectiveness of the model. It considers a case study from each of the first three generations of UDCs to assess their methods, challenges, successful outcomes and difficulties.

Review of first-generation UDCs

Although the task of both the original UDCs was to address the regeneration of docklands, they were in two very different cities. The LDDC also covered an area almost seven times that of MDC and had over 40,000 people resident within its boundaries. In contrast, only 450 people lived in the MDC's area. The chapter now looks at a case study of the LDDC experience as it was very influential on national policy and led to a debate about the role of urban regeneration.

London Docklands Development Corporation

In 1981, much of the London docklands had lain vacant for ten years. They were obsolete as the rise of modern large container ships had shifted docks to deeper water ports further down the estuary. It was an industrial wasteland of abandoned docks and derelict warehouses. Although adjacent to the international financial centre of the City of London, there was no prospect of redevelopment. It could be described as a prime example of property market failure.

LDDC was designated in 1981 to address the regeneration of 2070 hectares. In addition, part of the area, 195 hectares on the 'Isle of Dogs', at the same time had received Enterprise Zone (EZ) status as discussed in Chapter 21. This sub-area therefore benefitted from tax incentives that provided occupiers with exemption from local property tax, and developers were able to offset all of the costs against tax for up to ten years.

The first priority of the LDDC was to improve the basic infrastructure of the docklands, particularly to replace the narrow cobbled streets. A major new access road was built. It also had to improve the area's sewerage, gas and electricity capacity. The preparation of land for development was also a priority. In addition, it was important to change perceptions of the area to start to demonstrate its development potential. Part of this potential was based around its distinctive water features with its numerous docks and quays, but it was also its location near the heart of the city.

Despite its central location, the long-term scale of dereliction meant that there was no real estate market to build upon. There were also entrenched views about, for example, where offices should be, and no desire for business to move from these established locations. Land values over much of the area were initially negative with substantial costs required to be spent to prepare sites ready for development. The strategy of the LDDC was to attract new industries and housing.

To begin with, there was considerable success in attracting many newspaper printing presses to the area. The newspaper industry was introducing new technology. It was attracted to the new space available relatively close to its traditional home in the centre of London. These light industry developments were bespoke rather than speculative.

Initial speculative developments were on the EZ and took the form of business space. Developments were often mixed use, and many proposals did not come to full fruition. The uncertainty of the local real estate market led to many sites lying empty, in some cases for more than a decade. Nevertheless, the mid-1980s saw a construction boom in the docklands including significant demolitions of obsolete dock sheds and the conversion of existing buildings.

The building of a railway from east to west enabled the opening-up of the area for development. The Docklands Light Railway (DLR) was begun in 1984 and opened in 1987. It linked the area to the edge of the City of London and had a station at Canary Wharf in the middle of the EZ with its development tax incentives. It was a major step in the regeneration of the area.

New housing built for sale was a major innovation, as previously housing in the docklands area was nearly all socially owned terraced housing and flats. Other initial developments included an indoor sports arena. During these early years of the LDDC, there were small-scale developments of business space, which were let relatively quickly, but there was no discernible office market. Nevertheless, land values were rising by 1987.

The story of Canary Wharf as the centre of a major international financial district began in 1984. It was suggested that the shed that occupied it could be converted for 'back offices'. In the discussions that followed with LDDC during 1985, the proposal became a master plan for a 'mini-Manhattan'. It was a complete leap from the business park environment that hitherto had been expected for the Isle of Dogs, but was still in embryonic form.

There were still uncertainties about securing funding for such a large development in an unproven location. Eventually, a Canadian property developer committed to building the 50-storey office block, One Canada Square, the tallest building in the UK at the time. It was located in the EZ and so entitled to public subsidy towards its construction. The 20-acre site was purchased from the LDDC. Construction began in 1988. It was the transformative development that defined the LDDC.

The scale of the Canary Wharf proposals stimulated further development. By the beginning of 1988, a total of 2.7 million sq ft of development, primarily offices, was either committed or underway. Harbour Exchange office tower, comprising 482,950 sq ft, near to Canary Wharf, was an important development completed in 1989.

The recession of the late 1980s brought a sudden end to a national property boom, and the docklands did not escape the collapse of the real estate market. Canary Wharf was just one-third complete, but the recession halted work. Notwithstanding the economic downturn, the LDDC in 1991 could report on a decade of achievement with a claim that 27m sq ft of commercial and industrial development had been completed or was under construction.

Over the period, 1,500 new private houses had also been built and 600 hectares of land reclaimed. In terms of public infrastructure besides the DLR, 55 miles of new or improved roads had been built, and London City Airport had opened in the area. Over £140m had been spent on public utilities – gas, electricity, water and drainage.

The boom had created an overconfident sentiment in the property market everywhere and perhaps especially in the London docklands. Unfortunately, the onset of the recession diluted the demonstration effect of Canary Wharf and an anticipated second wave of development did not materialise. In fact, the effects of a downturn are always magnified in immature markets such as the docklands with half the offices empty in 1991 (see also Chapter 23).

The downturn was seen dramatically in the housing market when a local developer was one of the first in the UK to go into liquidation nationally, and house prices fell more than elsewhere in the country. No new houses were built for four years, and many existing owners were in negative equity.

The recession also threatened to shake the dockland's position as an emerging financial centre, not least because Canary Wharf was completed in November 1990 in the midst of the recession. The first tenants moved in August 1991, but there remained question marks over the location for financial services. Take-up of space in the building was minimal, and its developer went into administration in May 1992.

The problems facing the docklands were more than simply the property market fallout from the recession. The area's transport infrastructure could not cope with the size of its office developments, and access for commuters had become a major constraint on attracting tenants. Transport infrastructure had been seen as a major issue for the area from the beginning of the LDDC.

In the early 1990s, Canary Wharf in particular still suffered from the perception that the DLR would not be able to meet the travel demands of the workers located there. It had been designed for an area that was expected to be predominantly a mix of business space and residential. In 1991, the DLR was upgraded and extended to the heart of the City so that it was only ten minutes away. It was also ultimately upgraded by connections with the east of the docklands and to London City Airport.

Just as importantly, the area needed to be connected to the London Underground system. Tenants of Canary Wharf had been promised such a link, but it was delayed by funding problems. Eventually, the 16 km underground link began to be built in 1993, and the line was completed in in 1999 including a station at Canary Wharf.

The mid-1990s saw a return of confidence in the area. By 1995, Canary Wharf was 95% let. Major investment banks began to move into the area and establish headquarters in the latter half of the 1990s. The construction of a headquarters for Citibank began in 1997, the first office tower block in the area for nearly ten years. Total take-up of office space in 1995 was just over 1m sq ft, and this rose to 1.8m sq ft in 1997. In February 1999, Canary Wharf was 99.5% let and its working population was over 23,000.

The LDDC was wound up in 1998, and in its final annual report, it headlined its relevant achievements over its 17 years as follows:

- 25 m sq ft of office/industrial floorspace built
- 1,984 acres of derelict land reclaimed
- 24,046 new homes built
- 85,000 now at work in London docklands
- £1.86 billion in public sector investment
- £7.7 billion in private sector investment
- Docklands Light Railway
- London City Airport
- 144 km of new and improved roads

These are gross rather than net figures, and of course, some of these jobs have been 'poached' from central London.

The LDDC also oversaw a range of social infrastructure including shopping centres, a large exhibition centre and a university campus. It has also funded support for schools and vocational training centres. These educational facilities came late in the life of the LDDC as discussed below. By the final act of the LDDC, a critical mass for a new office market appeared to have been established, but it had taken most of the 17 years of its life. More than 20 years on the docklands is a well-established office centre.

The main lesson from the LDDC experience is the importance of transport infrastructure as the basis for transforming or regenerating an area ravaged by industrial decline to a major office centre. Some 44% of LDDC expenditure was associated with transport and access improvements.

Review of second-generation UDCs

The announcement of the second generation of UDCs was made by Mrs Thatcher, the prime minister, at a visit to a site on Teesside in 1987. The announcement was set against a backdrop of a derelict iron foundry to demonstrate the scale of the regeneration required. It was described in the popular newspapers as the 'Walk in the (urban) Wilderness'. The physical image of urban desolation captured by the announcement also demonstrated the scale of the economic regeneration required by these second-generation UDCs. Here, the chapter considers a case study of one, through the experience of Teesside Development Corporation (TDC).

Teesside Development Corporation

The task of the TDC was to regenerate former industrial land on both sides of the River Tees in the north-east of England. It bordered on the towns of Hartlepool, Stockton and Middlesbrough. The location can be characterised as the epitome of a neglected deindustrialised urban area in a peripheral region. Given the location, it was also the extreme challenge to attract private investment.

The core of TDC's activities focused on flagship regeneration policies on large individual derelict sites as symbols of physical change. Not all were successful. Its first major project was Teesside Park on the site of a disused racecourse with good access to nearby major roads. The first developments on the site were retail warehouses and a supermarket. It is now a fully fledged out-of-town shopping centre with a range of shops and restaurants, plus a cinema and bowling alley.

Many of its other initial efforts were in promoting locations for industrial units, managed workspace and the Tees Offshore base. This facility offers services for North Sea oil and gas activities including heavy lifting facilities.

The next successful flagship policy was Hartlepool Renaissance on the site of the town's south docks. A marina with 500 berths, bars, restaurants and 1,500 housing units were built on former dockland. To support the development, TDC removed and reclaimed contaminated land, refurbished marine walls and piers. There was a construction of a new bridge and road infrastructure to connect the site to Hartlepool.

The TDC also refurbished a dry dock that subsequently accommodated a historic yacht as part of a tourist attraction. The surrounding area has been rebuilt as a seventeenth-century maritime township with period shops and houses. It is now a visitor attraction called Hartlepool's Maritime Experience.

The final flagship policy was 'Teesdale' on the site of a former large engineering works. It was originally an area of 250 acres across the river from Stockton town centre. It needed new roads and bridges to open up the site to the town, including the Teesquay Millennium pedestrian bridge. A barrage was also built to control the tides and improve the visual appearance.

The site was ultimately developed as a mixed-use area. It encompassed a business park and a university campus together with some housing for sale and social rent. Residential nursing homes were also located in the area.

In the final analysis, the TDC achieved the following statistics:

* 1,295 acres of reclaimed land
* 1,4m sq ft of commercial floorspace
* 1,306 homes
* 22 miles of roads
* £4625m TDC expenditure
* £1089m private investment

The TDC was typical of second-generation UDCs and their approach to regeneration. It had probably the greatest test with the largest area within its boundaries and a national geographical location that was always going to be difficult to attract private investment. In relative terms, its area was twice the size of that of LDDC's and had none of that UDC's locational advantages.

As a reflection of the area's characteristics, TDC reclaimed more land than any other UDC. However, the private investment generated was less than any of the other second-generation UDCs. The number of new jobs established was less than half of these UDCs too. It did do relatively better in terms of the number of homes built.

Review of third-generation UDCs

The third generation represented a change of tack towards the edge of city centres and much smaller areas. As a result, their regeneration tasks were to be completed over a much shorter timescale. But there was also a slight difference in their overall goals with an objective to the removal of supply constraints on their respective city office markets (see Chapter 19). The proximity to office centres suggested a greater degree of growth potential than UDCs of the previous generations.

The basic strategy of UDCs remained unchanged, namely the encouragement of ambitious flagship developments. This was achieved through active intervention in the land and real estate markets, and by providing necessary infrastructure to support new developments. It also involved backing developments by funding, and where necessary, brokering deals by assembling land plots. The chapter now considers a case study of Leeds Development Corporation (LDC) that had the second largest area of the third-generation UDCs.

Leeds Development Corporation

Although the city had prospered as an office centre in the 1980s, the peripheral edges of the centre had not benefitted from this growth in economic activity. It was this fringe area that was covered by LDC. In fact, it was two separate and contrasting sub-areas. One to the immediate south of the centre was a former industrial area. The other, a north-west corridor from the centre along a river, suffered severe contamination stemming from its former use as a power station.

Almost half of LDC's expenditure was spent on land assembly, reclamation and site preparation. When spending on environmental improvements, infrastructure and grants to developers are taken into account, the total rises to three-quarters of its total expenditure on promoting real estate development.

An immediate successful high-profile demonstration development was the museum and visitor centre linked to an adjoining brewery. Later, the LDC was able to promote a further major tourist attraction, the Royal Armouries Museum. It relocated from London to a dock site within the area following site preparation works and expenditure on improved access.

Besides these tourist destinations, the LDC supported the development of a range of offices and business and industrial parks in the southern sub-area. A prominent development was Leeds City Office Park at Centre Gate built on a redundant gasworks. The development comprised three-storey buildings around pedestrianised space. The LDC also helped the refurbishment of leisure properties on the edge of the city centre.

Overall, the LDC produced the following statistics:

- 168 acres of reclaimed land
- 1,23m sq ft of commercial floorspace
- 571 homes
- 11.6 miles of roads
- £71m LDC expenditure
- £357m private investment

The gross employment created by LDC was 9066 new jobs. In comparison with other third-generation UDCs, its methods and results were comparable. Even though they were all off city centre location, individual strategies and outcomes were shaped by the characteristics and physical problems of their areas. For example, the Bristol Development Corporation identified a serious issue of its area's accessibility. It therefore placed great emphasis on building a direct link to the national road network.

Dependency on macroeconomy

UDCs have provoked considerable arguments about their achievements. There are questions about the emphasis on real estate development and its relationship to employment generation. Unfortunately, to answer this question, there is no counterfactual to compare. The evaluation in this chapter is more specifically linked to the outcomes of the regeneration process. In this section, the role of the UDC as a market-driven costless regeneration tool is discussed. In the following section, the overlapping issues of physical vs social regeneration, people vs place regeneration, and trickle-down effects are considered.

Market-driven costless solution

The LDDC was very much the leader and shaper of government policy on UDCs during the decade of the 1980s. It took an aggressive negative approach to planning with a stress on flexibility and bending to the needs of the market and private investors. Land was allocated to the most profitable use with little constraint on design. At least initially there was little or no reference to community facilities, social housing needs and local unemployment/training (see later).

The success was measured by the scale of new development but also the increase in land values. At the beginning of the regeneration process, land values had been zero in 1981 with the land lying derelict and unwanted. However, by the late 1980s residential land was valued at £3m per acre and land for offices at £4m. As a result, by 1992 LDDC had made a surplus of £141m on its land sales.

At the end of the 1980s, there were very optimistic projections about what could be achieved in terms of private investment and the ratio of private to public investment. It was being heralded as a *total success*, and the reason for the designation of the second- and third-generation UDCs. At the same time, the alternative EZ model was deemed too expensive (see Chapter 21). A UDC was seen as costless solution to urban regeneration. Essentially, the process would start with public expenditure on land assembly/reclamation/infrastructure. These physical improvements then increase land values, and the UDC can recoup its costs by land sales.

The trouble with this analysis was that it was predicated on a benign macroeconomy. Economic growth in the 1980s had contributed to a real estate boom that had supported the activities of LDDC. As noted earlier, the recession ultimately had a severe detrimental impact on new development in the London docklands.

The impact of the recession was far worse for the second- and third-generation UDCs set up in the late 1980s. They had to buy land at or near the peak of the market (whereas LDDC bought at the bottom in 1981). Increased land values ate into budgets allocated by central government and led to failure to meet acquisition and reclamation targets, which in turn fed through to regeneration outcomes.

The regeneration schemes also proved to be difficult because development was almost impossible even in prime locations during the recession as private investment stalled. It was subsequently difficult to sell land for some years, and so this often meant a loss on resale. Given the timing of the recession shortly after second- and third-generation UDCs had been established, they did not complete the task in their allotted time.

The pattern of development emphasises that real estate-led regeneration requires a sustained period of national economic growth to support it. Given the length of time required to bring such initiatives to fruition, there is almost an inevitability that a recession will disrupt the process. Real estate-led urban regeneration is not a quick fix. Indeed, this explains the failure of the Broadway Times Square regeneration plan noted above in New York that floundered when the macroeconomy turned down.

In the case of the docklands, the recession occurred once the fundamental developments had momentum. For the UDCs that were set up subsequently in UK provincial cities on the back of the 'success' of the LDDC, the positioning of the recession in the development time line was less rosy. The swift collapse of rents and capital values meant that flagship property developments did not materialise or were delayed or were pruned back to shadows of their original plans.

Regeneration of places or people

The operation of the LDDC created a debate about the nature of urban regeneration. A series of questions were raised:

- Was physical regeneration sufficient?
- Who benefits?
- Should regeneration create social polarisation?
- Regeneration of people versus places?

In fact, all these questions are linked and the issues raised overlap. These questions are considered in this section drawing on the experience of the London docklands.

Sufficiency of physical regeneration?

A heated argument began in the late 1980s with severe criticism of the emphasis on 'physical regeneration' or the transformation of places by UDCs. Physical regeneration was the first step of UDCs as a base for stimulating private investment. This investment would then bring a restructuring of the local economy.

Critics argued that social regeneration was more important; in other words, regeneration should focus on investment in social housing, education and community development. These goals were argued to be important because there is a need for a well-educated productive workforce.

There were a number of inquiries by (select) committees of the House of Commons, one of whom concluded,

> There needs to be a reasonable balance between physical development and the social needs of those living there.

The government countered by saying UDCs were exclusively physical but ultimately conceded the need for social regeneration, and their aims were adjusted. In its later years, LDDC developed social infrastructure programmes to support the community and other UDCs also had similar activities.

Who benefits?

A further criticism was that the short-term benefits of the LDDC's activities went predominantly to developers, financiers, 'back to the city professionals' and tourists. This reflected the fact that the new development in the LDDC area was concentrated in offices and expensive residential flats.

The counter-arguments focused on the trickle-down effect to the local community via employment and spending. The original local population certainly benefitted from the availability of new jobs, although they were mainly low-level service employment opportunities. The argument is partly about what could have been an alternative regeneration strategy for the area.

Social polarisation?

The concerns about social polarisation in the London docklands were twofold, linked to the housing and labour markets. The former issue stems from the fact that virtually all new housing in the area was built for sale and these were almost exclusively bought by in-migrants. The prices of these luxury apartments were well beyond the reach of the existing population. Existing residents continued to live in poor-quality social housing that was neglected, receiving no funds for modernisation.

In the labour market, local workers had been primarily employed in the docks before they moved downstream to deeper water to suit container ships. They lacked the qualifications to get most of the new jobs in the area. Furthermore, the increase in local services employment opportunities predominantly taken up by 'locals' were generally those of cleaners or receptionists. These jobs were typically taken by women.

Despite the number of jobs created in the LDDC area, local unemployment was higher in 1992 than in 1981. In 1981, there were 3533 (17.8%) registered unemployed and living in the area. Just before the LDDC was wound up in December 1997, there were still 2853 (7.2%) registered unemployed (definitions changed during the period). In other words, the numbers of the unemployed had fallen by just under 20% (but the numbers living in the area had increased significantly, see below).

Regenerating people or places?

The LDDC vividly demonstrated that there can be a conflict between regenerating places and people. There was antagonism from the beginning with little cooperation, even hostility, between LDDC and the local councils. Planning responsibilities had been taken away from the councils, and there were no local councillors initially on the board of LDDC. The local councils did not approve of the LDDC strategy and had no input into it. For ten years, there was no initiative aimed at local people and very limited funding to improve the social housing stock.

Over the lifetime of the LDDC, the area was dramatically transformed. In 1981, there were 14,743 households of whom 90% lived in social housing and 5% were owner-occupiers. In 1998, the population had more than doubled with 38,000 dwellings in total, and the tenure structure had been radically redrawn – now 45% of households were owner-occupiers, 44% social housing tenants and 11% private tenants.

The problem is that regeneration by definition requires the restructuring of the local economy and real estate market. The key task is how to manage this change. In the case of the LDDC and the community (representatives), they were at loggerheads from the beginning. Symbolically, the LDDC demolished local workshops on the Thames waterfront to make way for new flats.

This sounds an extreme policy, and the LDDC said these businesses were not viable and only occupying the premises at nominal rents until demolition. It was emblematic of the range of conflicts between the existing and new population in the area. The share of the benefits of the regeneration was predominantly biased towards the new residents and workers. This leads to the question 'Can an area be deemed to be regenerated if the community is not?'

LDDC was an extreme example, and the other UDCs did not follow the same path. Other UDCs had more local community buy-in, and the regeneration outcomes did not pose the same stark contrasts. In general, there was not the sharp divide between the original population and newcomers, because in most cases the former was minimal.

Summary

A UDC can have different definitions around the world. The New York State UDC that has operated since 1968 is similar to the UK model. Some UDCs are public agencies that act as direct developers to support economic development. Most UDCs have planning powers and bring together public resources with the expertise and dynamism of the private sector. The UK model is intended to nurture private development in urban regeneration areas.

Specifically, UDCs aim to remove real estate supply constraints in distressed urban areas. They are set within a public policy framework to address market failure by public intervention. The primary mechanism for UDCs is to assemble sites, and reclaim and service derelict land to bring it into effective use. They can also provide funding infrastructure and improve the environment to support private investment.

Each UDC is appointed by the government with a remit to develop a strategy to resolve the local physical and economic problems. Their main funding in the 1980s was derived from central government, but UDCs could also receive income from land and real estate sales. Virtually all of their expenditure was on land purchase, site assembly and preparation. There were also significant amounts spent on place marketing to attract investors.

Between 1981 and 1992, there were four generations of UDCs designated. Beginning in 1981, the LDDC and MDC were appointed to address the regeneration of derelict docklands. The second generation in 1987 primarily covered areas of former industrial wasteland in northern cities. Mini-UDCs were established in 1988 at the edge of provincial city centres. There were the two ad hoc UDCs announced in 1992.

The lifespan of these UDCs varied from six to seventeen years. The UDC's approach to regeneration did not finish with their closure but continued through URCs and then Mayoral Development Corporations in the 2010s. The primary differences between these different agencies are the funding and their relationships with the local community especially through planning.

The regeneration area of the London docklands had lain more or less vacant for ten years. The first priority of the LDDC was to improve the basic road and physical infrastructure of the docklands. There was an urgency to prepare sites for development. It was necessary to attract demand by changing perceptions of the area.

At first, there was success in promoting the area as a location for light industry and business space. The mid-1980s saw a construction boom in the area involving demolitions of obsolete dock sheds and the conversion of existing buildings. New housing for sale was also being built. A key event was the opening of the DLR in 1987. London City Airport also opened in the area.

The redevelopment of Canary Wharf with a 50-storey office block was the transformative development that defined the LDDC. It stimulated similar new developments and contributed to a boom. However, the recession at the end of the decade brought a collapse in the real estate market and developments stalled in the commercial and housing sectors. In fact, it raised questions about the area's future as a financial centre, especially as the developer of Canary Wharf went into administration.

The issues for the docklands went beyond the recession and centred on the ability of the transport infrastructure to cope with additional office developments. Eventually, the DLR was upgraded and extended, and an underground rail link was built. Confidence in the area eventually began to recover by the mid-1990s, and new development returned bringing a critical mass for a financial centre.

The mission of TDC was to regenerate vast areas of former industrial wasteland in the north-east of England. Its activities emphasised flagship regeneration projects on high-profile individual derelict sites as symbols of substantial physical change. These new developments encompassed a shopping centre on a former racecourse, and marinas and tourist attractions that could take advantage of waterside locations.

One large project is illustrative of the process. Increased access to a nearby town centre of a site of a former large engineering works enabled it to be developed. A water barrage was also built to enhance the physical environment. The site was regenerated as mixed-use area, comprising a business park, a university campus and housing.

The challenge for TDC was that it had a huge area to regenerate in a locality that was difficult to attract private development. Reflecting this test, it reclaimed more land than any other UDC, but its market outcomes in terms of jobs and private investment were relatively less than other second-generation UDCs.

The role of the third generation shifted to regenerating the edge of city centres with the objective of reducing supply constraints on their office markets. The basic strategy of UDCs remained unchanged through the fostering of ambitious flagship developments. It continued to involve intervention in real estate markets and infrastructure provision.

In Leeds, the LDC was tasked with regenerating two separate areas near the city centre. The focus of its expenditure was predominantly on land assembly, reclamation and site preparation. There were prominent museum developments together with a range of offices business and industrial parks. A notable development was Leeds City Office Park on a redundant gasworks.

The UDC programme of the 1980s provoked considerable dispute about its achievements. The government promoted UDCs as a marker-driven costless solution to urban regeneration. This policy was derived from the experience of LDDC. The model was essentially that initial physical improvement expenditure could be recouped by land sales when values rose as a result of the success of the regeneration.

This model was based on the LDDC's success during the 1980s real estate boom. However, the recession at the end of the decade queried the whole basis of the model. Unfortunately by then the second and third generations of UDCs had been designated on its understanding. The recession seriously hampered the activities of these later UDCs. They bought land when prices were high and struggled to promote development in the aftermath of the recession. Real estate-led regeneration depends on a positive platform of national economic growth.

The operation of the LDDC created a debate about the nature of urban regeneration. Its emphasis on physical regeneration was queried by critics that saw social regeneration as a higher priority. Eventually, the government conceded that there needed to be also social regeneration.

An additional criticism was that the fruits of the LDDC's activities went to 'outsiders' rather the local community. The counter-argument was that the new offices and housing meant that there were new employment opportunities. However, the new jobs for the original residents were mainly low-level service employment opportunities, and focused on women. There was also social polarisation as the prices of the newly built apartments were too high for the existing population.

The LDDC strikingly showed that there can be a clash between regenerating places and people. Over the lifetime of the LDDC, the area was dramatically transformed by new housing and workplaces. However, for the most part the new housing and job opportunities passed the original community by. These divisions were not apparent in other UDCs because most had only a small initial population within their boundaries.

Learning outcomes

Variants in UDCs are applied around the world.

UDCs have planning powers and bring together public resources with the expertise and dynamism of the private sector.

UDCs in the UK regenerate distressed urban areas by assembling sites and bringing derelict land into effective use. To do so, they can also provide funding to support private investment.

The main funding of UDCs in the 1980s was derived from central government, but they could also receive income from land and real estate sales.

Between 1981 and 1992, there were four generations of UDCs designated. The first three generations each had distinct characteristics.

The lifespan of the original UDCs varied from six to seventeen years.

The UDC's approach to regeneration has continued first through URCs and then through Mayoral Development Corporations.

The first priority of the LDDC was to improve the basic road and physical infrastructure of the docklands to provide the base for new development.

The London docklands initially attracted new housing, light industry and business space. The redevelopment of Canary Wharf with a 50-storey office tower was a game-changer for the area.

The recession at the end of the decade brought a collapse in the docklands' real estate market, and developments stalled in the commercial and housing sectors.

The hiatus in development following the recession raised issues about whether the transport infrastructure in the docklands could deal with additional office developments.

During the 1990s, the public transport infrastructure in the docklands was upgraded supporting the establishment of the area as a financial centre.

Like other second-generation UDCs, TDC was charged with regenerating substantial areas of urban wasteland. Its strategy was to focus on transforming prominent derelict sites into new uses and demonstrate physical change.

The focus of the third generation of UDCs was to regenerate the edge of city centres and relax supply constraints on their office markets.

The financial model for the 1980s UDC programme was that initial physical improvement expenditure could be recouped by land sales as values rose with the success of the regeneration.

The recession at the end of the 1980s undermined the basis of the model. The fall in property values and hence development activity had a detrimental effect on the success of later UDCs.

Real estate-led regeneration is underpinned by national economic growth.

The focus of LDDC on physical regeneration was queried by critics that saw social regeneration as a higher priority. Ultimately, the government accepted that both were needed.

The LDDC's activities were also criticised as favouring 'outsiders' rather the local community. The jobs available for original residents were mainly low-level service employment opportunities for women.

Social polarisation occurred in the dockland area because the prices of the newly built often luxury apartments were too high for the existing population.

There can be a clash between regenerating places and people as the LDDC experience demonstrated.

Bibliography

Deas I, Robson B and Bradford M (2000) Re-thinking the Urban Development Corporation 'experiment': The case of Central Manchester, Leeds and Bristol, *Progress in Planning*, 54, 1–72.

Dodman D (2008) Developers in the public interest? The Role of Urban Development Corporations in the Anglophone Caribbean, *The Geographical Journal*, 174, 1, 30–44.

Jones C (2013) *Office Markets and Public Policy*, Wiley-Blackwell, Chichester.

Robinson F, Lawrence M and Shaw K (1993) More than Bricks or Mortar? Tyne and Wear and Teesside Development Corporations: A Mid-term Report, University of Durham, Durham.

23 Development of sustainable markets

Objectives

The fundamentals of real estate-led urban regeneration are to directly intervene in the market to support and attract private property development and investment. The last two chapters on enterprise zones (EZs) and urban development corporations (UDCs) have considered a range of issues that arise by looking at the individualities of these particular initiatives.

In this third chapter on the topic, the focus is on the role of grants to developers for urban regeneration that is today often referred to as gap funding. In doing so, we explore the internal logic of addressing market failure that has been a dominant paradigm for urban policy. In particular, the chapter considers evaluation measures of success of real estate-led regeneration.

The use of short-term measures to evaluate projects is queried, and the chapter considers longer term perspectives on achievement. In particular, it considers the formation of new local land use markets brought about by regeneration expenditures. To do so, the chapter examines the dynamics involved in the eventual establishment of sustainable markets for publicly supported new land uses in distressed areas.

The chapter has the following structure:

* Logic of grants to developers
* A long-term perspective
* Case studies
* Real estate initiatives in a wider context

Logic of grants to developers

The basic idea of funding developers to undertake urban regeneration in the UK was imported from the United States as explained in Chapter 18. When first introduced in 1982, the financial support was called Urban Development Grants in England. Later, they were renamed Urban Regeneration Grants and eventually City Grants in 1988. There were variants of these grants in Scotland, Wales and Northern Ireland, but all shared the same fundamentals. It should be remembered that UDCs also provided the same grant support, and the discussion below also applies to that activity.

Since 1993, these grants have lost a formal name and are referred to as gap funding that reflects their purpose. The concept of gap funding is that it is a grant to a developer for a regeneration project that otherwise would not be profitable. In this way, the

DOI: 10.1201/9781003027515-26

grant would attract private investment into, say, an inner-city area. A grant was theoretically designed to offer the minimum public sector contribution to allow a project to go ahead. Beyond this essential principle, the precise rules about the suitability of regeneration projects vary.

Clearly, a government is not going to fund all regeneration development or refurbishment projects that are not viable. There need to be criteria for selection, and this implies meeting of urban regeneration goals. Obvious examples of goals include the following:

- Strengthen local economy/rebuild confidence, and
- Bring derelict or disused property back into use.

In the UK, in the 1980s the regional variants noted above applied marginally different priorities to these basics in the choice of projects.

Besides these outcomes, potential projects are also inevitably judged on the basis of value for money, comparing the amount of public investment with the private sector investment/development outcomes. Following this logic, prospective development projects are usually compared/assessed on the following criteria:

- Leverage – the ratio of private investment to public expenditure,
- Cost of jobs generated, or
- Cost of housing units produced.

These types of evaluations originated for Urban Development Action Grant in the United States. However, the calculation of these ratios is not straightforward and was subject to considerable dispute in the United States.

Employment generation

The first problem is that these ratios need to be estimated prior to the start of the development. Looking first at jobs, the employment generation in the form of the net additional jobs created will need to be forecast. There is an inevitable tendency to overestimate by those who are proposing the development. To judge the employment impacts, there needs to be a post-evaluation.

Judging urban regeneration development projects simply on the employment potential is also arguably overly restrictive. Different types of property/land use projects will inevitably lead to variations in the numbers of jobs (and types of jobs). Some land uses are more labour-intensive such as retail, leisure and hotels. Location of the development will also have an impact on employment.

There are potential displacement effects that need to be taken into account in forecasting net employment. For example, a new shopping centre may simply displace jobs, but a decision will also depend on the location and form of existing retail facilities. Some jobs are part-time, so how are these numbers balanced against full-time opportunities? Similarly, how do you weigh up the quality of jobs generated in different regeneration projects?

The problem is not just the forecast of jobs but also their cost. One study in the UK suggests hotels cost most per job, followed by retail, offices and industrial development. A whole gamut of issues arises. There is the problem of trade-offs between the criteria. The same study that finds the cost of jobs is cheapest for industrial development found that it scores relatively poorly on the private investment leverage ratio.

It is not just a matter of creating jobs or the type of jobs, but there is usually a need to target take-up of jobs by those living in residential areas with high unemployment. As previous chapters have shown, generating jobs in a particular area does not necessarily mean that they are taken up by local residents. Jobs, more often than not, go to better qualified people from elsewhere within the urban labour market.

Projects may be required to guarantee 'local' employment opportunities, for example by ensuring interviews for local people who are registered as unemployed. While there is a clear logic in targeting jobs on the most deprived groups in cities, there is also a conundrum. The overall aim is to attract private investment to inner-city locations experiencing physical and economic distress. It needs a beneficial climate to induce business into those locations, and adding employment constraints could act as a deterrence.

Leveraging private investment

The use of leverage ratios is also not straightforward as a metric for comparing projects/areas to support. At its basic, the criterion suggests the greater amount of private investment generated for every £, dollar or euro the better. But it is not that simple.

First, projects that have a high ratio of private investment generated may not actually need public support. If the public sector is only contributing 5% of the total project cost giving a very high leverage ratio, then the project could probably go ahead without that contribution. On the other hand, a low leverage ratio such that £1 of public money generates £5 of private money sounds less attractive to the public purse. However, without the 16.6% public sector contribution in this example, it is probable that the project is not viable.

Second, using leverage ratios to compare the suitability of projects between areas is also problematic. Chapter 22 notes the difficulties of attracting private investment into the urban wasteland of Teesside in a peripheral region of the north of England. Promoting investment in the London docklands was a much easier task. Comparing leverage ratios between Teesside and London would always favour the latter, because of the relative attractions of the localities to the private sector. Leverage ratios are always going to be less for the most risky distressed urban areas.

Third, there can be similar question marks about the use of public money to support certain projects. The most obvious controversial case is luxury apartments that could generate a high leverage ratio. Public financial support for the development of luxury apartments has been justified in terms of rebuilding confidence in an area and bringing derelict land/buildings back into use (see above). Critics might argue that the money could achieve equivalent goals with social housing.

Fourth, leverage ratios vary depending on the stage in the development of an 'area initiative'. Private investment should logically rise through the life of such a project. This process is recognised in the workings of the UDCs as explained in Chapter 22. Initial public sector investment on infrastructure provides a platform for subsequent private investment. The leverage ratios therefore increase over time.

A long-term perspective

The basis of these evaluations or justification for individual projects is arguably too narrow. There is a need for criteria to judge the suitability of regeneration projects to fund. It is debatable about whether decisions can be based simply on the (likely) induced scale of private investment. Further, it is not possible to comparatively assess projects on the number and cost of job generation in areas with distinctly different levels of unemployment.

A wider perspective sees the success of projects as their contribution to the broader regeneration of a neighbourhood or area. Projects should be demonstrations of the future possibilities for an area together with positive social and community impacts. These latter impacts might encompass greater tenure choice, transforming the physical environment and an improvement in civic pride. But these effects are dependent on regeneration schemes being successful and the new uses of land promoted having a long-term future.

A long-term approach goes back to the basics of what real estate-led urban regeneration is attempting to achieve. The promotion of private investment is not the end of the story. The underlying theory of government policy is to resolve market failure in the long term. To achieve this goal, ultimately there needs to be the establishment of new real estate markets in regeneration areas. There needs to be more than the physical regeneration of the area.

Gap funding is a means to restore market confidence via financial support. More generally, UK Government regeneration policies have been aimed at restoring private sector confidence within a limited timescale before the public sector withdraws. As shown in some EZs in Chapter 21, this does not always work. The long-term aim to create a sustainable real estate market was not necessarily achieved. To understand why, it is useful to consider the fundamentals of the policy and its interaction with real estate markets.

Real estate-led urban regeneration initiatives are based on the idea of public money to pump-prime the private sector until private development is just viable in its own right. A necessary condition for success is that rents/capital values rise to the long-run average rent/value so that this would make development in theory viable without public support.

The challenge is that urban regeneration schemes are in secondary/tertiary locations where mainstream private investment has been limited in the recent past, possibly for decades. These projects are fostering new or embryonic markets. They are often innovative development schemes with multi-uses or mixed uses. Sometimes they are large scale projects that can only be completed over a long timescale.

The nature of these projects, developing new concepts coming to fruition at some time in the future, creates valuation problems in the development appraisals. For example, how do you value an apartment to be built in a converted mill in an industrial area where there is no housing at present and it will take three years to build?

Professional valuers will inevitably be conservative. Regeneration schemes therefore often promote new land uses in secondary locations where the outcomes are difficult to evaluate. The initial purchasers of these properties are having to buy into the project vision, so valuations are a problem not just for the developer but also for subsequent owner/investor.

Regeneration projects therefore encompass significant risks in terms of the values that can be achieved on completion. There is also great scope for unexpected development costs from often building on former industrial sites that adds to the risks. Nevertheless, research shows that returns from regeneration projects can be at least as comparable with the mainstream.

Simply viewing an urban regeneration project as a pump-priming private investment task is too naïve. The analysis is incomplete because it ignores the fragility of 'new' markets and so markets need to mature to convince investors/developers that they should continue to participate. In other words, sustainable markets for the new use(s) are achieved, and this requires two conditions:

1   Private sector development is viable in its own right, and
2   Market has matured through a period of market activity.

The first condition is a necessary condition for market viability, but it is insufficient without the latter one. The latter condition requires that a resale/reletting market has been established for the new uses. In turn, it will enable the credibility of the products as investments to be generally recognised. It also implies the achievement of a critical mass to ensure its credibility. A parallel process is considered in Chapter 13 with regard to the acceptance of new real estate forms by financial institutions.

Regeneration case studies

The chapter now considers a number of regeneration case studies that draw out the nature of the processes by which the establishment of new uses and markets evolves. They relate to housing developments in provincial cities of the UK. These retrospective studies over the period 1980–2000 flesh out the issues discussed in the previous section.

Merchant City in Glasgow

The area currently known as the 'Merchant City' is an area on the edge of Glasgow city centre that was in apparent terminal decline at the end of the 1970s. The area was originally an extension of the city of Glasgow in the eighteenth century as tobacco and sugar trades flourished. It contained warehouses and the mansions of the merchants from that time. After the demise of these trades, the buildings continued to be used as warehouses for other goods.

During the late twentieth century, this original warehousing function of the area ceased as the occupiers moved to more decentralised locations. The area spiralled into decline with buildings abandoned. By 1980, about a third of the property in the area was vacant. Many of the properties were in poor repair and had been gifted to the city council to avoid the expenditure to resolve the problems. As a result, the council owned two-thirds of the vacant property.

The city examined options for the regeneration of the area and decided to refurbish the buildings and promote it as a private residential area. It was a bold step because there had been no housing built in the city centre for nearly a hundred years. As part of the strategy, it gave the area the Merchant City name in recognition of its past and added it to all the street signs.

The essential policy problem was that as there was no owner-occupied housing in the area, there was no market to build on. A small private scheme was initiated in 1982 to convert 23 flats with grants from the council to the developer, and a guarantee from council to buy the flats if necessary. This first small development was completed in 1983 and the flats sold very quickly.

A number of other schemes followed funded with grants to developers, and by 1986, more than 300 housing units had been created. In parallel with these housing developments, a number of buildings were restored and converted to leisure uses. The housing conversions also often included retail units on the ground floor.

By the beginning of 1994, over 1,000 units had been converted/built incrementally in 18 developments in the Merchant City. The majority of these were small flats aimed at and occupied by young adults in their twenties in non-manual occupations. More than half of these buyers were migrants into the city.

This implied picture of a successful regeneration initiative ignores how and to what extent a market was established for these flats. Despite the success of the first development, there were still obstacles to the promotion of the area as a residential location. The agents for the first large development felt it necessary to organise an introductory session for building societies/banks and their surveyors to convince them to finance purchases.

By 1986, when the first phase of this large development was marketed, there were long queues of prospective buyers wishing to make reservations months before likely completion. Sales of new properties in the Merchant City grew rapidly. Between 1986 and 1989, they represented around 9% of new sales in the city. The latter half of the 1980s saw estate agencies set up to serve this developing market.

This period in retrospect was the heyday for Merchant City developments. New sales decline from 1990 onwards, partly reflecting the number of new developments being marketed with the national recession. However, the growth of resales in the Merchant City from 1988 partially offsets this trend.

These sales trends hide significant patterns in prices. In the initial developments, flat prices were very cheap to attract buyers to an unproven location. Three years later, the average price had risen by 50%. Average new prices fell for the first time in 1992 in the aftermath of the recession, with this trend continuing into 1993.

This fall in prices combined with the severe decrease in new sales perhaps set the first seeds of doubt in the success of the initiative. New developments in the pipeline failed to materialise. There was a loss of market confidence in the area, and this was reflected in comparative house price trends elsewhere in the city.

There was a general upward trend in house prices in Glasgow during 1992, and established residential areas had not been so susceptible to falling prices from the late 1980s. In fact, the Merchant City had had its own distinctive pattern. After the initial surge in house prices up to 1985, average prices in the Merchant City did not rise by as much as the citywide average over the subsequent eight years.

The Merchant City therefore suffered relatively worse from the national recession of the late 1980s than the city's housing market as a whole. This was probably in part the result of the characteristics of demand from young professionals employed in service industries that were particularly badly hit by the recession. The combination of naturally high turnover and forced sales sowed doubt in the investment potential of housing in the area. New development stopped and potential buyers began to question the permanence of the market.

The Merchant City housing market that had bloomed through the 1980s now faced a serious question mark about its future and its sustainability. However, as discussed in Chapter 9, the Merchant City can be seen as part of a wider revival of inner-city living by middle-income households. After more than ten years from the inception of the initiative, the Merchant City eventually became established as a niche in the local housing market.

In the latter half of the 1990s, new development returns to the area, and now without public subsidy. In terms of the criteria for a sustainable market outlined above, it can be seen that both conditions have been achieved. Private development was now viable in the area, and market activity had demonstrated long-term demand. The time taken to achieve a sustainable market was approximately 15 years.

Inner-city estates in Liverpool

In the mid-1970s, Liverpool city council sold off land cheaply in three inner areas to builders to construct houses for sale. Much of this land was adjacent to social housing. The completed housings were predominantly semi-detached houses that were offered to households on the waiting list for social housing.

The new estates were easily sold as there was excess demand with prices set low to attract buyers. They were very popular to begin with, and prices on resale rose quickly. However, by the early 1980s one estate began to suffer from riots nearby. Another of the three estates endured a high proportion of foreclosures with sales by auction bringing the introduction of renting from private landlords. A study in the late 1990s found that prices were very low on both of these two estates and properties are difficult to sell.

The success of the other estate appears to be as a result of the neighbourhood location given there is little difference in the characteristics of the housing built. Twenty years later, prices had more than kept pace with city prices overall. The estate was well kept and many of the houses have extensions, porches, double glazing and improved driveways. Further estates were built nearby to compete in the same submarket. On this evidence, only one of the three estates achieved a sustainable market.

Refurbished flats in Salford, Manchester

At the beginning of the 1980s, an estate of five-storey walk-up flats at Regent Park owned by the local council was converted for owner occupation. The flats were originally built between 1939 and 1949 and were suffering from high vacancies and turnover. Although structurally sound, they were in a poor state of repair and were very unpopular with tenants. The council decided to sell them to a developer.

The reformulation of the estate involved selective demolition, and refurbishment of the remaining. The refurbishments included private balconies and new closed entrances with entry phones. A perimeter wall was built with one entrance creating a gated community, approximately a mile from Manchester city centre.

The refurbished flats were sold quickly and cheaply in 1984 to one- or two-person households primarily in their twenties who were likely to move on relatively quickly. From the beginning, the price of resales did not tend to rise even in nominal terms with inflation. A further reason may have also been because the original sale prices included kitchens with fitted white goods such as fridges that quickly lost their value.

Part of the reason for the weak resale prices was also the high rate of repossessions, with the fallout from the recession. Banks/building societies then resold foreclosed properties at low prices deflating prices of other market sales. There were also many properties on the market from nearby substitute developments.

Resale prices slid and never fully recover. As a consequence, relative prices for Regent Park fell from two-thirds to one-half of local average prices by 1997. The poor state of the estate's environment meant that the flats found it difficult to compete with warehouse conversions in the city centre. Many flats were bought by landlords. The lack of success of the development meant that there was no local demonstration development impact.

New housing for sale on a social housing estate in Glasgow

Easterhouse is a large peripheral social housing estate in Glasgow. It covers a large area and in reality is a number of communities, rather than one estate. The city council decided to build new detached, semi-detached and terraced houses for sale within the estate as well as improving some existing housing flats for sale to diversify the area's tenure.

The introduction of this new housing for sale effectively represented a new tenure to the area. The renovated flats for sale were initially three storey blocks with security entry systems and communal gardens and were sold from 1985. The new housing was also marketed in stages in clusters or different areas from 1985 through to 1992.

The extent that a market for these houses has been subsequently established can be assessed by reference to the number of resales and any rise in house prices. As with the other case studies, there is the familiar pattern of an initial rise in resale prices that can be attributed to the low original prices. Foreclosures then impact on the resale values of the refurbished flats with the onset of the recession at the end of the 1980s.

In the 1990s, the numbers of resales are variable from one year to the next, but there is continuing demand. There is evidence therefore of market activity, but price trends are not stable. Resales in two of the three clusters/areas in Easterhouse have negative or static changes in prices over the 1990s. The pattern for the other cluster is inconclusive with the resale prices of those housing units formerly improved for sale falling. On the other hand, resales of the new houses in this area demonstrated a continuing, if at times faltering, upward trend.

By the millennium, there is no evidence of sustainable markets for housing for sale developing in Easterhouse. All the developments that had occurred had received public subsidy. A resale market has been established, but second-hand house prices had not risen to a level that made private development viable without subsidy. A sustainable market had not been achieved at the time of the study over ten years on from the first developments for sale.

Commentary on case studies

All of these case studies of regeneration are about creating a market for housing to buy where one did not exist. The markets in these case studies are initially very fragile as they are introducing a new use to an area. As they develop, these embryonic markets are subject to internal and external shocks. In one inner Liverpool estate, the internal shock caused by a local riot had a terminal impact on building a successful local market.

The most important external factor was the national economy and notably the negative consequences of a recession. As the Merchant City case study demonstrates, a new and still to some extent unproven market is more precarious in weathering such a storm than established neighbourhoods in the city. Many of the other case studies suffered significantly from foreclosures following a recession.

The establishment of a mature/sustainable market takes a relatively long time. Even if a second-hand market is established, prices may or may not rise to make further development viable without a grant. Success in these terms, if it occurs, is not overnight, and the Merchant City took 15 years. The London docklands covered in Chapter 22 arguably took almost 20 years.

The chapter has outlined arguments that government support for development in regeneration areas should be judged in the long term via the concept of market sustainability. The case studies demonstrate the importance of market sustainability but also raise a fundamental question about measurement of success.

Regeneration schemes that appear successful in the short term by reference to the investment generated or jobs created may not be viable in the long run. Time paths to success or failure are convoluted. There is an identification problem of which regeneration projects will ultimately succeed that cannot necessarily be determined at the outset.

Real estate initiatives in a wider context

The book has chronicled a whole range of real estate-led local economic development and regeneration initiatives. In the UK, probably the first real estate-led local economic development policy was the building of industrial estates in peripheral regions during the 1930s as part of regional policy. Advance factories subsequently became almost a ubiquitous tool of local economic development although the theory behind the policy was rarely spelt out.

Real estate-led policies became more sophisticated in the late 1970s/early 1980s and encompassed housing and commercial uses. Besides the policies we have discussed in the last few chapters, public agencies have also provided land for development (affordable housing) and undertook master planning of private development.

Since around 1980, real estate-led strategies have often explicitly been set a task of addressing market failure. But there is an identification problem (see above) because the absence of property forms in specific locations may be because they are not viable. In other words, the market is not failing, but rather there is simply a market gap.

How do we decide between a market gap and a market failure? There is no answer in the short term. The establishment of markets is a long-term process for all real estate-led initiatives discussed. However, the market may not respond because of market failure but because the type of development will never be profitable. This means it might take 20 years to decide if there is a market gap rather than market failure. If it is a market gap, then the property could require continuing subsidy as rents will never reach market viability.

The real estate market failure perspective therefore has severe limitations as a practical tool of urban regeneration/local economic development. An alternative perspective is to simply accept that the real estate market is inefficient with the potential for widespread market failure. It can mean that the local economy is not maximising its potential because of real estate market constraints as discussed in Chapter 19. It can also be because of local infrastructure constraints on development.

From this perspective, where there are, for example, large areas of derelict and underused land, then the conclusion is that the local economy is not functioning efficiently. Intervention in the real estate market, including new supporting infrastructure, can be justified by expanding the production possibilities frontier of the city. It will still be only a platform for developing new local land use markets.

This task may involve diverting demand by making/building properties available in certain areas. This justification potentially raises the issue of whether regeneration is aimed at a city or a local community. Restructuring the land use structure of part of a city may stimulate the economy of the whole city, say repurposing vacant building, but may not directly benefit the immediate local community.

This issue is highlighted by the diverting of demand to create a new office centre at Canary Wharf. This almost certainly expanded the output/production possibilities frontier of London but, as we noted in Chapter 22, brought questionable benefits to the local community. Which 'community's' interest is paramount?

Summary

Providing grants to developers to promote regeneration began around 1980, and in the UK it is now simply known as gap funding. It is a mechanism to support regeneration projects that otherwise would not be profitable. The funding should logically be the minimum public sector contribution to allow a project to proceed.

The choice of projects to fund in this way will be based partly on their scope to achieve physical and economic urban regeneration. Value for money is also relevant based on the amount of public expenditure and the amount of private investment, jobs or housing units that could be generated. However, the calculation of these outputs is difficult with any certainty at the design stage.

Assessing urban regeneration projects on the basis of net employment benefits has a number of challenges. Some types of land use inevitably are more labour-intensive, and numbers are a function of the type of location. Forecasts also need to take account of the balance between part-time and full-time employment and the quality of the jobs created. The cost to the public purse has also to be factored in.

Creating jobs is only part of the urban regeneration equation, they also often need to ideally go to residents in areas of high unemployment. One way to solve this is to require an occupier of a completed development to offer interviews to local people, but in doing so, this policy may deter private investment.

The use of leverage ratios also has a number of practical issues. Projects that generate high levels of private investment would probably go ahead anyway. In effect, the state is wasting its money that should be spent on projects in more distressed areas. The more deprived the area, the less private investment is likely to be induced for every £ of public money. Leverage ratios are also lower at the beginning of a regeneration initiative.

Judging individual projects on jobs or investment to be generated is a limited mindset. They need to be seen within a long-term lens of what is possible for an area, and the potential of new uses. In particular, the economic goal is the establishment of new real estate markets in regeneration areas.

Real estate-led urban regeneration initiatives use public money to pump-prime the private sector in theory until private development is just viable in its own right. However, the challenge is that urban regeneration schemes are in secondary/tertiary locations where mainstream private investment has been limited in the past, possibly for decades. These projects are also fostering embryonic markets, and there is the potential for unforeseen development costs.

To take account of these factors, an assessment of long-term success should be by reference to whether a sustainable market for the new use(s) has been achieved. It is based on whether further private sector development is viable without public support, and there has been a period of market activity to establish its credibility.

The Merchant City case study examines how an area of obsolescent warehouses on the edge of Glasgow city centre is converted into a residential area over 20 years. With no owner-occupied housing in the area, there was no market to build on. A small subsidised development tested the waters in 1983 and the flats sold very quickly. It was followed by other schemes supported by grants.

By the start of 1994, 18 developments had produced over 1,000 housing units in the area. Nevertheless, there had been market obstacles to overcome. Mortgage banks had to be convinced to finance purchases. Strong initial demand was eventually stemmed by the impact of the recession at the end of the 1980s.

Flat prices initially were very cheap, but rose quickly before increases levelled off. Average new prices fell for the first in time in 1992 in the aftermath of the recession. With prices rising elsewhere in Glasgow, and combined with the number of new sales slumping, it generated the first seeds of doubt in the success of the initiative. New development stopped.

Eventually, after more than ten years from the inception of the initiative, the Merchant City became an established niche in the city's housing market. In the latter half of the 1990s, new development returns to the area, and now without public subsidy. A sustainable market had been achieved after approximately 15 years.

The case study of new houses on three estates built in the mid-1970s in the inner city of Liverpool reveals a more mixed picture of success. The new estates were easily sold as there was excess demand with prices set low to attract buyers. They were very popular to begin with, and prices on resale rose quickly. However, in the early 1980s two estates began to suffer from a loss of demand. A study in the late 1990s found that prices were very low in both estates, and properties were difficult to sell. Only one of the three estates achieved a sustainable market.

The case study of modernised social housing flats converted for owner occupation is also a precautionary tale. The estate was refashioned as a gated community, approximately a mile from Manchester city centre. The refurbished flats were sold quickly and cheaply in 1984, but the price of resales did not maintain their real value. Resale prices were dampened particularly by the high rate of repossessions exacerbated by the recession of the late 1980s. Ultimately, the estate found it difficult to compete with equivalent more central city developments.

The case study in Easterhouse, a large social housing estate in Glasgow, involves new houses for sale together with the improvement of some flats for sale. They represent the first new housing for sale on the market in the area. An initial rise in resale prices is followed by foreclosures among the refurbished flats bringing a negative impact on their resale values.

In the 1990s, there is continuing turnover of resales but despite this market activity prices are stable or falling, and at best showing a faltering, upward trend. While a resale market has been established by the millennium, second-hand house prices had not risen sufficiently to make private development viable without subsidy. By the end of the decade, there is no sustainable market for housing for sale.

In all these case studies, the success of the developments is dependent on creating a market for owner-occupied housing where one did not exist. Over time, these markets are buffered by the impact of a recession and changing local circumstances. As embryonic markets, they are particularly susceptible to recessions. As a result, not all initiatives are successful.

The emergence of a mature/sustainable market takes some time to nurture, certainly more than a decade. In one case study, a second-hand market was established, but price levels were insufficient to enable further profitable development. There is an identification problem of which regeneration projects will ultimately succeed that cannot necessarily be determined at the outset.

The long-term nature of regeneration processes means that it is difficult to identify a market failure to be addressed. Unfortunately, the absence of property forms or land uses in specific localities may simply be because they are not viable. Rather than a market failure, it may mean that the market has correctly identified a lack of viability; in other words, there is a rational market gap.

An alternative standpoint is to simply intervene in the inefficient real estate market to improve the physical environment and promote local economic potential. Such intervention could include new infrastructure to remove locational constraints. The new opportunities permitted may act as a springboard for developing new local land use markets. This restructuring of land uses is designed to enhance the city's economy but may not directly benefit the immediate local community.

Learning outcomes

Gap funding to developers supports regeneration projects that otherwise would not be profitable.

The choice of projects to fund in this way is based on a combination of their scope to achieve physical and economic urban regeneration and value for money.

The assessment of urban regeneration projects on the basis of net employment benefits is difficult as it needs to take account of the type of jobs established and who will take them.

The use of private investment leverage ratios for regeneration projects can be misleading as a measure of success.

The ultimate goal of regeneration projects is the establishment of new real estate markets in these localities.

Real estate-led urban regeneration initiatives involve schemes in secondary/tertiary locations devoid of recent private investment.

The long-term success of a regeneration project should be judged by whether a sustainable market for the new use(s) has been achieved.

The Merchant City was an area of obsolescent warehouses on the edge of Glasgow city centre. Developers were given grants to convert buildings into a residential area. After weathering the impact of a recession, it established itself as a niche in the housing market, and eventually after 15 years, development occurred without public subsidy.

The case study of new houses on three estates built in the mid-1970s in the inner city of Liverpool reveals a more mixed picture of success. Despite initial popularity in the early 1980s, two estates began to suffer from a loss of demand, and only one of the estates achieved a sustainable market.

The modernisation of social housing flats converted for owner occupation near Manchester city centre in 1984 also had a chequered market history. Resale prices were dampened particularly by the high rate of repossessions exacerbated by the recession of the late 1980s. It never fully recovered.

New houses for sale were built in Easterhouse, a large social housing estate in Glasgow in the mid-1980s and early 1990s. Some social housing flats were also improved and then sold. Subsequent resale market activity demonstrated an ongoing demand for the properties, but by the millennium, prices had not risen sufficiently to make private development viable without subsidy.

The emergence of a mature/sustainable market takes some time to nurture, certainly more than a decade. The long-term nature of these regeneration processes means that it is difficult to identify at the outset of a project whether there is market failure that can be addressed.

Real estate-led regeneration can be conceived as intervening to improve the physical environment and promote local economic potential. Such intervention could include new infrastructure to remove locational constraints and develop new land use markets.

Bibliography

Barnekov T and Hart D (1993) The changing nature of US urban policy evaluation: The case of Urban Development Action Grant, *Urban Studies*, 30, 9, 1469–1483.

Jones C and Watkins C (1996) Urban regeneration and sustainable markets, *Urban Studies*, 33, 1129–1140.

Jones C and Brown J (2002) The establishment of markets for owner occupation in public sector communities, *European Journal of Housing Policy*, 2, 3, 265–292.

Index

Printed in Great Britain
by Amazon